EUROPEAN CITIES IN THE KNOWLEDGE ECONOMY

The European Institute for Comparative Urban Research, EURICUR, was founded in 1988 and has its seat with Erasmus University Rotterdam. EURICUR is the heart and pulse of an extensive network of European cities and universities. EURICUR's principal objective is to stimulate fundamental international comparative research into matters that are of interest to cities. To that end, EURICUR coordinates, initiates and carries out studies of subjects of strategic value for urban management today and in the future. Through its network EURICUR has privileged access to crucial information regarding urban development in Europe and North America and to key persons at all levels, working in different public and private organizations active in metropolitan areas. EURICUR closely cooperates with the Eurocities Association, representing more than 100 large European cities.

As a scientific institution, one of EURICUR's core activities is to respond to the increasing need for information that broadens and deepens the insight into the complex process of urban development, among others by disseminating the results of its investigations by international book publications. These publications are especially valuable for city governments, supranational, national and regional authorities, chambers of commerce, real estate developers and investors, academics and students, and others with an interest in urban affairs.

EURICUR website: http://www.euricur.nl

This book is one of a series to be published by Ashgate under the auspices of EURICUR, the European Institute for Comparative Urban Research, Erasmus University Rotterdam. Titles in the series are:

Metropolitan Organising Capacity
Leo van den Berg, Erik Braun and Jan van der Meer

National Urban Policies in the European Union
Leo van den Berg, Erik Braun and Jan van der Meer

The European High-Speed Train and Urban Development
Leo van den Berg and Peter M.J. Pol

Growth Clusters in European Metropolitan Cities
Leo van den Berg, Erik Braun and Willem van Winden

Information and Communications Technology as Potential Catalyst for Sustainable Urban Development
Leo van den Berg and Willem van Winden

Sports and City Marketing in European Cities
Leo van den Berg, Erik Braun and Alexander H.J. Otgaar

Social Challenges and Organising Capacity in Cities
Leo van den Berg, Jan van der Meer and Peter M.J. Pol

City and Enterprise
Leo van den Berg, Erik Braun and Alexander H.J. Otgaar

The Student City
Leo van den Berg and Antonio P. Russo

European Cities in the Knowledge Economy

The Cases of Amsterdam, Dortmund, Eindhoven, Helsinki, Manchester, Munich, Münster, Rotterdam and Zaragoza

LEO VAN DEN BERG
PETER M.J. POL
WILLEM VAN WINDEN
PAULUS WOETS

European Institute for Comparative Urban Research
Erasmus University Rotterdam
The Netherlands
www.euricur.nl

LONDON AND NEW YORK

First published 2005 by Ashgate Publishing

Published 2018 by Routledge
2 Park Square, Milton Park, Abingdon, Oxon OX14 4RN
605 Third Avenue, New York, NY 10017

First issued in paperback 2021

Routledge is an imprint of the Taylor & Francis Group, an informa business

Publisher's Note
The publisher has gone to great lengths to ensure the quality of this reprint but points out that some imperfections in the original copies may be apparent.

A Library of Congress record exists under LC control number: 2005924973

ISBN 13: 978-0-367-78713-4 (pbk)
ISBN 13: 978-0-8153-8889-0 (hbk)

Contents

List of Figures

List of Tables

Preface

Much has been written and spoken in recent years about the 'knowledge economy'. The Western economies have for some time found themselves in a transition phase, with the accent shifting in the sense that knowledge is steadily gaining weight as a production factor. The development towards a knowledge economy is manifest not only in the growing service sector, but also in such 'traditional' sectors as agriculture and manufacturing industry. The core activity in most sectors is no longer the physical manufacture of a product, but the development of new products and production processes, the generation of new knowledge, and the devising of marketing concepts.

What does this shift mean for cities and for urban policies? Many cities struggle with this question, but only a few studies have addressed it in a comparative and integrative way. Therefore, it seemed appropriate to conduct an international comparative analysis. Nine European cities (Amsterdam, Dortmund, Eindhoven, Helsinki, Manchester, Munich, Münster, Rotterdam and Zaragoza) shared this view, and together with them we embarked on a comparative research project. This volume is the final result of this research.

We want to express our very special thanks to four of the key initiators of this study: Erik van Merrienboer (City of Eindhoven), Jan Smeekens (NV Rede), the present Mayor of Eindhoven, Mr Sakkers, and the previous Mayor of Eindhoven, Mr Welschen. Their support has greatly contributed to this study. Furthermore, we would like to thank our contact people from the participating cities. Their help and input has been of great value in preparing and organising our visits to the cities. And, perhaps most importantly, it was very pleasant to work with them. Therefore, many thanks to Carine van Oosteren (Amsterdam), Dieter Steemann (Dortmund), Asta Manninen (Helsinki), Dave Carter and Sarah Adolph (Manchester), Christian Sorge (Münster), Hans Peter Heidebach and Thomas Fischer (Munich), Jan van 't Verlaat (Rotterdam) and Jose-Carlos Arnal (Zaragoza).

We also want to thank all the people we interviewed in the nine cities who were willing to dedicate some of their precious time to our study. They were our main source of information and new insights. But, of course, the authors take full responsibility for the content of this study.

Finally, we thank our colleagues Paolo Russo and Giuliano Mingardo for their contributions in the initial stage of the project, and our secretary, Ankimon Vernède, for her administrative support.

Leo van den Berg
Peter M.J. Pol
Willem van Winden
Paulus Moets
Rotterdam, July 2005

Chapter 1

Research Framework

1.1 Introduction

The economies of the developed world have for some years found themselves in a transition phase, with the accent shifting in the sense that 'knowledge' is steadily gaining weight as a production factor. The development towards a knowledge economy is manifest not only in the growing service sector, but also in such 'traditional' sectors as agriculture and manufacturing industry. The core activity in all sectors is no longer the physical manufacture of a product, but the development of new products and production processes, the generation of new knowledge, and the devising of marketing concepts. The knowledge economy also entails new location factors. Knowledge-intensive activities require highly-educated workers. Countries and regions that know how to attract and keep such workers have an advantage. Such matters as the quality of the living environment, cultural provisions and the level of all manner of other services carry ever more weight. In addition, the quality of the local knowledge and educational infrastructure is gaining ground as a location factor. The growing internationalisation of education offers new chances for regions to attract knowledge but may, on the other hand, give rise to leakage effects. The knowledge economy is also a network economy. Specialised companies need to cooperate to bring forth new products, techniques and concepts. That is why, in a knowledge economy, the 'organising capacity' of actors has become such a vital factor for a region's or country's enduring economic prosperity. It is not only networks of companies that are needed, but in particular also strategic alliances between companies and institutions of education and research. The paradox of the knowledge economy is that on the one hand, business companies have become more flexible with the virtual abolition of frontiers and the internationalisation of the economy, while on the other, they are more than ever tied to a location because they depend on highly-educated staff and on their integration into local knowledge networks.

The report which follows addresses the local dimension of the knowledge economy and policy options. It is the result of a comparative urban study of nine cities: Amsterdam, Dortmund, Eindhoven, Helsinki, Manchester, Munich, Münster, Rotterdam and Zaragoza.

This report is organised as follows. In the remainder of this chapter, we elaborate on the research questions and the methodology (1.2), and present an overview of recent literature about cities and urban policy in the knowledge economy (1.3). Based on that, we present a research framework that forms the basis for the case studies (1.4). After this, the case studies are described: Chapters 2 to 10 contain the nine case studies. In Chapter 11 we draw conclusions and present a synthesis.

1.2 Research Questions and Methodology

The first research question concerns the *transition towards a knowledge economy.* There is no such thing as a univocal transition path. However, we will try to identify some typical elements by charting and analysing the development paths of a number of cities. The focus will be on those cities that in the last couple of decades have evolved into a high-grade local knowledge economy. Another point to explore is how far specialisation or, conversely, diversification of the urban structure has laid the foundation for 'success'.

A second, closely related question concerns the local government. *What can the local government do to upgrade the local economy and guide it towards greater knowledge intensity?* Points of interest to explore are: what are the cities doing and planning; what policy options are they choosing; and what is their policy orientation? Do cities focus explicitly on certain sectors? Should cities just create the proper conditions and leave the rest to market forces? The last option is sometimes recommended in the literature, the argument being that the local government cannot be expected adequately to foresee rapid developments in the market and respond with the appropriate policy measures.

We also deal with the process aspects of policy. None of the policy options can be attained by the local government on its own; good cooperation with all stakeholders is imperative. In other words, the network capacity of the city will largely determine the achievement of ideas and plans, the generation of funds, etc. We shall therefore also take into account the city's organising capacity.

For each participating city we have investigated what the local government has done in the past few years to raise knowledge intensity, and how the stakeholders have been involved. In that way we have identified some effective strategies. We must keep in mind the element of contingency: what has worked well in one city may be totally unsuitable for another. Cities differ in many respects, and the (inter)national context is undoubtedly important. A sound 'context analysis' is therefore part of the synthesis, which puts the experiences of all the cities into perspective and presents some recommendations.

The research has been carried out in two stages. Desk research was central to the first stage. Much is already known about the knowledge economy and the role of cities in it. An extensive literature study has enabled us to put together a survey of the 'state of the art', and to produce the building blocks for a theoretical framework.

At the second stage we probed the subject more deeply in nine European case studies: Amsterdam, Dortmund, Eindhoven, Helsinki, Manchester, Munich, Münster, Rotterdam and Zaragoza. In each city we held between eight and 16 in-depth interviews with key people from government and enterprise. For each city we made a generic analysis of the development of its knowledge economy in the last decade. How knowledge-intensive is the local economy? How can that be measured? What are the city's prominent assets and resources? What does the economic structure look like? What generic economic/spatial policy has been undertaken?

In the next section the literature review of our study is presented. Important questions to be answered are: what are characteristics of the knowledge economy; what is the role of cities in the knowledge economy, and what determines their success; and what is the role of urban management/policy in the knowledge economy? In Section 1.3.4, conclusions of the literature review are given. Then, in Section 1.4, our research framework is presented. This consists of two parts: the foundations and the activities of cities in the knowledge economy. This framework will form the structure of the case studies.

1.3 Literature Review[1]

1.3.1 What are the Characteristics of the Knowledge Economy?

1 It is not easy to define which cities are successful in the knowledge economy, because it is difficult to define what falls within its ambit. It can be argued that all capitalistic economy is a knowledge economy, because today, to varying degrees, production systems are increasingly moving their processes away from the material and tangible, putting more emphasis on information as added value (e.g. logistics, web services, e-culture, information and communication technologies (ICT), etc.), and depending on knowledge as crucial input.

2 In discussions about the knowledge economy, distinctions are made between various types of knowledge and information. The difference between codified knowledge and tacit knowledge is very important. Codified knowledge is information that is widely available through ICTs (especially Internet) and other media. It is accessible to everyone and confers no competitive advantage. Tacit knowledge is available only to limited numbers of contacts and often has to be passed on face-to-face. It thus tends to benefit those locations where there is most access and contact – namely the largest cities (Lever, 2002). Lambooy (2002) discerns three levels of complexity: data (unstructured facts); information (structured facts); and knowledge (the competence of individuals to judge and evaluate, to use data and information, and to reformulate and solve problems).

3 In the knowledge economy, knowledge and information are the main inputs and outputs. In the words of Stiglitz (1999), 'knowledge and information is being produced today like cars and steel were produced a hundred years ago. Those who know how to produce knowledge and information better than others reap the rewards ...' (p. 1). The growing class of 'knowledge workers' does not produce any tangible product, but is continuously transforming knowledge and information into new knowledge and information for which there is a market. They make up the vast majority of the workforce in advanced economies. Reich (1991) invented a term for this class of people: symbolic analysts. These people tend to be well educated; they earn high salaries, and subsequently put high demands on the quality of their living environment.

4 It can be argued that, due to globalisation and the new ICTs, the diffusion speed of information and knowledge has increased dramatically. New knowledge or innovations that formerly took months to spread are now globally available in seconds. This speeds up the process of new knowledge creation. Also, because so much information and knowledge is available nowadays, it has become a crucial ability to *select* and *interpret* new information and knowledge, and to turn it into profitable activities (see Castells, 2000; Howells, 2002).

5 The knowledge economy is a network economy. The rapid developments in knowledge and information mean that no single person or company can master all disciplines, or monitor all the latest developments. For companies, it is increasingly crucial to engage in strategic networks, in order to 'tap' into complementary knowledge resources in a flexible way. Networks enable faster responses to rapidly changing markets and

technologies, and they are conducive to creativity, thus producing new combinations. Every 'node' (which can be a company or a person, but also a city) in the knowledge network has to develop its own specialisation (Castells, 2000).

6 In the knowledge economy, there is a high premium on entrepreneurship and innovation. The shift away from relatively stable and standardised mass manufacturing to highly volatile niche markets, combined with the rapid changes in technology, opens many new opportunities for entrepreneurs who are able to find the niches and mobilise resources. Many argue that the 'entrepreneurial climate' in a city or country is therefore an important determinant of success in the knowledge economy. Schumpeter (1980) distinguished five kinds of innovation: products, processes, new markets, new inputs (like new materials), and new organisational forms. In the dynamic process of competition three kinds of competences are mentioned: cognitive, innovative, and organisational. Besides these entrepreneurial competences, several external structural characteristics can also influence the outcome of the competitive process.

7 The knowledge economy is very volatile. Companies can grow very quickly, but also decline very quickly. This can be explained by various factors. On the one hand, technology and markets change very rapidly, which means that companies can win (or lose) market share if they make the right (or wrong) decisions. On the other hand, in many sectors in the knowledge economy, it is very expensive to develop a new product (e.g., a movie or a software program), but once it has been developed, the reproduction costs are almost zero. Thus, if the product is a 'hit' the company makes a lot of money and can expand, but if it is a flop it loses out. For cities, this has consequences. Firstly, a strong dependence on a single company (think of Nokia in Helsinki) is dangerous. Secondly, fostering innovative start-ups might have a high premium, if they happen to grow large.

8 In the knowledge economy, the old distinction between manufacturing and services is less useful. Manufacturing and services are no longer two totally separate matters: increasingly manufacturing uses service activities and service companies become involved in manufacturing. From this perspective, Daniels and Bryson (2002) argue that the terminology needs to be replaced by an emphasis on two sets of related issues. Firstly, we need a focus on knowledge and information flows that will identify and explain the complex web of connections that exist within and between companies. Such a focus would identify the structure of an economic sector rather

than service or manufacturing components. Secondly, they argue that it is time to develop a service-informed understanding of the manufacturing sector. This would highlight the service aspects of manufacturing and at the same time reveal the difficulties of continuing to classify activities as either services or manufacturing.

9 There is no single and unambiguous way for countries to make the transition to a knowledge economy. Different countries, with their own traditions, culture, history and background, have taken very different paths. Castells and Himanen (2002) make a distinction between three rather extreme models: the US model, the Singapore model, and the Finnish model. These three countries ranked at the top of most competitive 'new economies' by the end of the 1990s. All of them had observable strengths in infrastructure, production, and knowledge of information technology. They have the highest penetration rates of new technologies such as the Internet and mobile phones. Also, in terms of competitiveness and innovative capacity, all scored very high. At the same time, these countries are very different from each other in a number of respects. Finland is very different from the others with respect to its large welfare state and commitment to egalitarism. This country combines very low poverty and social exclusion with high economic dynamics. Singapore is an example of a country that is very competitive but lacks democratic governance. The US stands out for its extremely flexible labour markets and the high rewards on entrepreneurship, but has also high levels of inequality and social exclusion.

10 The Finnish case raises the question of whether, in the knowledge economy, the textbook trade-off between equity and equality still holds. More inequality (large income differences) increases incentives to engage in successful activities. However, at the same time it can be argued that higher equality leads to more demand for high tech and innovative products. The argument runs as follows: in consumer markets, there is a shift away from standardised 'mass' markets to more specialised and sometimes even interactive demand. It is commonly agreed that the budget share of basic goods declines with rising income. With more equal incomes, more people have means to purchase innovative goods, which can spur the local or national innovative industry (Zweimüller, 2000). The Finnish experience with mobile phones supports this claim. Recent research has suggested that, at least for countries, the 'traditional' thesis that greater equality leads to less efficiency is not proven (Atkinson et al., 1995). In particular, if transfer payments are used for education, growth rates can actually rise.

11 As stated earlier, the knowledge economy is gaining importance throughout the world. However, this does not mean that every kind of information and knowledge becomes more relevant. An essential question is: information and knowledge on what, for what? For economic development it is mainly those types of information and knowledge that can be transformed into new, competitive business that seem important. Many countries and regions struggle with this aspect of the knowledge economy: e.g., in the UK it was found that regions perform strongly on the qualifications of the workforce whereas they have poor performances on output indicators such as regional GDP and productivity. A possible cause might be the building of a 'paper knowledge economy' based on a well performing qualifications industry, instead of a demand-driven economy that has different needs concerning business, skills and education (Hepworth and Spencer, 2003).

12 Armstrong (2001) is critical of the knowledge economy ideas: many policy makers look at the task of stimulating science-based industries through the lens of enterprise ideology. They assume that scientific fundamentals are already in position and that applications in economic sectors can take place by stimulating entrepreneurship. However, they disregard that the mechanisms active in the knowledge economy are much more complex. It is not clear what attributes and behaviours are important for successful enterprise. Furthermore, the relevance of risk-taking behaviour in entrepreneurship is questionable; risk has not been a prevailing characteristic of new business creation in general, and science-based start-up companies in particular. Risk-taking does not seem to fill the gap between basic science and applied technology. Nor is there significant proof that new product development on the basis of existing technologies has depended on a mentality of risk-taking.

1.3.2 What is the Role of Cities in the Knowledge Economy, and What Determines their Success?

1 Cities are the focal points of the knowledge economy, because it is mainly in cities that knowledge is produced, processed, exchanged, and marketed. Cities are best endowed with knowledge infrastructure (universities, other educational institutes, etc.); they tend to have higher than average shares of well-educated people (the 'symbolic analysts'); they are best endowed with electronic infrastructure; they are well connected to the global economy through airports; they have a function as a place where knowledge is exchanged, and as breeding ground for talent and new combinations. It

is therefore no coincidence that many larger cities have experienced a remarkable revival in the 1990s.

2 The creation of new knowledge mainly takes place in cities. Cities with a strong knowledge base seem to gain in the knowledge economy. In Finland, Helsinki and other Finnish university cities were the main drivers of economic growth during the second half of the 1990s (OECD, 2002). Mathiessen et al. (2002) analysed the scientific output of 40 cities. They assumed that a solid knowledge base is reflected in the economic life of a city and that it is, therefore, of increasing importance for urban economic growth and change. Can cities do without producing knowledge themselves? According to Castells and Hall (1994), this is problematic: in their view no region can prosper without some level of linkage to sources of innovation and knowledge production.

3 The urban knowledge economy thrives on talented people who create new knowledge and ideas. From this perspective, Florida (2000) studied the location behaviour of 'talent'. Among other things, he found that quality of life in a place is a key determinant to attracting and retaining these people. 'Talented people do not simply select a place to work based on the highest salary, they are typically concerned with a whole series of place-based characteristics' (Florida, 2000, p. 6). Florida (2000) empirically found that cultural activities and amenities[2] are increasingly central determinants of urban competitiveness. Talented people are attracted by places where they can enjoy life (Castells, 2000). There is also a cumulative effect: as Florida puts it '... talent tends to attract talent' (Florida, 2000, p. 15). Van den Berg (1987) also finds that quality of life in urban areas is a location factor of growing importance: highly skilled people will live in cities that offer high quality of life, and companies will follow. Many empirical studies confirm the link between human capital and urban economic growth (see for instance Glaeser, Sheinkman and Sheifer, 1995; Simon, 1998).

4 Urban diversity promotes creativity. The scale of cities and the diversity of their inhabitants create the interactions that generate new ideas. Diversity is a measure of the degree of system openness. It can be said that the places that attract diverse groups of people (by ethnicity, nationality, gender and sexual orientation) have an environment that is easy to plug into; such places can be said to have low entry barriers for talent (Florida, 2000).

5 To a large extent, in the knowledge economy big is beautiful. The size of the city matters as an attraction factor both for companies and for knowledge workers. For companies, in a larger city is it easier to find specialised staff. Glaeser (2000) found that it is the need to access common pools of labour

rather than access to suppliers and customers that drives the tendency of firms to cluster together in cities. For the knowledge workers, being in a large metropolitan area increases the variety of jobs available. This is especially relevant for households with two knowledge workers. Larger cities tend to have bigger airports through which more destinations can be reached, and many of them are nodes on a high-speed rail network; larger metropolitan areas are relatively attractive for foreign direct investment as well. Their scale offers scope for international subcultures and amenities such as international schools.

6 In the knowledge economy, one of the drivers of innovation is the exchange of tacit knowledge among actors. Cities can be good environments for this type of exchange. In 1890, Marshall was already describing the powerful dynamics in industrial districts, where geographically concentrated groupings of firms, large and small, interact with each other via subcontracting, joint ventures or other collaborative means, gaining external economies of scale in doing so (Cooke, 1995), thus deriving international competitiveness from local sources. Porter (1990) describes how clusters of densely networked firms serve global markets while deriving their strength from a regional basis. He discernes four conditions as essential in that development: factor conditions (quality of labour, capital, knowledge available); demand conditions (scale and quality of the regional home market); supplier industries (globally competitive suppliers, specialised services); and business strategy (not only rivalry between local firms, but also willingness to cooperate in research, sales and marketing). In particular, the interplay of competition and cooperation is fundamental. Too much competition may be destructive, but the same holds for too much cooperation when it degenerates into the formation of cartels (Cooke, 1995; Harrison, 1994). Lazonick (1992) and Boekholt (1994) stress that in clusters, a major role is played by other than interfirm linkages: links with government-supported scientific institutes, ties with the scientific community and professional associations are important factors in a cluster's performance.

7 The question remains, though, as to why proximity still seems to matter in networks, where modern communication technology theoretically permits spatial dispersion. Several reasons are put forward. Firstly, face-to-face contacts appear to be very important as sources of (technological) information and in the exchange of tacit knowledge (Leonard-Barton, 1982; Malmberg et al., 1996). Spatial proximity greatly enhances the possibility of such contacts. In this respect, Howells (2002) argues that

knowledge is codified at global level, while it is tacit at local level. Secondly, cooperation between actors requires mutual trust. This holds particularly when sensitive and valuable information is exchanged, for instance in a joint innovation project. Several authors (for instance, Piore and Sabel, 1984) argue that cultural proximity, i.e. the sharing of the same norms and values, is an important factor in that respect, since cooperation is a human phenomenon. A very relevant issue concerning the spatial dimension of clusters is how local networks relate to global networks. In the local-global interplay, transnational companies (TNCs) play a special role. Malmberg et al. (1996) stress that if a TNC is rooted and integrated ('fledged') in the region and engaging in regional networks, it can act as an important disseminator of new knowledge, information and innovation from abroad into the region. This is particularly relevant for research and development activities: knowledge flows are facilitated by personal relationships and mobility of employees or spinouts from the large firm.

8 The economic base of a city is a determinant of its success in the knowledge economy. In general, cities that used to specialise in traditional industry and port activities do less well than cities with a more diverse economic base. They tend to have a less well educated population, lower quality of life and housing stock, and often suffer from a bad image. The assets that counted in the industrial age (proximity to raw materials, seaports) have lost much of their value. In Europe and the US, many industries have turned away from the old and polluted industrial centres, and moved to areas in the 'sunbelt' – less polluted and crowded suburban areas – or to smaller towns (see, for example, van den Berg, 1987). These structural factors are hard to change over time, although some cities have managed remarkably well.

9 Simmie (2002) states that innovative activity is highly concentrated in some metropolitan and regional capital cities. He reviewed the local knowledge spillovers hypothesis as an explanation for the geography of innovation, arguing that it offers only a partial explanation. The reasons for this are, firstly, that there are many city regions that possess universities and industrial R&D (research and development) facilities but have yet to join the top-ranking cities as centres of innovation; secondly, innovation is not just driven by technology push factors, but also by demand pulls (often by clients based in other advanced countries). It is argued that international contacts and networks conducted by face-to-face contacts which are facilitated by international hub airports are critical factors for international knowledge transfer. Finally, he argues that the most successful cities are

those that are able to combine both rich local knowledge spillovers and international best practice in the design and specification of innovations; because of the need for face-to-face meetings to achieve many of these exchanges of experience, the geography of innovation is a function of both physical and time proximity.

10 In the knowledge economy, there is a strong relation between economic performance and equity, and poverty and social exclusion. The literature suggests a two-way relationship. On the one hand, economic growth in knowledge intensive sectors helps to reduce poverty: new jobs are created in personal services, hotel and catering industry and retail, which often require low educational qualifications. From this perspective many cities increasingly regard economic development policy as a key instrument in reducing poverty and inequality. On the other hand, reducing inequality, poverty and social exclusion may also affect growth. If these issues are neglected or overlooked, in the long run the consequences could be dramatic. As Hall and Pfeiffer (2000, p. 23) put it: 'a city that prospers economically but fails to distribute the wealth with some degree of equity runs the clear risk that it disintegrates into civil war between the haves and have-nots, a war in which both sides are losers'. The challenge will not simply be to redistribute money from the rich to the poor, but rather 'reinstating them (the socially excluded) into the mainstream social fabric'.

1.3.3 What is the Role of Urban Management/Policy in the Knowledge Economy?

1 How should urban regions be governed in order to thrive in the knowledge economy? The developments described in the last sections make it clear that traditional top-down models of urban management are no longer adequate. Kearns and Paddison (2000) argue that, to a large extent, urban governments have lost control over urban economies and societies because of economic globalisation and the mobile capital investments associated with it. The old top-down model does not fit well in a world of urban competition, which has become more intense and universal than ever; cities and regions compete for inward investment, visitors, real estate developments and inhabitants. This logic of competition asks for a more entrepreneurial form of urban governance. Furthermore, the traditional model is insufficiently able to deal with changes of all varieties. The hierarchical decision-making structure means that it takes a very long

time before decisions are taken and translated into policy measures; new ideas and initiatives hardly get a chance. However, the increased pace of economic and technological change and consumer preferences (so typical of the knowledge economy) demands types of urban management that are better able to signal new developments and translate them into adequate policies.

2 Due to all these factors, during the 1990s the term 'urban management' gave way to 'urban governance', reflecting the idea that not everything can be managed from the top down. As Kearns and Paddison (2000, p. 847) put it: 'urban governance is not an attempt to regain control so much as an attempt to manage and regulate difference and to be creative in urban arenas which are themselves experiencing considerable change'. Stoker (1996) argues that governance is about the capacity to get things done in the face of complexity, conflict and social change. Urban government needs to empower itself by using resources and skills from other organisations. In other words, urban governance can provide new ways to achieve strength, creativity and resources. Public-public and public-private partnerships are key elements of governance (Stoker, 1995) as vehicles for resource sharing, resolving conflicts and discovering mutual interests. Conflict management is key in achieving urban sustainability, as we have seen that there are several conflicts between the objectives of economic development, mobility, quality of life and social inclusion in urban areas.

3 Public-private partnership does not only increase public resources. Recent research (van den Berg, Braun and Otgaar, 2000) suggests that, increasingly, companies are willing to contribute to different aspects of sustainable urban development in order to improve their image, to attract labour potential, or to improve the accessibility of their estates, among other things. The key challenge for urban management is to mobilise these resources, and deal with the business community in new ways.

4 Similar to the 'urban governance' approaches, van den Berg et al. (1997, p. 253) introduced the concept of urban organising capacity, which they define as 'the ability to enlist all actors involved, and with their help to generate new ideas and develop and implement a policy designed to respond to fundamental developments and create conditions for sustainable development'. They state that this capacity is influenced by a number of factors: the adequacy of the administrative organisation of the region; the quality of strategic networks; the spatial-economic conditions that prevail in the region; the presence of leadership, a vision; and the levels of political and societal support for policies.

5 What can cities do to strengthen their knowledge base? Stiglitz (1999, p. 22) argues that cities should stimulate the externality-generating knowledge; 'the objective of the government is not to pick winners, but to identify externality-generating innovations … many applied research projects generate large externalities. The object of government policy is to identify winning projects with large externalities'. Also, he argues that for developing countries success in the knowledge economy requires a change in culture (ibid., p. 8): 'Development is about the transformation of societies which ultimately involves people changing how they think … but people cannot be forced to change their minds, so they have to do it by themselves' (ibid., p. 6).

6 Lever (2002) says that the (local) government should achieve the right balance between basic and applied research, and should encourage the production and transfer of tacit as well as codified knowledge. In particular, basic research should be encouraged, because 'it is basic research from which innovation ultimately flows' (ibid., p. 862), and several studies have found that market failure leads to an underproduction of basic research (ibid., pp. 861–2).

7 Van den Berg, Braun and van Winden (2001) discuss the role of policy in developing knowledge intensive growth clusters. Firstly, they argue that a well-defined and shared vision and a strategy on the development possibilities of a cluster is indispensable for an efficient allocation of resources and efforts to stimulate the cluster. Political and societal support are also necessary conditions for a cluster policy. Political support helps to bring about positive collaboration at the local level. Proper presentation and communication of policies are of paramount importance to achieve results. Societal support is important for the acceptance of policies aimed at growth clusters. Finally, public-private cooperation – on the strategic, tactical and operational levels – is very important for a successful cluster policy. An essential factor for success is the early involvement of the private sector in the development of locations, the attraction of companies, etc. (see also Knight, 1995). The knowledge, expertise and involvement of the private sector can be very valuable to the decision-making process and can considerably enhance the chance of success. In addition, government can act as network broker, stimulating the formation of inter- and intra-sectoral networks by bringing people and firms together. Local and/or regional government can engage in public-private partnerships directed at the stimulation of the growth cluster, for example by providing facilities or specific education.

8 Van den Berg and Russo (2003) discuss how cities can make more out of their universities and attract/retain students – future knowledge workers – to the city. They found that city governments and other organisations have only recently started to realise that they should have distinct policies to attract and accommodate different segments: freshers, graduates, visiting fellows, and young members of staff. The basic assumption of this research is that an explicit strategy targeting the student community is necessary for the well-being and the competitiveness of the city. The student community needs to be fully integrated into the local socioeconomic environment, culturally and socially active, and motivated to stay after the completion of studies; the education facilities required by a competitive city are flexible, spatially balanced, and well connected. Van den Berg and Russo provide four guidelines for local government action: 1) use excellence in education to promote the city as a business location; 2) create 'excellence centres' for higher education; 3) act as a catalyst to fund locally-oriented research; and 4) fund 'chairs' at local higher educational institutes dealing with local issues.

1.3.4 Conclusions: Literature Review

1 This overview shows there is a rich and expanding literature on the knowledge economy in general, and on the implications for cities.
2 Much of the literature (theoretical and empirical) focuses on 'formal' basic scientific knowledge as the key resource for the knowledge economy, on both the local and the national levels. There is much less attention on other aspects; for instance the creative capacity of people in strongly growing sectors such as culture, arts and design, or the role of the market and organisational knowledge that is embedded in the urban service sectors. In our approach, we have explicitly taken these types of knowledge and creativity into account.
3 The empirical literature has a strong bias towards large cities. In particular the dynamics in 'world cities' are well documented. Much less is written about the role and development paths of medium-sized cities in the knowledge economy. This raises fundamental questions about the assumed role of scale, diversity and specialisation, and the role of cities in urban networks. Cities may compensate for a lack of scale and diversity by being part of an effective network, in which they develop a functional specialisation. New research is needed to learn more about these aspects in the light of the knowledge economy.

4 In general, little has been written about adequate policy responses at the local level for cities in the knowledge economy. The recommendations in the literature tend to be either very general or directed towards specific aspects (i.e., cluster development, attracting students/talent). An integrative approach seems to be lacking. Our study should therefore develop tools to give more explicit and integral policy recommendations.

1.4 The Research Framework of this Study

In this section, we will present and elaborate our research framework on a knowledge city. How can we judge the position of urban regions in the knowledge economy? Processes of network interaction and knowledge exchange are fundamental in the knowledge economy. We distinguish the foundations and the activities of a knowledge city (see Figure 1.1). The former pay a lot of attention to measuring the levels of activity (numbers of people employed, numbers of businesses in various sectors, etc). While the foundations analyse whether sets of activities (which might be functionally related) are found in the case studies, the sections on the five types of knowledge activities investigate whether organisations in functionally related activities do really interact (commercially). Such interaction between activities does not necessarily have to be the case; this might prove to be something that can be improved in certain cities. Put another way, the foundations mainly deal with the structure, whereas the activities focus on the process in the context of a knowledge economy.

We assume that seven foundations have to be present in a city to become and to remain successful in the knowledge economy. In Section 1.4.2 we will explain the activities of a knowledge city. There, we will go into which foundations are important for these knowledge activities.

1.4.1 The Foundations of a Knowledge Economy

1 Knowledge base The first foundation stone for a knowledge city is its knowledge base. According to Lever (2002) this includes tacit knowledge, codified knowledge and knowledge infrastructure. As stated in Section 1.3, the creation of knowledge mainly takes place in cities. The quality, quantity and diversity of the universities, other education institutes and R&D activities determine to a large extent the starting position of a city in the knowledge economy. Moreover, a city can also have a kind of creative knowledge base.

Florida (2002) refers in this respect to the presence of a class of creative people who write software, songs and stories, create designs and discover new ways to combine elements.

2 Economic base The second foundation stone, the economic base, largely determines not only the economic possibilities but also the difficulties for an urban region, within the knowledge economy. Urban regions with an economy dominated by service activities often have a better starting position in the knowledge economy than those specialising in manufacturing and port industries. Moreover, cities with a diversified economy are less vulnerable in rapidly changing economic circumstances. Urban economies that are dependent on one economic sector can be confronted with huge socioeconomic problems when its competitive position is weakening. Besides, cities with a diversified economy can become incubation places for new developments and economic innovation. Jacobs (1961, 1984) states that, in particular, cities that, in the course of tens or sometimes hundreds of years, have built up a richly diversified economy may count on a favourable future economic development.

3 Quality of life As van den Berg (1987), Castells (2000) and Florida (2002) states, quality of life is a key determinant to attracting and retaining knowledge

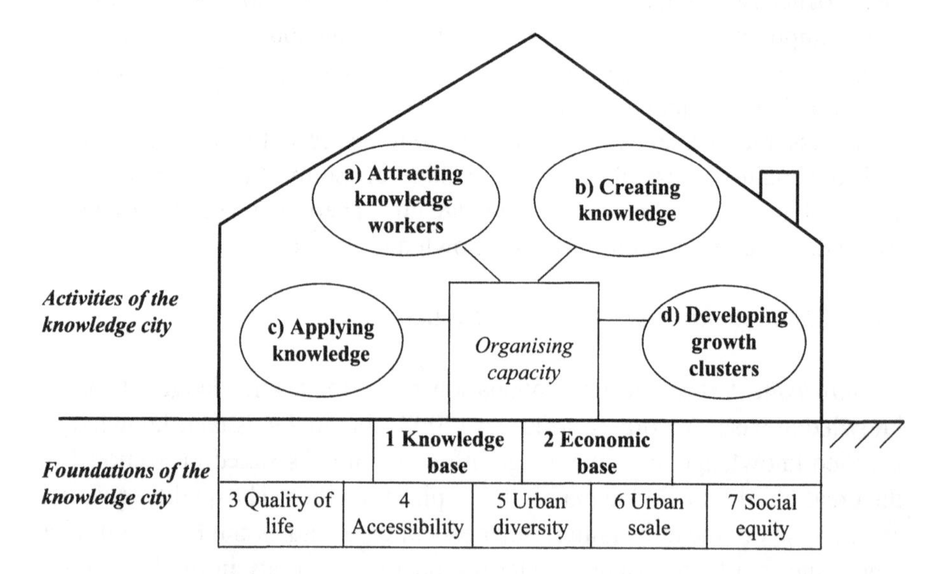

Figure 1.1 The foundations and activities of a knowledge city

workers, the symbolic analysts. Important in this respect are an attractively built environment, high-quality houses, attractive city parks, attractive natural surroundings, and a rich variety of cultural institutions. Moreover, there have to be good facilities, such as high quality hospitals and (international) schools. Besides, to generate a good urban quality of life, traffic systems – such as highways and airports – should not generate too much air and noise pollution.

4 Accessibility The knowledge economy is a global economy and a network economy. A good international, regional and multimodal accessibility is therefore crucial for successful knowledge cities. They have to have good and fast access to international airports and high-speed-train stations, but also good regional linkages to other urban knowledge centres (in particular by rail and road), and an efficient local infrastructure network to accommodate face-to-face contacts. Naturally, knowledge cities have to have high-quality electronic infrastructure for vast and swift global communication.

5 Urban diversity As stated, urban diversity promotes creativity. To Jacobs (1961) and to Florida (2000) the diversified city is *the* key to a socioeconomic successful *and* attractive metropolitan area. Only the urban areas that are sufficiently diversified will continually attract inhabitants, companies and visitors. Several empirical studies found that diversity fosters growth in cities (Glaeser et al., 1992) or at least in their most innovative sectors (Henderson et al., 1995, in Duranton and Puga, 2000). Looking more directly at innovation, Feldman and Audretsch (1999), Harrison et al. (1996) and Kelley and Helper (1999) show that diversity fosters innovation in cities, while narrow specialisation hinders it (Duranton and Puga, 2000). Urban diversity is on the one hand a source of economic opportunities, but on the other hand it can be a source of problems because tensions between groups of inhabitants may arise. Jacobs (1961) analyses at micro-level which requirements vital city districts must fulfil. She distinguished four basic preconditions for city diversity (need for primary mixed uses, small blocks, mix of buildings of different ages and concentration of people). They are not just personal preferences; she legitimised these preconditions by economic arguments.

6 Urban scale Knowledge-intensive activities take place in particular in medium-large and large cities. In larger cities there are more scale economies for knowledge activities, there is a larger market for specialised services and there is a larger common pool of symbolic analysts. Moreover, larger cities normally have better international transport infrastructure, offer scope for international subcultures and all kinds of international amenities. Besides, creative workers prefer inspiring cities with a thriving cultural life, and international orientation and high levels of diversity. Larger metropolitan areas are much more likely to attract these types of workers than remoter, smaller places. Smaller-sized cities located in or close to (and well connected to) relatively large metropolitan areas can benefit from these scale advantages and can thus also play a role in the knowledge economy.

7 Social equity In order to aim for sustainable urban growth it is important to reduce poverty and inequality. This is considered a fundamental precondition for cities to prevent tensions between 'haves' and 'have-nots' and between different ethnic groups. Such tensions may have an impact on the safety perception of urban inhabitants, workers and visitors. Urban safety has become a basic precondition for urban economic growth. Or, to put it negatively, insecurity, caused by whatever factors, can seriously hamper it. Increasingly, the perceived safety of a place is becoming an important location factor for companies and citizens (van den Berg, Pol and van Winden, 2002).

1.4.2 Activities of a Knowledge City

We distinguish four core activities of a knowledge city: a) attracting and retaining knowledge workers; b) creating new knowledge; c) applying new knowledge and making new combinations; d) developing new growth clusters. In the previous section we introduced the foundations for a knowledge city. The quality and quantity of these foundations have to be sufficient (balanced) for a city to be successful in knowledge activities. For some of these activities certain foundations are more important than for others. This is indicated in Figure 1.2 and will be explained and elaborated on below. The activities of a knowledge city can be compared to dishes and the foundations to ingredients. You can specialise in preparing one or more dishes, but you always need to have the basic ingredients. This is the same for cities with knowledge activities and foundations. An urban region can specialise in one or more activities of a knowledge city, but it will always need the basic foundations to have a balanced growth.

As can be seen in Figure 1.2, to develop one or more knowledge activities there has to be sufficient *organising capacity* within an urban region. Organising capacity is understood as the ability of those responsible for solving a problem to convene all concerned partners (public and private, internal and external), in order to jointly generate new ideas and formulate and implement a policy that responds to fundamental developments and creates conditions for sustainable economic growth. Organising capacity refers to the entire process from the identification of certain needs, through the development of strategies and policy, to the implementation of the policy and the monitoring of the results (van den Berg, Braun and van der Meer, 1997). Key tools for organising capacity are vision, strategic networks, leadership, political support, societal support and communication. As the knowledge economy is a network economy, investing in strategic networks appears to be crucial for knowledge cities.

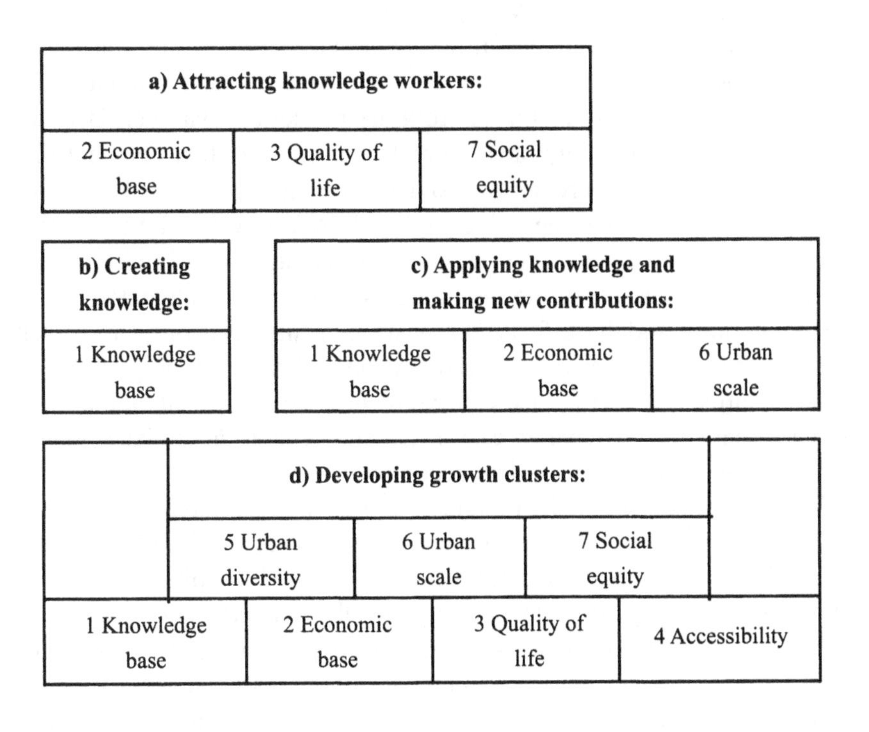

Figure 1.2 Activities of a knowledge city and the necessary 'building blocks'

(a) Attracting and retaining knowledge workers The most important foundations of a knowledge city for attracting and retaining knowledge workers appear to be its economic base, its quality of life and social equity. In principle, the economic base, of which the labour market is part, is the

most important element for attracting knowledge workers. Logically, in the first instance, these symbolic analysts are attracted by the available jobs in an urban region. However, with the increasing competition between cities and the increasing welfare it can be expected that in particular those urban regions that offer a high quality of life will prosper in the knowledge economy. With their increasing welfare, knowledge workers attach more and more value to a high quality of life. The third important foundation is sufficient social equity, because too much social difference may lead to social exclusion, divided urban areas, tensions between population groups, and a perception of lower urban safety and attractiveness.

Regarding (potential) knowledge workers, a distinction can be made between students and graduates. It turns out that factors that are relevant in attracting students as undergraduates are not the same where graduates – the potential skilled workers of the future – are concerned (van den Berg and Russo, 2003). Students may have a great time in a city that does not offer much to families, entrepreneurs or other social groups, provided student life is well organised and the university facilities are self-sufficient. This is the case in cities with a strong 'technical' orientation like Eindhoven and Birmingham. In these cases, though, students are not really integrated into the local community, and this affects their post-university decisions. Conversely, a city might be very attractive for visitors and a pleasant, high-quality place to live for citizens, as is the case with Munich, Lyon or Utrecht, but this does not mean that students have an easy life there; in fact, these places may be the hardest ones to find convenient accommodation.

Cities like Lyon, Lille, Birmingham and Eindhoven have a problem of 'brain drain': being second (or secondary) cities in their national urban systems, they suffer from competition from attractive areas like capital cities (Paris, London) or other large urban agglomerations (the Randstad). Other cities like Utrecht, Venice, Munich and Helsinki have the opposite problem of being 'too attractive' so that they cannot accommodate all the potential newcomers.

The following questions are relevant to get an indication of the capacity of a city to attract and retain knowledge workers:

- What is the size of the relevant labour market?
- How many students does the city have, and where do they come from?
- Why do they come to the city?
- After their graduation, where do they work, and why?
- Soes the city have or is it considering policies both to attract students and to keep them in the city?

- What is the proportion of well-educated people in the labour force?
- Why is the city popular/unpopular with knowledge workers as a place to live?
- Are there any explicit (marketing) campaigns or implicit strategies (housing policy, amenities) to attract and/or retain the higher-educated people?

(b) Creating new knowledge The most important foundation for creating new knowledge is naturally its knowledge base. The better its institutional infrastructure to produce, process and exchange knowledge, the more successful its activity to create new knowledge will be. The following questions are relevant to get an indication of the capacity of a city to create new knowledge:

- What are the cities' strengths in academic research?
- What is the quality, quantity and diversity of other education institutes and R&D activities?
- How have these strengths evolved over the last ten years? Have there been major changes?
- Does the city have any policy for promoting certain types of research?
- What are the cities' strengths in the creation of non-academic knowledge (arts, design, etc)?
- What is done with the newly acquired knowledge: is it applied within the region, or sold/transferred elsewhere?

(c) Applying new knowledge and making new combinations The most important foundations for a city in applying new knowledge are its knowledge base and its economic base. Moreover, to translate knowledge into business it is crucial that there are good networks within and between knowledge institutes and the business community. In addition, it is also possible to apply knowledge that comes from outside the region. For this activity, the urban scale is relevant, because there can be scale economies in applying new knowledge. In large cities, universities and research centres are more likely to come into contact with the industry and develop innovative products. That is why among 'top-research' rankings we find many small cities, but among 'patent locations' we only find large metropolitan areas.

The embeddedness of educational institutes (in particular universities) proves to be very fundamental for knowledge cities. The 'knowledge economy' is evidently reflected in an increase in university–firm interaction (van den Berg, Braun and van Winden, 2001). In the top part of Figure 1.3, several degrees of strategic interaction between the business community and the educational institutes are illustrated. Universities are not only important as education centres but also as research locations. For research, just as for education, a pyramid can be drawn up (bottom part of Figure 1.3) that indicates the level of strategic interaction between the business community and research units.

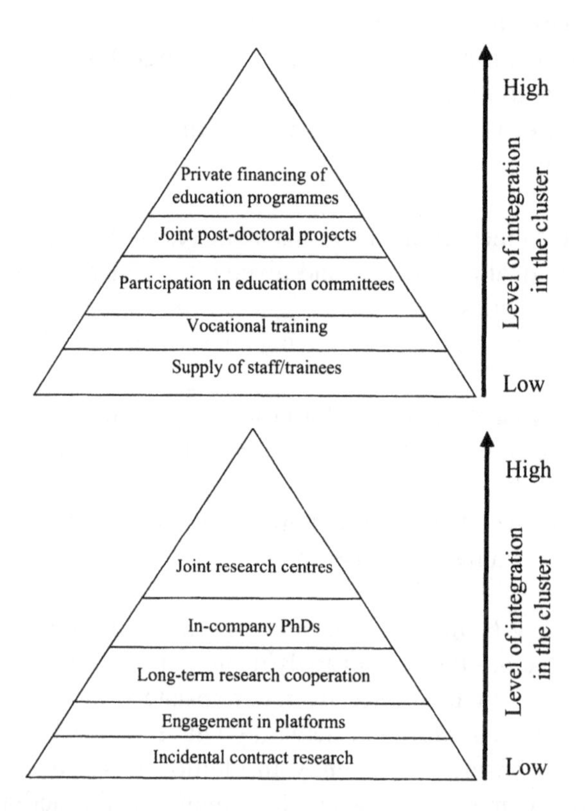

Figure 1.3 Embeddedness of education and research in the local economy

Source: van den Berg, Braun and van Winden, 2001.

New combinations can be expected in particular in those cities that have a good institutional knowledge infrastructure, good local networks and sufficient urban diversity. Urban diversity normally stimulates the combination of activities, processes, products, markets, inputs and organisational forms. Important preconditions for this are, however, mutual trust and some cultural proximity. Making new combinations is particularly challenging for old manufacturing and port industries. Their historical economic base could be the foundation for new modern industries.

An indication of the success of creating new combinations can be obtained by looking at the level of new firm creation. Young firms are often dynamic and innovative, and generate jobs; they can be important for large firms as partners in innovation or as suppliers. They may help to tie young talent to the region, particularly when new firms are strongly linked in the cluster, for instance by strategic relations with local universities or large firms. New firm creation in European cities generally lags behind the figures of the USA, particularly of high-tech starters (van den Berg, Braun and van Winden, 2001). Appropriate public-private structures to guide starting firms are assumed to be a very important factor in the degree to which people are inclined to start businesses, but cultural elements (the level of 'entrepreneurial spirit') are also likely to play a role.

New firms create dynamics: they offer employment, create added value, and may act as useful suppliers for existing firms in a cluster. Particularly when active in expanding markets, new firms may grow very rapidly. New firms are started from several sources: from existing firms, universities (researchers who commercialise a scientific discovery), or other educational institutes.

The following questions are relevant to get an indication of the capacity of a city to apply new knowledge and make new combinations:

- How is the interaction between knowledge institutes and local business organised (examples: 1) local industry is involved in the setup of education and research programmes; 2) local industry finances professor positions or PhDs; 3) incubators; 4) venture capital funds)?
- Why is the interaction good/bad?
- What has been/will be the role of local government in this interaction?
- Over the last ten years, how well have 'traditional' dominant economic sectors in the region (if they exist) transformed into more knowledge-intensive activities (i.e., from mainport to 'brainport', or transforming industrial heritage into ICT business locations, or using old ports for tourist purposes, etc.)?
- Has there been any explicit local 'industrial policy' to help companies become more knowledge intensive?
- What is the level of new firm creation?
- Does the city contain multinational companies? What are their links within the region? Do they play an important role in transferring knowledge?

(d) Developing new growth clusters All foundations of a knowledge city are important for developing new growth clusters: its knowledge base, its economic base, its quality of life, its accessibility, its urban diversity, its urban scale, and social equity.

The quality of life – the attractiveness of a city (in terms of housing, cultural and leisure facilities) – proves a fundamental factor in cluster development as a means to attract and retain highly skilled people to the region. Sufficient social equity is considered to support this quality of life but the economic and knowledge base of an urban region is also relevant.

An important aspect of a cluster's functioning is its size and its development level. Does the cluster possess 'critical mass'? How many companies and educational and research institutions are active in the cluster? Critical mass, and thus urban scale, is important for various reasons. Firstly, it ensures a market large enough to support the (specialist) activities in the cluster. A second advantage is that the presence of many companies may ignite keen competition and thus force companies to operate efficiently and effectively. Thirdly, the chance of fast penetration of many types of innovation is greater if the cluster is larger. Fourthly, regional cooperation is easier to accomplish within a large cluster, as it is easier to find a complementary partner within the region. Finally, scale offers prospects for the sharing of resources, the benefits of a shared pool of specialised labour, and the scope for a cluster 'superstructure' like joint education facilities (van den Berg, Braun and van Winden, 2001).

Often there is a self-reinforcing development: a large-scale cluster entails division of labour and specialisation; the large, specialised job market generates knowledge transfer; this permits further sophistication of the 'cluster product': that, in turn, may activate more demand; next, the increase in demand stimulates firms to expand, induces cluster-specific new firm creation and attracts more firms to the cluster, so that the economies of scale increase further. See Figure 1.4 for a graphical representation of this 'virtuous circle'. Nevertheless, the circle is by no means an automatism. The potential danger is that success could at the same time induce sluggishness and conservatism in (key) players in the cluster.

Next, the quality of the cluster actors is a relevant factor. Quality may refer to the degree of international competitiveness of firms, the technological sophistication of their output, the standing of a university, etc.

The presence of one or more engines in a region – which may be large multinational firms, but can be other actors as well – is also considered to be a determinant of a cluster's functioning, in their role of spider in global and local webs, or as 'flagships' of the cluster as a whole.

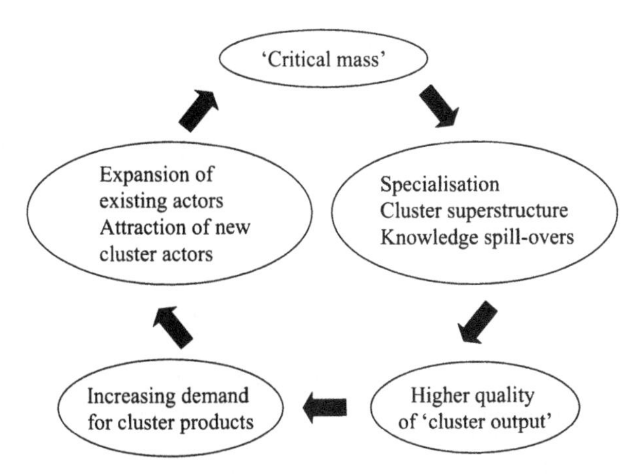

Figure 1.4 The 'virtuous circle' of cluster development

Source: van den Berg, Braun and van Winden, 2001

Besides its scale and quality, the degree of strategic interaction is assumed to be largely decisive for a cluster's performance. Strategic interaction implies long-term relations other than strictly financial, between organisations. Within the region such interaction can be achieved on various levels: among companies, between companies and institutions of education or research, among educational institutions, etc. As indicated in the previous section, strategic interaction can serve a variety of purposes: to create scale, to use one another's knowledge (of markets, technology, organisation), to make use of one another's networks, to solve common problems together, or to enhance flexibility. Fundamental conditions for interaction are that actors involved should know one another and trust each other. Next, the parties need to some extent be complementary. That last aspect is also associated with the scale. In a large and varied cluster, the chance of finding a suitable partner is considerably greater than in a small one.

The following questions are relevant to get an indication of the capacity of a city to develop new growth clusters:

- Can we observe the emergence of relatively new growth clusters in the city (i.e. biotechnology, arts/design/music, architecture, ICT)?
- Why have these clusters emerged here?
- Is there a link between the new clusters and the traditional economic structure of the region?

- What are the ambitions and policies of local government concerning the new clusters?
- What is the relevant region of emerging clusters?
- Do you consider neighbouring cities as competitors or partners?

Notes

1 The authors thank Drs Giuliano Mingardo and Dr Paolo Russo for their valuable contributions to and comments on this chapter.

2 The relation between amenities and economic development is two-way. The quality of amenities may strengthen the local economy because it attracts firms and inhabitants. At the same time, many amenities are a function of the urban economy. This holds particularly for market-oriented amenities, for instance, the retail sector and the hotels and restaurants sector. Also, stronger urban economies generate more tax revenue, which in principle can be used by city managers to improve the quality of public amenities.

References

Armstrong, P. (2001), 'Science, Enterprise and Profit: Ideology in the knowledge-driven economy', *Economy and Society*, Vol. 30, No. 4, November, pp. 524–52.

Atkinson, A.B., L. Rainwater and T.M. Smeeding (1995), *Income Distribution in OECD Countries: Evidence from the Luxembourg Income Study*, Paris: OECD Publications.

Berg, L. van den (1987), *Urban Systems in a Dynamic Society*, Aldershot: Gower.

Berg, L. van den and A.P. Russo (2003), *The Student City: Strategic planning for students' communities in EU cities*, Rotterdam: Euricur.

Berg, L. van den, E. Braun and J. van der Meer (1997), *Metropolitan Organising Capacity; Experiences with organising major projects in European cities*, Aldershot: Euricur Series, Ashgate.

Berg, L. van den, E. Braun and A.H.J. Otgaar (2000), *City and Enterprise – Corporate Social Responsibility in European and US Cities*, Rotterdam: Euricur.

Berg, L. van den, E. Braun and W. van Winden (2001), *Growth Clusters in European Metropolitan Cities*, Aldershot: Euricur Series, Ashgate.

Boekholt, P. (1994), Methodology to identify regional clusters of firms and their needs, Paper for Sprint-RITTS workshop, Luxemburg.

Castells, M. (2000), 'The Information City, the New Economy, and the Network Society', in *People, Cities and the New Information Economy*, materials from an International Conference in Helsinki, 14–15 December.

Castells, M. and P. Hall (1994), *Technopoles of the World*, London: Routledge.

Castells, M. and P. Himanen (2002), *The Information Society and the Welfare State*, Oxford: Oxford University Press.

Cooke, P. (1995), *The Rise of the Rustbelt*, London: UCL Press.

Daniels, P.W. and J.R. Bryson (2002), 'Manufacturing Services and Servicing Manufacturing: Knowledge-based cities and changing forms of production', *Urban Studies*, Vol. 39, Nos 5–6, pp. 929–45.

Duranton, G. and D. Puga (2000), 'Diversity and Specialisation in Cities: Why, where and when does it matters?', *Urban Studies*, Vol. 37, No. 3, pp. 533–55.

Feldman, M. and D. Audretsch (1999), 'Innovation in Cities: Implications for innovation', *European Economic Review*, No. 43, pp. 409–29.

Florida, R. (2000), *The Economic Geography of Talent*, Pittsburgh: Carnegie Mellon University, September.

Florida, R. (2002), *The Rise of the Creative Classs*, New York: Basic Book.

Glaeser, E.L. (2000), 'The New Economics of Urban and Regional Growth', in G. Clark, M. Gertler and M. Feldman (eds), *The Oxford Handbook of Economic Geography*, Oxford: Oxford University Press, pp. 83–98.

Glaeser, E.L., J.A. Sheinkman and A. Sheifer (1995), 'Economic Growth in a Cross-section of Cities', *Journal of Monetary Economics*, No. 36, pp. 117–43.

Hall, P. and U. Pfeiffer (2000), *Urban Future 21: A global agenda for twenty-first century cities*, London: E&FN Spon.

Harrison, G.J. (1994), *Europe and the United States: Competition and cooperation in the 1990s*, Armonk, NY: Sharpe.

Hepworth, M. and G. Spencer (2003), 'A Regional Perspective on the Knowledge Economy in Great Britain – Report for the Department of Trade and Industry', London: The Local Futures Group, http://www.localfutures.com.

Howells, J.R.L. (2002), 'Tacit Knowledge, Innovation and Economic Geography', *Urban Studies*, Vol. 39, Nos 5–6, pp. 871–84.

Jacobs, J. (1961), *The Death and Life of Great American Cities*, New York: Random House.

Jacobs, J. (1984), *Cities and the Wealth of Nations: Principles of economic life*, New York: Random House.

Kearns, A. and R. Paddison (2000), 'New Challenges for Urban Governance', *Urban Studies*, Vol. 37, No. 5, pp. 845–50.

Kelley, M.R. and S. Helper (1999), 'Firm Size and Capabilities, Regional Agglomeration, and the Adoption of New Technology', *Economics of Innovation and New Technology*, Vol. 8, No. 1, pp. 79–103.

Knight, R.V. (1995), 'Knowledge-based Development: Policy and planning implications for cities', *Urban Studies*, Vol. 32, No. 2 pp. 225–60.

Lambooy, J.G. (2002), 'Knowledge and Urban Economic Development: An evolutionary perspective', *Urban Studies*, Vol. 39, Nos 5–6, pp. 1019–35.

Lazonick, W. (1992), *Organization and Technology in Capitalist Development*, Aldershot: Elgar.

Leonard-Barton, D. (1982), *Swedish Entrepreneurs in Manufacturing and their Sources of Information*, Boston: Center for Policy Application, MIT.

Lever, W.F. (2002), 'Correlating the Knowledge-base of Cities with Economic Growth', *Urban Studies*, Vol. 39, Nos 5–6, pp. 859–70.

Malmberg, A., O. Sölvell and I. Zander (1996), 'Spatial Clustering, Local Accumulation of Knowledge and Firm Competitiveness', *Geografiska Annaler*, 78 B(2), pp. 85–97.

Marshall, A. (1890), *Principles of Economics – Volume 1*, London: Macmillan.

Matthiessen, C.W., A.W. Schwarz and S. Find (2002), 'The Top-level Global Research System, 1997–99: Centers, networks and nodality. An analysis based on bibliometric indicators', *Urban Studies*, Vol. 39, Nos 5–6, pp. 903–27.

OECD (2002), *Territorial Review of Helsinki*, Paris: OECD.

Piore, M.J. and C.F. Sabel (1984), *The Second Industrial Divide: Possibilities for prosperity*, New York: Basic Books.

Porter, M.E. (1990), *The Competitive Advantage of Nations*, New York: Free Press.
Reich, R. (1991), *The Work of Nations*, New York: Brandon House.
Schumpeter, J. (1934/1980), *The Theory of Economic Development*, Oxford: Oxford University Press.
Simmie, J. (2002), 'Knowledge Spillovers and Reasons for the Concentration of Innovative SMEs', *Urban Studies*, Vol. 39, Nos 5–6, pp. 885–902.
Simon, C. (1998), 'Human Capital and Metropolitan Employment Growth', *Journal of Urban Economics*, No. 43, pp. 223–43.
Stiglitz, J.E. (1999), 'Public Policy for a Knowledge Economy', London: Department for Trade and Industry and Centre for Economic Policy Research, 27 January.
Stoker, G. (1995), 'Public Private Partnerships and Urban Governance', paper presented to the Housing Studies Association Conference, Edinburgh, July, Dept of Government, University of Strathclyde, Glasgow.
Stoker, G. (1996), 'Governance as Theory: Five propositions', paper presented to Enjeux des Débats sur la governance, Université de Lausanne, November, Dept of Government, University of Strathclyde, Glasgow.
Zweimüller, J. (2000), 'Schumpeterian Entrepreneurs Meet Engel's Law: The impact of inequality on innovation-driven growth', *Journal of Economic Growth*, 5 (June), pp. 185–206.

Chapter 2

Amsterdam

2.1 Introduction

The City of Amsterdam (736,000 inhabitants in 2003) is the capital of The Netherlands. During the 1990s its economy developed into one of the most successful local economies in the Dutch economy. In the 1980s the urban economy in general stagnated, whereas the region of Amsterdam as a whole was developing very well. From the second half of the 1990s onwards in particular, the city did very well. It attracted almost 100 foreign companies per year, amongst which were companies' European headquarters and European distribution centres as well as IT companies. The proximity of the national airport Schiphol, the international orientation, the liberal attitude, the quality of life and the strong tradition in knowledge intensive sectors, like finance, publishing, marketing, IT and telecommunication, all contribute to the favourable conditions. The rate of growth of employment – on average over 3 per cent per year – was higher than the national average, the Dutch economy being one of the most successful in the European Union. In particular, IT and new media activities have been growing at a very rapid pace. Since 2001 the economy of the city has shared the stagnation of both the national economy and of other economies in Western Europe. Moreover, the international level playing field is changing. New competitors have appeared, both in Eastern Europe and in Asia.

In Section 2.2, we will explore the knowledge economy strategies of the local authorities. Then in Section 2.3, we will analyse the foundations of the knowledge economy for Amsterdam. In Section 2.4, we will go into the four knowledge economy activities for Amsterdam. Finally, Section 2.5 presents some conclusions and perspectives.

2.2 Knowledge Economy Strategy of the Urban Region

2.2.1 Knowledge Economy Strategy of the City of Amsterdam

The Economic Development Department considers knowledge infrastructure, knowledge organisations and their interrelations to be part of Amsterdam's

Figure 2.1 Amsterdam in The Netherlands

competitive advantages. It acknowledges, however, that Amsterdam's potential in this field can be used much better. Economic policy frameworks pay a lot of attention to the knowledge economy. The Amsterdam strategy towards the information society takes the form of a *strategy towards a diverse economy, with emphasis on the knowledge-based sectors*. The approach is more bottom-up than top-down and local government focuses on general conditions. A sector-specific approach is therefore not generally set out in writing, although policies lay emphasis on specific activities like IT and life sciences. Generally speaking, the approach is rather to stimulate the actors in the knowledge infrastructure and the business sectors to envisage the relevant developments, to anticipate their future and to take up their relevant roles (van der Meer, 2003). However, since the beginning of the 1990s the Economic Development Department has been actively involved in projects stimulating economic spin-offs from the knowledge infrastructure and in fostering innovations through the dissemination of new technologies among small and medium-sized enterprises (SMEs). A successful example is Amsterdam Sciencepark (ASP). This is a public-private partnership between the City of Amsterdam, the University of Amsterdam, the National Scientific Organisation and the Rabobank. Since its inception in 1992 hundreds of innovative new companies have been established as spin-offs from local knowledge institutions.

Another example is Knowledge Network Amsterdam (Kenniskring Amsterdam, KKA), also established in the first half of the 1990s, its founding fathers being local government, enterprises and universities. An interesting new development is the formation of a new strategy by the KKA, the Amsterdam Innovation Motor (AIM).

2.2.2 The Knowledge Network Amsterdam (KKA)

The KKA aims at knowledge transfer between research, education, enterprises and government. Since its establishment in January 1994, the KKA has developed into a broad platform of individuals from the sciences, education, trade and industry as well as from local and provincial government.

The initiative to start the KKA was taken when studies revealed that a good deal of knowledge is available in the Amsterdam region, but that only a limited part of this knowledge finds it way into product and market development. KKA establishes cooperation and promotes effective use of the existing knowledge infrastructure.

The KKA operates with the commitment of more than 125 local organisations. Depending on their degree of involvement and/or capacity for support, these organisations make financial contributions. Seventeen organisations comprise the core donors (van der Meer, 2003).

Until recently, the activities of the KKA have been focused on networking and information exchange rather than on concrete actions. However, on several occasions in the past the KKA has given support to specific projects. Nowadays the KKA is focusing on broad themes, like information and communications technologies, life sciences and sustainability. There is an awareness that the KKA has to transfer to a more operational framework in order to increase its effectiveness. Some steps within the KKA are now being undertaken to change this situation. KKA thus wants to transform itself into a kind of regional innovation platform that operates in a more proactive way.

In the first instance, KKA did not have enough of a managerial character, while it was focused too much on polytechnics and SMEs. Currently, CEOs from large business companies and academic actors are involved in the organisation. Represented on the daily board of the KKA now are the mayor of Amsterdam, CEOs of the ING bank, the chair of the board of the University of Amsterdam (UvA), the president of the Chamber of Commerce Amsterdam and the Dean of the Faculty of Earth and Life sciences of the Free University of Amsterdam.

2.2.3 The Amsterdam Innovation Motor (AIM)

Inspired by Finnish experiences, Amsterdam is developing the AIM in the hope of accelerating the knowledge economy in the Amsterdam region. The global goals and aims are (KKA, 2003):

- to create an innovative (inter-)national competitive knowledge economy with a matching location climate;
- to start up promising clusters within the region based on excellent knowledge;
- to disseminate the achievements to outsiders who are potentially of interest for the Amsterdam region.

The AIM is still in its infancy. One of the first steps will therefore be the formulation of a challenging mission and a clear vision. Then, the starting position of Amsterdam has to be assessed: what are the qualities of education, research, innovation capacity and the ICT infrastructure in the Amsterdam region? Furthermore, an action programme has to be launched, together with an institutional framework. The level of cooperation should be raised: perhaps courses in team building (as the national funding institute Tekes does in Finland) have to be organised.

2.3 Knowledge Foundations

In this section, the knowledge foundations – as discussed in the theoretical chapter – will be assessed for Amsterdam. They include the knowledge base, the economic base, the quality of life, accessibility, urban diversity, urban scale, and the city's social equity.

2.3.1 Knowledge Base

Within Amsterdam there are two 'broad' universities: the University of Amsterdam (UvA) and the Free University (VU, *Vrije Universiteit*). 'Broad' means that both universities undertake research and education in a wide range of humanities, sciences, social and behavioural sciences. There are also two academic hospitals, the Academic Medical Centre (AMC) and the Academic Hospital VU, related respectively to the UvA and the VU.

Moreover, there are nine polytechnics (*hogescholen*) located in the Amsterdam region, amongst which are two specialising in arts: the *Amsterdamse Hogeschool voor de Kunsten* and the *Gerrit Rietveld Academy* (see Section 2.4.1). They contribute to the creative knowledge base of the Amsterdam region. Both universities merged administratively with a polytechnic.

Table 2.1 shows the number of students in Amsterdam. One quarter of the population is student, of which around 35 per cent follows high-grade education (polytechnic or university). It also shows that there has been a relatively high increase of high-grade students, comparing 2001/02 with 1998/99 (respectively 5 and 7 per cent).

2.3.2 Economic Base

Amsterdam has long been an important pillar of the Dutch economy. Amsterdam is the unchallenged financial-economic, cultural and tourist centre of The Netherlands, and offers a suitable environment for the location of international firms that scores well in the various international place rankings. Economically, the region of Amsterdam is a prominent decision-making centre for The Netherlands, boasting valuable trump cards such as the historical inner city, Schiphol Airport, and the new central business district, the *Zuidas* (Southern Axis).

The most important economic clusters in Amsterdam are (City of Amsterdam, 2002):

- finance and business services. Banks, insurance companies, consultancy and stock exchange companies generate about 150,000 jobs, almost one-third of the employment in Amsterdam. The head offices of two global players in this field, ABN-AMRO and the ING bank, are located at the Southern Axis. To this cluster also belongs a strong concentration of foreign international headquarters;
- port and related industry. Including supplying companies, this cluster generates about 55,000 jobs. Many shipped products are also processed in the port area, generating quite some added value;
- Schiphol Airport and related activities generate about 50,000 jobs;
- ICT and new media generate about 40,000 jobs. Amsterdam has a strong publishing branch. Publishing companies such as Wegener, Elsevier,

Table 2.1 Number of students in Amsterdam (in 1998/99–2001/02)

	1998/99 Absolute	2000/01 Absolute	2001/02 Absolute	2001/02 % of population of Amsterdam	Increase 1998/99–2001/02 Absolute	Increase 1998/99–2001/02 %
Primary education	57,050	57,014	57,105	8.0	55	0.0
Secondary education (avo, vmbo, havo, vwo)	32,416	32,703	32,981	4.0	565	2.0
Vocational training (bol/mbo)	12,822	13,336	13,224	2.0	402	3.0
Polytechnics	30,315	31,530	31,956	4.0	1,641	5.0
Universities	33,968	35,270	36,237	5.0	2,269	7.0
Miscellaneous		13,191	13,256	2.0		
Total	172,653	183,044	184,759	25.0	12,106	7.0

Source: Amsterdam Statistical Department (www.onstat.amsterdam.nl).

Kluwer and VNU are located in the Amsterdam region. Moreover, hundreds of small content-focused new-media companies are located in the inner city;

- tourism and conference facilities generate about 45,000 jobs. The city aims to attract more 'high-grade tourism', directed at cultural events. As a conference city, Amsterdam ranks eighth in the world;
- retail and small-sized companies generate about one quarter of urban employment.

2.3.3 *Quality of Life*

The supply of leisure activities is broad and widely appreciated, and quality housing is available. Thus, it comes as no surprise that Amsterdam is a favourable living location, resulting in high prices for housing. For these reasons, there is a shortage of housing facilities. The relatively expensive housing can be a barrier for attracting students and entry-level knowledge workers. This cannot be easily solved, as expansion options are limited by building constraints. In general, the quality of the living environment in Amsterdam, including the cultural climate and the nightlife, is quite high and an asset in the international competition among cities for top quality business.

As stated, the Southern Axis is becoming the economic centre of the Amsterdam region. According to city plans, this area will also become a major transport hub after becoming the high-speed-train (HST) stopping-place and putting the main infrastructure underground. The process of putting the infrastructure underground and building a substantial share of houses and metropolitan amenities is considered crucial for the quality of life in this area. Moreover, to guarantee sufficient liveliness in this urban area, the city also plans to locate other urban functions such as museums, education institutes and sporting facilities.

2.3.4 *Accessibility*

Amsterdam will be connected to the new high-speed line to Paris by 2007. For the Dutch capital that is a reason to invest in the line's junctions. There will be three HST stations in the Amsterdam region, with divergent development prospects. First, there is the station at Schiphol Airport, which has attracted considerable investment in recent decades. The second HST station is the Central Station (CS), in which much is currently being invested to increase in particular its capacity and quality as a main transport junction. The third is Amsterdam-Zuid (South-Amsterdam station). The area around this future HST station, the Southern Axis, is the economic development location for the region of Amsterdam, where many investments have been planned.

The principal new investment in complementary access to the junction concerns a city railway, the north-south line, which is to connect the CS with Schiphol near the *Zuidas*. Because the area immediately adjoins the ring road, car access is potentially excellent. Nevertheless, some people are concerned about the accessibility if the capacity of the ring road remains as it is now: the A-10 highway ring road as well as in the inner city are regularly quite congested.

The AMS-IX (Amsterdam Internet Exchange: http://www.ams-ix.net/) plays an important role for Amsterdam. An Internet Exchange is a place where Internet Service Providers (ISPs) can interconnect their independent networks and exchange Internet traffic with each other. AMS-IX is one of the largest Internet nodes in the world. The presence of AMX-IX – in combination with a dense fibre network which connects most business locations – is an important location factor for companies operating in the ICT cluster. This is, however, predominantly a marketing advantage, because there is no technical reason for companies to be located in the immediate vicinity of this Internet Exchange.

2.3.5 Urban Diversity

According to Richard Florida (2002), the knowledge economy is dependent on three factors: 1) technology, 2) talent, and 3) tolerance. Amsterdam is doing quite well in attracting talent from all parts of The Netherlands and from abroad. Amsterdam also has a positive track record in the field of tolerance: comparatively, Amsterdam appreciates a diversity of inhabitants. The inner city of Amsterdam is considered by many to be an inspiring environment for knowledge exchange. However, some centres of knowledge-intensive activities are located at less inspiring locations: the Southern Axis and the Sciencepark Amsterdam. However, both areas have plans to attract a wider variety of urban functions, thus potentially making these locations more vibrant.

Almost 48 per cent of the present population in the city of Amsterdam originates from another country (see Table 2.2); much more than the average migrant population for The Netherlands as a whole (10 per cent). They come from 170 different countries, with many different native languages. This imposes issues of social and economic integration as a key challenge for urban development. For instance, the need to improve the quality of the education increases when a growing part of the population is of foreign descent. A substantial part of this group has an education shortfall, and thus has a need for additional education. Participation of these groups in higher education is at a low level, but growing.

A particular characteristic of the Amsterdam demographic development is the immigration of people from industrialised countries (mostly the EU and the western hemisphere). From 1997 to 2001 the population of Amsterdam grew by 2.7 per cent. At the same time the number of inhabitants originating from these industrialised countries grew by 5 per cent. On a yearly basis this part of the population increases with 1,000 people (see Table 2.2). They appear to be attracted by the creative culture of the city, and mainly find employment in the services, banking and ICT sectors (van der Meer, 2003). The increase of these inhabitants outweighs more or less the number of Dutch inhabitants leaving Amsterdam.

The highest number of immigrants, however, come from non-industrialised countries. Moroccans, Turcs and Antilleans are the fastest growing population groups, with growth rates of respectively 20, 16 and 14 per cent. Besides the Dutch, the largest ethnic group in Amsterdam is the Surinamese. With almost 72,000 inhabitants, they form one-tenth of Amsterdam's population.

Table 2.2 Ethnic composition of Amsterdam 1998–2003

Ethnic groups	1998	2003	2003	Increase 1998–2003	
	Absolute	Absolute	%	Absolute	%
Surinamese	70,379	71,984	10.0	1,605	2.0
Antilleans	10,718	12,207	2.0	1,489	14.0
Turks	31,822	36,834	5.0	5,012	16.0
Moroccans	50,853	61,146	8.0	10,293	20.0
Southern Europeans	16,372	17,551	2.0	1,179	7.0
Non-industrialised countries	64,647	83,053	11.0	18,406	28.0
Industrialised countries	67,986	71,116	10.0	3,130	5.0
Dutch	405,398	382,154	52.0	–23,244	–6.0
Total	718,175	736,045	100.0	17,870	2.0

Source: Amsterdam Statistical Department (www.onstat.amsterdam.nl).

The numbers above illustrate that Amsterdam is a multicultural society, a city of many religions and many languages. However, non-nationals are as yet poorly represented in innovation clubs. Nowadays, middle-aged

Dutch-born men dominate these clubs. It is expected that within a decade the majority of the Amsterdam population will be non-nationals. It is in this respect considered to be important to involve them much more in all kinds of knowledge-intensive activities. In addition, it was also stated that women are under-represented in innovation clubs (our list of discussion partners is illustrative in this respect) and that it would be desirable to involve them more in such organisations as well.

2.3.6 Urban Scale

Amsterdam cannot be considered a very large city, having 730,000 inhabitants within the municipality and 1,400,000 inhabitants in the region. The city is, however, a major node within a densely populated urban field of seven million inhabitants, namely the Randstad. The number of interactions between the cities in the Randstad is increasing. Currently, many of Amsterdam's interactions take place within the northern part of the Randstad (Amsterdam-Utrecht). After realisation of the HSL-Southline in 2007, travel time between the centres of Amsterdam and Rotterdam will be reduced to only 35 minutes. It is expected that as a result the number of interactions between these two cities will increase rapidly. Consequently, the Randstad is more and more becoming one coherent region with many interactions between its subcentres. The urban scale of this new relevant region can be considered as large enough to exceed demand thresholds for a large number of amenities and infrastructure which are relevant in the knowledge economy. Most universities and national R&D institutes, as well as the greater part of all business in The Netherlands, is concentrated within the Randstad. It is expected that better connections between the urban regions will foster the further cooperation between knowledge institutions and between knowledge institutes and business.

2.3.7 Social Equity

During the 1980s, the unemployment level in Amsterdam was relatively high. In 1987, it was 15.5 per cent (as against 8.5 per cent in The Netherlands). In 1990, it decreased to 10 per cent, but in 1994 it rose again to 14 per cent (NEI-Kolpron, 2001). Although at the time of writing unemployment in Amsterdam is still at a lower level than it was in 2001 (10 per cent), there has been an increase since the start of 2003. In January 2003 unemployment in Amsterdam was 8.3 per cent and in October 2003, 9.2 per cent (http://www.onstat.amsterdam.nl). This rising unemployment is naturally a threat to social equity in the Amsterdam region, and can be an incentive to take appropriate measures to improve the match between supply and demand for labour. The most important policy step in this respect is considered to be the improvement of the quality of education, since most of the unemployed inhabitants have a low level of education.

Table 2.3 Social benefit percentages in Amsterdam

	2000	2002	2003	Share of population Amsterdam (2003) (%)	Change 2000–03	
					Absolute	Relative (%)
Social security benefits	56,125	49,082	45,038	6.0	−11,087	–20.0
Disabled persons' allowances	9,766	12,567	13,097	2.0	3,331	34.0

Source: Amsterdam Statistical Department (www.onstat.amsterdam.nl).

Table 2.3 shows that 6 per cent of the Amsterdam population depends on social security benefits. In the period 2000–2003, the number of people receiving these benefits decreased dramatically. This can be explained predominantly by a change of local policy to be more stringent in assessing who is entitled to these benefits.

2.4 Knowledge Activities

In the last section the foundations of Amsterdam's knowledge economy have been discussed. Now, we turn to knowledge activities and ask how well the city (and the actors in it) manage to: 1) attract and retain talented people; 2) create new knowledge; 3) apply new knowledge and make new combinations; and 4) develop new growth clusters.

2.4.1 Attracting and Retaining Knowledge Workers

Amsterdam is very attractive to knowledge workers because of its high metropolitan quality of life, its international atmosphere and its good international accessibility, in spite of drawbacks such as the shortage of residences and the relatively high costs of housing compared with other Dutch cities. One of the strengths of Amsterdam is its creative, intellectual and tolerant atmosphere. This is very attractive to knowledge workers, particularly those working in creative industries.

The large variety of internationally-oriented education institutes contributes to the capacity of Amsterdam to attract students easily. There are two universities in Amsterdam, offering a wide variety of disciplines, and nine polytechnics. More than 36,000 students follow education at an Amsterdam university and almost 32,000 students at a polytechnic (see Table 2.1). The universities will be described in the next section on knowledge creation. The relatively broad areas in which education operates are related to the research fields mentioned in that section. Moreover, Amsterdam has five independent international schools, contributing to an international location environment.

The large variety of institutes is also considered by some to be a weak point of Amsterdam as a knowledge region, because as yet Amsterdam does not promote itself in a specific knowledge field. Considering the importance of a multidisciplinary approach in R&D and the very promising future of, for instance, bio-informatics and other combinations of knowledge fields, such as e-science, others claim this variety to be a strong point. Within Amsterdam there is, however, a growing acknowledgement of a lack of contact between regional knowledge institutes. It is therefore considered important to develop good city marketing. This will be done under the name 'Amsterdam Partners'. The common denominator will be 'creativity, commercial spirit and innovation'. Amsterdam has a strong brand name and has many good connotations (City of Amsterdam, 2003). The challenge will be to create a good association between this brand name ('Amsterdam') and several knowledge clusters.

Some education institutes that contribute to profiling Amsterdam as an international knowledge region and to attract knowledge workers are The International School for Humanities and Social Sciences, the Gerrit Rietveld Academy and The Network University. They are described below.

Box 2.1 The International School for Humanities and Social Sciences (ISHSS)

The ISHSS is an independent school within the University of Amsterdam (UvA). In 1984–85 the UvA started cooperating with a university in New York; from then on, some education was given in English. In 1992, an international Masters programme was started. About 600 students take part in this programme. Almost all students come from abroad; hardly any Dutch students take part. ISHSS offers education at MA level. The education programme is pretty tight and rapid: there is an entry selection, a progress check and a graduation deadline. Each year about 900 people apply for ISHSS; 25 per cent of the applicants are accepted. Some 33 per cent come from Anglo-Saxon countries (UK, USA), 33 per cent from continental Europe (including Russia) and 33 per cent from other parts of the world.

ISHSS contributes to attracting (potential) knowledge workers to Amsterdam. The school tries to attract excellent students. By selecting students entering the school, it has a good opportunity to do this. Moreover, its tight study schedule stimulates students to excel. In order to retain these knowledge workers in the Amsterdam region, it will be important to stimulate apprenticeships in regionally-based (business) organisations. Currently, apprenticeships are mostly short-term and often arranged on the basis of personal acquaintance. By better institutionalising them, apprenticeships could be made more productive.

Box 2.2 The Gerrit Rietveld Academy

The Gerrit Rietveld Academy (GRA) is a university of professional education in art and design (see: www.gerritrietveldacademie.nl). The institute has some 750 students and 160 teachers. About 48 per cent of the students come from abroad, with 60 different nationalities represented. The foreign students mostly come from the upper social classes in their country of origin. GRA has a separate programme for some asylum seekers. More recently, GRA has also started focusing on attracting Moroccan, Antillean and African students. As a result of the big number of foreign students the main language in GRA is English. Most of the teachers work for GRA one day a week; the rest of the time they work as professionals in art and design.

Currently, GRA offers two educational programmes, one at BA and one at MA level: autonomous visual arts and design. The institution is in the process of starting a programme of 'Art and Economics' together with 'De Baak' (a management centre run by the employers organisation VNO-NCW). Furthermore, GRA cooperates with the UvA on the subject 'art in the public domain'. This programme is expected to include the organisation of symposia, inviting guest speakers for UvA and GRA and initiating some publications. The Southern Axis area in Amsterdam, in which the renovated building of GRA is located, is the primary research area of the art institute. In this respect there is close cooperation with people from the project office Southern Axis.

GRA wanted to move to a new location in Amsterdam, but the municipality has convinced the institute to stay in the southern part of the city on favourable conditions that will be fulfilled by the municipality. The municipality will pay for the renovation and new buildings.

Box 2.3 The Network University

Network University (TNU: www.netuni.nl) started some five years ago. TNU offers virtual education globally. The courses are taught at 'real' universities and other organisations. The 'real' universities have to make considerable efforts to keep up with the current society, but with virtual systems education programmes can be adapted much more quickly. Firms, NGOs, ministries, etc. are involved in TNU. TNU's roots are in the UvA: many international students want to come to The Netherlands, but the number of education places for these students is limited. The current TNU focus is on postgraduate students; they themselves contribute to improving educational programmes. An example of a TNU course is 'Transforming Civil Conflicts', in which students from around the world participate. Another course, 'Mapping the Future', mainly consists of Dutch students; therefore it is taught in a non-virtual way.

Plans for a media institute

There are plans for a new education institute that offers a programme based on new demands in the knowledge economy: the Amsterdam New Media Institute. This is intended to offer courses at undergraduate level for new media. In this school several relevant disciplines, such as marketing, arts, law, and computer programming, should be combined. The graduates are expected to find work in the ICT sector, the new media, advertising, radio and television, publishing companies, etc. The idea is to start with a Summer School in the summer of 2004.

2.4.2 Creating Knowledge

The number of research institutes is relatively large in Amsterdam. There is therefore quite a large potential for creating new knowledge, which can partly be applied for business purposes. Both Amsterdam universities score well internationally regarding their citation impact score. There are 22 European universities that rank above the world average. Both of Amsterdam's universities are in this list (see Table 2.4); the University of Amsterdam ranks fifteenth, and the Free University ranks twentieth. In addition, most of the universities (7) on this list are from The Netherlands. Thus, a good score for a relatively small country.

The Free University of Amsterdam (VU) has 12 faculties with a wide variety of disciplines, ranging from humanities, sciences, social and behavioural sciences to medicines. To give some more insight into its specific strengths, the VU has formulated seven knowledge areas: (1) System Earth: 'nature, space and raw materials'; (2) Life Sciences: 'from molecule towards life'; (3) Health and Disease; (4) Informatics; (5) Communication, Text and Culture; (6) Juridical and Administrative Issues; and (7) Economy and Society (www.vu.nl). In those seven areas the VU wants to promote itself in research and education and wants to strengthen the relations with the external environment. For instance, regarding life sciences the VU wants to develop further the network with business companies and other research institutes.

Like the VU, the University of Amsterdam (UvA) also offers a broad range of knowledge disciplines. The UvA has seven faculties: (1) Economics and Econometrics; (2) Humanities; (3) Medicine (in collaboration with the AMC-hospital); (4) Social and Behavioural Sciences; (5) Sciences, Mathematics and Informatics; (6) Law; and (7) Dentistry (www.uva.nl). Within those seven faculties there is education and research. On the one hand, the UvA considers its broad education programmes and multidisciplinarity as an important strength. On the other hand, it admits the weakness of not having a clear focus as a knowledge institute. The UvA is therefore aiming for a clearer organisation structure for the students, employees and external relations.

Other research institutes In Amsterdam, there are three research institutes in sciences which are international leaders: CWI, NIKHEF and AMOLF, all located in the Sciencepark Amsterdam (see Section 2.4.3). The Centre for Mathematics and Informatics (CWI) is the National Research Institute for Mathematics and Computer Science in The Netherlands. With 130 researchers, CWI undertakes research in mathematics and computer science and transfers new knowledge in these fields to society in general and trade and industry in particular. CWI is 70 per cent funded by NWO, the National Organisation for Scientific Research. The remaining 30 per cent is obtained through national and international programmes and contract research commissioned by industry (www.cwi.nl).

In the National Institute for Nuclear Physics and High Energy Physics (NIKHEF) research in the physics of elementary particles and their interactions is undertaken by 120 academics. NIKHEF participates with others in the

Table 2.4 European universities with the highest impacts

Universities	Impact score
University of Cambridge	1.55
University of Oxford	1.48
Eindhoven University of Technology	1.40
Technical University of Munich	1.40
University of Edinburgh	1.35
University of Freiburg	1.34
University of Karlsruhe	1.34
University of Twente	1.34
Erasmus University Rotterdam	1.32
University of Heidelberg	1.32
University of Strasbourg 1	1.32
Catholic University of Louvain-la-Neuve	1.30
University of Helsinki	1.29
University of London	1.29
University of Amsterdam	1.25
Leiden University	1.25
Delft University of Technology	1.24
Technical University of Denmark	1.24
University of Stuttgart	1.24
Free University of Amsterdam	1.22
Karolinska Institute	1.22
University of Antwerp	1.20

Source: Third European Report on S&T Indicators, 2003.

preparation of experiments at the Large Hadron Collider at CERN in Geneva. More than half of the academic staff are PhD students and postdoctoral fellows. Technical support is provided by well-equipped mechanical, electronic and information technology departments with a total staff of about 100. NIKHEF collaborates with four Dutch universities (UvA, VU, Nijmegen and Utrecht) (www.nikhef.nl).

In the Institute for Atomic and Molecular Physics (AMOLF), research is focused on the understanding and manipulating of the physical attributes of material. It presently focuses an important part of its research on the physics of living matter. AMOLF is an institute of approximately 100 scientists. An important goal of the institute is the training of scientists and technical engineers for advanced research (www.amolf.nl). Both NIKHEF and AMOLF are active in the area of nanotechnology.

Another important institute for Amsterdam is SARA, the National High Performance Computer and Networking Center. SARA is an independent organisation with offices in Amsterdam and Almere, and collaborations with many partners. The most eye-catching activities and facilities of SARA are the national supercomputers TERAS and ASTER, the Amsterdam Internet Exchange (AMS-IX), the academic high-bandwidth network SURFnet5 and the virtual reality facility, the CAVE. An important addition to the available facilities is SARA's expertise and the resulting consultancy and support activities.

Moreover, in North-Amsterdam there is the Research and Technology Centre of the multinational company Shell which is engaged in oil processing, the petrochemical industry and new technologies for sustainable energy. There are also plans to develop a second Sciencepark focused on all aspects of sustainability, the proposed partners being both universities, ECN (national R&D institute for energy) and NUON, a large energy company.

An important institute in Amsterdam for creating new knowledge in life sciences is Sanquin. This is a not-for-profit organisation, affiliated with the Amsterdam Medical Centre (AMC). It was set up in 1998 after a merger of the Blood Banks and the Central Laboratory of the Dutch Red Cross Blood Transfusion Centre. Sanquin's research programme is focused on its mission to ensure the Dutch blood supply and to advance transfusion medicine meeting high standards in quality, safety and efficiency. The organisation has a yearly turnover of €130–140m. Some 400–500 people work in diagnostics and research; this involves both academic and entrepreneurial activities. The financial base of Sanquin consists of supplying blood. The profits made on this activity are surplus value. The institute also obtains money from NWO and the Cancer Foundation; these subsidies are won through open competition. About five years ago Sanquin started facilitating some start-up firms. Sanquin wanted a science park, but the biotech market collapsed. Therefore, the construction of 6,000 m^2 in laboratory space has been put on hold.

Sanquin is located next to the NKI; The Netherlands Cancer Institute. This is a large knowledge centre that belongs to the top organisations in the world. Just like Sanquin, the NKI also combines science with entrepreneurial activities. Sanquin and NKI together form a kind of bio-cluster on the west side of the city. Research in the NKI covers all major areas of cancer research, with special emphasis on mouse models, mouse (reverse) genetics, cell biology, immunology, translational research requiring close collaboration between clinical and basic scientists and epidemiology and psychosocial cancer research.

2.4.3 Applying Knowledge/Making New Combinations

To be and remain competitive in the globalising economy, business companies in the Amsterdam region should provide products with high added value to compensate for high wages. These should predominantly be products that are close to the end-user and products for which users are willing to pay a relatively price. Good examples are products associated with health: for instance, the butter brand 'Pro Active' instead of 'Becel'. New products are expected from life sciences particularly which will be associated with health and well-being.

In order to be more successful in applying knowledge to new business products, the entrepreneurial attitude in The Netherlands, including the Amsterdam region, has to change. For some decades the general attitude of most scientists was favourable towards fundamental research, but less so towards applied research. There was too much risk aversion and there are too few entrepreneurs. Education institutes, amongst others, should pay more attention to developing entrepreneurial skills. The awareness of this problem has been growing for several years and action has been taken to improve attitudes and help scientists to develop entrepreneurial skills. Moreover, there is a need for financial incentives to speed up cooperation. Though there is a widespread awareness that broadly supported plans for strategic cooperations with clear economic perspectives function better than artificial cooperations mainly based on subsidies, there is still a need for incentives to stimulate new initiatives for mutual cooperation. This can be compared with the success of the VOC company during the seventeenth century. This success was particularly due to the willingness of the involved actors to *invest* in new activities. Not all activities were successful; but the operation as a whole was very profitable. Currently, a situation in which public as well as private actors are willing to invest in promising knowledge-intensive activities and are not afraid of some inevitable failures is also desirable.

Links between organisations It appears that there are too few links between large and small companies, though there is a large (potential) need for these relations. The creativity of small companies is needed by larger firms to helpt them to stay successful, while small companies need larger ones to help them tomarket products worldwide. Thus, in order to realise new market combinations, interactions between small firms and multinational firms could be stimulated.

A good example of close cooperation between scientific organisations and business is a programme on IT, new media and life sciences. These R&D

institutes are located at Sciencepark Amsterdam and work together with large firms like IBM, KPN, VNU and AKZO. The programme has been co-financed by central government (ICES-KIS, BSIK). However, generally speaking there are too few links between companies and education, though here also there is a great (potential) necessity for these links. First of all, there is considerable need for a better matching of regional education and regional business activities. Secondly, there is a need for internships at education institutes as well as in business organisations. Thirdly, there is a need for education maintenance programmes (life-long learning) and post-Masters education.

Sciencepark Amsterdam The Sciencepark Amsterdam is a cooperative initiative among different scientific organisations, university research and teaching institutions, entrepreneurs and multinationals in the fields of ICT and biotechnology. It was initiated in 1989 by the University of Amsterdam, the NWO, the Rabobank and the Economic Department of the City of Amsterdam in a public private partnership.

The Sciencepark Amsterdam offers an academic and research infrastructure, as well as state-of-the-art ICT infrastructure. The area brings together research in physics, chemistry, astronomy, geography, biology, informatics and computer science. The Sciencepark Amsterdam has the goal of exchanging knowledge among the scientific institutions on the one hand, and with the business and industrial community on the other. It thus has to contribute to the application of knowledge within the Amsterdam region. The joint approach strengthens the existing knowledge infrastructure and produces knowledge-intensive products and activities.

ASP is an initiative of the city of Amsterdam, the Dutch organisation for Scientific Research (NWO), a bank (Rabobank) and the University of Amsterdam (UvA). It manages and leases the Matrix businesss complexes. Since its foundation in 1989 five business complexes have been realised:

- Matrix 1: the largest building. It was the first accommodation for starting entrepreneurs;
- Matrix 2: the building for starting and growing businesses;
- Matrix 3: larger companies can be found in this building;
- Matrix 4: the University of Amsterdam is accommodated in this building;
- Matrix 5: the Biopartner Center houses innovative businesses, mainly in the biotechnology sector.

Start-up policy Start-up policy is one of the activities of the Economic Development Department. Start-ups are considered important for the Amsterdam economy because they create employment, stimulate innovations and refresh the composition of business life in Amsterdam. However, the local government is aware of the danger of distortion of competition. This is the main reason why support is only offered during the start-up period. Maturing companies are supposed to be able to help themselves.

2.4.4 Developing New Growth Clusters

Geographical clusters The Economic Development Department of the City of Amsterdam distinguishes five knowledge clusters in the region which should be further developed: 1) the University of Amsterdam (UvA) – NWO – Sciencepark Amsterdam; 2) The Academic Medical Centre (AMC); 3) The Netherlands Cancer Institute (NKI)/Sanquin; 4) The VU University/ Academic Hospital; and 5) Shell Research – sustainability (within the Shell area, Shell, ECN and the university are planning to invest in a business park for sustainability) (City of Amsterdam, 2002).

It must be stated that the clusters above are first of all *physical* clusters. They can be considered as nodes on the A10-highway ring around Amsterdam: number 1 is on the east side, 2 on the south-east side, 3 on the west side, 4 on the south side and 5 on the north side. The clusters above are historically developed in island-like economic clusters. It might, however, be desirable and even necessary to define mutually dependent *functional* clusters, which can operate in a competitive way in the international market.

A strong knowledge-intensive cluster at the time of writing is finance and business services (about 150,000 jobs). This cluster is particularly regionally and nationally important, but internationally it has lost its prominent position. Although there are the head offices of two important global players (the ABN-AMRO and the ING bank) and a broad variety of foreign banks, Amsterdam has lost position vis-à-vis strong European financial clusters in London and Frankfurt. Other important economic clusters are the port and related industry (55,000 jobs, Schiphol Airport (50,000 jobs), tourism and conferences (45,000 jobs) and IT and new media (40,000 jobs) (City of Amsterdam, 2002).

Within the AIM cooperation choices are made regarding those *functional* knowledge-intensive clusters which should be further strengthened in the Amsterdam region, namely life sciences, ICT and new media, and sustainability. These categories of economic activities are, however, quite broad. Many urban regions want to promote themselves in these fields;

the current definitions appear not to be sufficiently distinctive in the newly emerging playing field of knowledge cities. Therefore, one may argue that a better focus seems to be desirable. However, according to the Economic Development Department specialisation does not contribute to innovation as much as does a broad scientific orientation. It is argued that most innovations in fact are a combination of insights from different fields of investigation. On the other hand, life sciences include quite a large number of subfields (see description below), and sustainability includes even more. For this *knowledge domain* the following themes have been mentioned: water, energy/mobility, corporate governance and risk reducing.[1] Much is expected from a new business park on sustainability in Amsterdam. But the plans for this park are not yet developed far enough to assess its opportunities. The life sciences and ICT and new media clusters are described below.

ICT and new media One of the main factors driving the ICT cluster in Amsterdam is the presence of a large number of customers in the different sectors of the economy. One could say that the conglomerate of financial and commercial headquarters in Amsterdam constitutes the local engine. The concentration of knowledge and infrastructure in the Sciencepark Amsterdam gives fuel to the engine. Another factor is the highly concentrated urban environment, which is a favourable place for small firms in ICT/new media, whose productivity depends on a high level of interaction with other firms in the ICT sector (for co-production) and a high level of interaction with firms in the other sectors (e.g., as customers). As many as 3,000 such small firms have been counted in the most densely built inner city district of Amsterdam (van der Meer, 2003).

In 2000 Amsterdam succeeded in attracting the large ICT company CISCO. In addition, many of the large service enterprises in the region (e.g., Schiphol Airport, KLM airlines, ABN/AMRO banking) have in-house ICT facilities, which in some cases have more than 1,000 employees (van der Meer, 2003).

The new media cluster is rather sensitive to the economic climate. About 90 per cent of the people work on a freelance basis. The new media sector can be characterised as quite international, small scale, and with a liberal attitude. These firms are attracted particularly by the Amsterdam image. New media companies are developing everywhere in The Netherlands, but their growth is highest in Amsterdam. There are several network organisations for new media entrepreneurs, such as the Amsterdam New Media Association (ANMA), the Interactive Professionals Association Netherlands (IPAN) and the Internet Society (ISOC, an international organisation). They play a role in better profiling this economic cluster of the Amsterdam region.

Life sciences cluster Life sciences can be defined as the science of molecular, cellular and development processes that create and determine a living organism (Universiteit van Amsterdam, 2003b). The life sciences cluster in Amsterdam is the biggest one in The Netherlands: some 4,000 people work in the cluster, about 6,000 students are registered in life sciences. Organisations active in life sciences are, amongst others, the VU, the UvA, the two academic medical centres, the Biosciences Incubator in the Amsterdam Sciencepark, The Netherlands Cancer Institute, the Central Laboratory of Netherlands Red Cross, and the physics and informatics institutes AMOLF and CWI. Important life sciences disciplines in the Amsterdam region are Molecular and Cellular Biology, Biomedical Sciences, Biotechnology, Genomics and bio-informatics, and Computer Sciences and image analysis.

A key institute for stimulating new life sciences companies is Biopartner, a public foundation aiming to support starting biotech companies. Biopartner was launched in 2000 by the Dutch Ministry of Economic Affairs, initially for the period 2000–2004. In total, €45.4m was allocated for stimulating entrepreneurship and the entrepreneurial climate for bio-starters and bio-businesses in The Netherlands. It is, however, unclear whether there will be a follow-up programme for Biopartner.

Amsterdam does not have a strong life sciences *image*, notwithstanding the large variety of life sciences institutes. Major reasons for this appear to be the insufficient promotion of innovations that come out of R&D within the region and, until recently, the limited number of spin-off companies. There used to be a lack of synergy between organisations working in life sciences but this is starting to change due to the efforts of the mayor of Amsterdam. Until now the region has lacked a large firm in this sector. However, within the wider region there are some relevant large companies, such as Crucell in Leiden, Sovay in Weesp and food-processing companies in the Zaanstreek. The relevant region of Amsterdam in which daily economic activities take place becomes larger and larger and broadly exceeds the administrative borders of the city (see also the section on *urban scale)*. The above-mentioned areas can be considered as becoming part of the relevant region of Amsterdam, considering the good transport connections between Leiden, Weesp and Zaanstreek and Amsterdam.

Recently Crucell bought a UvA-related company, ChromaGenics, which researches gentechnologies. This company was created in 2000 with the help of, amongst others, Biopartner.[2] Crucell paid more than €4m for the company, because it considers the developed technologies of the company highly promising. Eventual new patents from ChromaGenics will go to Crucell. This is not considered as a large problem for the UvA, as patents are not seen as an aim themselves, but rather as a means for attracting companies to invest. This transaction can be considered as a good example of the interest of business companies in creating knowledge in life sciences at universities.

The actors involved in the life sciences cluster in Amsterdam are quite positive regarding the high market potential of the life sciences cluster, provided that there is more interaction, and more attention on commercial applications and image building.

2.5 Conclusions and Perspectives

2.5.1 Knowledge Economy Strategy

Although Amsterdam has a strong economic position within The Netherlands the city is facing increasingly fierce international competition in knowledge-intensive activities. A growing awareness of the importance of stimulating the knowledge economy can be noticed. But there are some doubts whether there is a sufficient *sense of urgency* in the Amsterdam region to take the appropriate necessary steps in order to *really* improve its competitiveness in the globalising economy.

Until now, the city has not had a clear vision of how Amsterdam has to develop in the knowledge economy. There is no clear *leadership* in the process towards a knowledge economy and risk-avoiding behaviour can be noticed. In addition, there appears to be a culture in the Amsterdam region of not cooperating coupled with a relatively large distance between the private and the public sectors.

As yet, there is no explicit codified knowledge economy strategy in Amsterdam. An important institution in the field of the knowledge economy could be the Knowledge Network Amsterdam (KKA). Until now it has operated mainly as a discussion and exchange forum. The KKA now intends to become more operational. However, an institutional approach rather than an entrepreneurial approach can still be observed. This is meeting with a considerable amount of criticism.

2.5.2 Foundations of the Knowledge Economy

The *knowledge base* of Amsterdam can be assessed as being good (see Table 2.5). There are many (potential) strong assets: two universities, two academic hospitals and a relatively large number of research institutes. The available potential could be further strengthened by more cooperation and by focusing on research fields with the potential to excel. The *economic base* of Amsterdam can be described as good. There are some important economic clusters, such as finance and business services, ICT and new media. *Quality of life* in Amsterdam is very good. The region is very attractive to knowledge workers. Areas such as the inner city, old south and IJ-banks are considered very attractive locations. *Accessibility* can be qualified as very good. Strong trump cards are Schiphol Airport and the presence of AMS-IX (Internet exchange), with new accessibility opportunities offered by the connection to the high-speed line to Paris and the construction of the north-south metro line. Amsterdam is considered to be a *well-diversified* city, with its mix of many business activities, education, arts and music and inhabitants. Almost half of the Amsterdam population is of foreign descent; the city is a multicultural society with many religions and languages. Currently the *urban scale* of Amsterdam is not very large. But the *Randstad* is transforming gradually into one coherent region. This level is very attractive for accommodating all kinds of facilities and amenities relevant for knowledge economy activities. *Social equity* is qualified as moderate. Unemployment is rising and could become a threat to stability in the city. As in many other large European cities, in Amsterdam there are considerable differences in welfare and opportunities between nationals and non-nationals. It is considered to be one of the major challenges to involve non-nationals better in knowledge-intensive activities (and in creative activities). Women are also still under-represented in these activities.

2.5.3 Activities of the Knowledge Economy

The urban region of Amsterdam is relatively strong in *attracting and retaining knowledge workers* (see Table 2.6). One of the strategic strengths of Amsterdam is its creative, intellectual and tolerant atmosphere. The city has an enormous attraction for (creative) knowledge workers from within The Netherlands as well as from abroad. Consequently, in Amsterdam there is quite an international environment. There are many education institutes in the region which play an important role in attracting new (potential) knowledge workers. The large

Table 2.5 An indication of the value of knowledge economy foundations in Amsterdam

	Foundations	Score in Amsterdam
1	Knowledge base	++
2	Economic base	+
3	Quality of life	++
4	Accessibility	++
5	Urban diversity	++
6	Urban scale	+
7	Social equity	□

Note: -- = very weak; – = weak; □ = moderate; + = good; ++ = very good.

Table 2.6 An indication of the value of knowledge economy activities in Amsterdam

	Activities	Score in Amsterdam
1	Attracting knowledge workers	++
2	Creating knowledge	+
3	Applying knowledge/making new combinations	□
4	Developing new growth clusters	□

Note: -- = very weak; – = weak; □ = moderate; + = good; ++ = very good

number and variety of knowledge institutes makes it difficult to formulate a clear focus for the Amsterdam knowledge economy. It is therefore difficult for Amsterdam to represent itself internationally as a knowledge city. A well-developed city-marketing strategy appears therefore to be desirable. Some first steps in this field have been taken.

The capacity of creating knowledge within Amsterdam is judged as being good, considering the large number and quality of research institutes. Both universities rank in the top 20 of European universities (citation impact score). Both are aiming for a clearer focus for their research and education activities. They could thus contribute to the desired regional focus. Within Amsterdam there are some internationally leading institutes in sciences, IT and life sciences. They are considered to form a good base for the Amsterdam knowledge economy. However, an important precondition will be good cooperation with relevant (market) actors.

Amsterdam appears not to have a well-developed environment for *applying knowledge*, as the culture of cooperation appears to be inadequate. Although there are a limited number of new combinations of practical market-driven activities and research activities (like the Sciencepark Amsterdam), more cooperation between relevant partners could lead to better and more innovative products. In particular, there could be better cooperation between small and large companies and between business companies and education and research institutes. It was stated that one of the preconditions for better cooperation is a more efficient and effective functioning of the city administration. Moreover, there should be a more entrepreneurial attitude and there is a need for investing in promising knowledge economy activities.

In Amsterdam, there is a need for developing some clusters with knowledge economy activities which *excel*. Promising *growth clusters* appear to be ICT and new media and life sciences. For the latter cluster, a large industry that could function as an economic engine is lacking, although there are some relevant companies within the wider region. The clusters defined by KKA, ICT and new media, life sciences and sustainability are relatively broad. A better focus on the promotion of specific strong fields within these clusters appears to be desirable in order to promote the region of Amsterdam better in relation to competing urban regions.

2.5.4 Perspectives

At the moment, Amsterdam's position is still rather good, particularly within the Dutch economy, but if the city does not take the proper action it could well fall in the knowledge economy rankings. The sense of urgency to undertake appropriate steps for approving the position of the Dutch capital city is growing, but there still appear to be countervailing powers.

Within the current international competition it is no longer sufficient to be good at something; it is necessary to excel in a certain field. Amsterdam is, for instance, good in sciences, music, art and finances. But it is not at the top in those fields. Amsterdam should thus develop one or more fields in which it can position itself in the top league. This should be done after close examination in which fields other competing cities (can) excel. This should lead to a better focus of the current choices of knowledge-intensive economic clusters.

In addition, it will be important that a sufficient variety of promising knowledge activities remains within the urban region. The actors in the knowledge economy debate cannot choose all the winning sectors in advance, because such developments cannot be (fully) predicted. They should, however, create the right conditions for new knowledge economy activities to excel. One of the most important of these conditions is (continuously) improving the quality of the regional education institutes. This concerns also, and urgently, the quality of the primary and secondary schools.

One of the promising clusters is the life science cluster. Several actions can be undertaken to strengthen this cluster. In the near future, the interaction between research groups has to be increased, the image has to be improved, and a large life sciences company could be attracted (or actors in the Amsterdam region could form such a firm themselves). Within life sciences, a strategic cooperation partnership could be set up related to a certain field. Important fields could be cancer, brain research, infection research and bio-informatics. There should be a good self-organising capacity to achieve such a partnership.

An important actor in the transition process of Amsterdam towards the knowledge economy could be the Knowledge Network Amsterdam. In that case, it should act more proactive instead of mainly exchanging information. The Knowledge Network could, for instance, operate as a *broker*, bringing knowledge parties together. This should focus on actions to bring knowledge-creating actors together (such as university organisations or relatively small young companies) and actors who are able to market knowledge on global scale (such as multinational companies). Moreover, there should be more awareness of the relevance of good public relations. Successes should receive more attention. To achieve all this, there is an urgent need for strong leadership to accelerate the process of Amsterdam towards the knowledge economy.

Notes

1 Source: Het Amsterdamse Innovatie Model (AIM); de startnotitie.
2 NRC Handelsblad, 30 March 2003, 'Geld en prestige in ruil voor een goed idee, Crucell koop biotechbedrijfje van Universiteit van Amsterdam'.

Discussion Partners

Mr Henk Eppink, Director Kenniskring Amsterdam.
Mr Frank de Graaf, Policy Advice and Regional Stimulation, Chamber of Commerce.
Mr Tijmen van Grootheest, President, Gerrit Rietveld Academy.
Mr Rob Hagendijk, Dean, International School for Humanities and Social Sciences, ISHSS.
Mr Frits Hermans, General Manager, Shell Research and Technology Centre, Amsterdam.
Mr Walter Hoogland, Director, ASP and Dean, Faculty of Physics, Informatics and Mathematics FNWI, University of Amsterdam (UvA).

Mr Gerd Junne, The Network University, Professor of International Relations, University of Amsterdam (UvA).
Mr Willem Kleyn, Deputy Director and Head of Research, Economic Development Department, City of Amsterdam.
Mr Frank Miedema, Sanquin Research and Education.
Mr Frans Nauta, Chair, Knowledge Land Foundation (Stichting Nederland Kennisland).
Mr Peter Odermatt, Head of Regional Economic Affairs, Chamber of Commerce.
Mr Appie A.J. Reuver, Director of External Relations, IBM Netherlands.
Mr Peter Rijntjes, former Director of NEMO.
Mr Arjen J. van Tunen, Director, Faculty of Science, Swammerdam Institute for Life Sciences, board member of Biopartner network.
Mr Ton Veldhuis, private advisor (Multimediair), Amsterdam New Media Association.

References

Bureau Bartels B.V. (1996), *Kennis als Strategische Kracht; Een Beleidsvisie voor Amsterdam*.
City of Amsterdam (2002), *Hermez, Het economisch resultaat moet er zijn, Beleidsprogramma 2002–2006*, Amsterdam.
City of Amsterdam (2003), *Choosing Amsterdam:Brand, concept and organisation of the city marketing*, Amsterdam.
Florida, R. (2002), *The Rise of the Creative Class*, New York: Basic Books.
Kamer van Koophandel en Fabrieken Amsterdam (2003), *Ondernemen met Kennis, De topkennisagenda van het bedrijfsleven in de regio Amsterdam*.
Meer, A. van der (2003), 'The Amsterdam Region ICT Cluster', in A. van der Meer, W. van Winden and P. Woets (eds), *ICT Clusters in European Cities during the 1990s – Development Patterns and Policy Lessons*, Rotterdam: Euricur.
NEI-Kolpron (2001), *Naar een ruimtelijk-economische beleidsvisie voor Amsterdam*, Rotterdam.
Sanquin (2003), *Scientific Report 2002*, Amsterdam: Spinhex and Industrie.
Stichting Nederland Kennisland (2003), *Plan van Aanpak – Amsterdams Innovatie Model*.
Universiteit van Amsterdam (2003a), International School for Humanities and Social Sciences (IHSS), Masters and Certificate Programmes, Amsterdam.
Universiteit van Amsterdam, Swammerdam Institute for Life Sciences (2003b), *Annual Report 2002*, Amsterdam.

Chapter 3

Dortmund

3.1 Introduction

Dortmund is located in the industrial Ruhr region. It is situated in the state of North-Rhine Westphalia (NRW) and is the economic and cultural capital of the Westphalia region. The city is famous for its historic background, which includes coal and steel manufacturing and several famous beer breweries. The city has a population of 588,000 people (Stadt Dortmund, 2004a). Dortmund is the sixth largest city in Germany; it is the second largest city (after Cologne) of the metropolitan Rhine-Ruhr agglomeration, which has 11.5 million inhabitants. Other large cities in the Rhine-Ruhr region are Bonn, Essen, Düsseldorf and Duisburg. The region is a typical polycentric area with no apparent hierarchy of cities.[1] The 'Ruhrgebiet' (5.4 million inhabitants) covers almost half of the Rhine-Ruhr region. The area is covered by the KVR: the Association of Local Authorities in the Ruhr Region. This association is a public body, which develops and promotes regional initiatives. The more narrowly defined Dortmund region is made up of Dortmund, Unna and Hamm, because this area coincides with the district of the Dortmund Chamber of Industry and Commerce. Most of Dortmund's cross-border economic activities are orientated as much towards the smaller towns in the more rural east, e.g. Unna and Hamm, as they are towards the cities of Bochum and Hagen, which are also close to Dortmund (Brödner and Rücker, 2004).

Dortmund's unemployment level is relatively high compared to other major cities in Germany (see Figure 3.2). The main reason is the demise of the coal and steel industry in the city which has resulted in a large decrease in employment. Between 1980 and 1997 the industrial sector lost some 60,000 jobs. In the same period the number of jobs in the tertiary economic sector (services) increased with 23,300 jobs. The number of unemployed people increased (Wirtschafts- und Beschäftigungsförderung Dortmund, 1999).

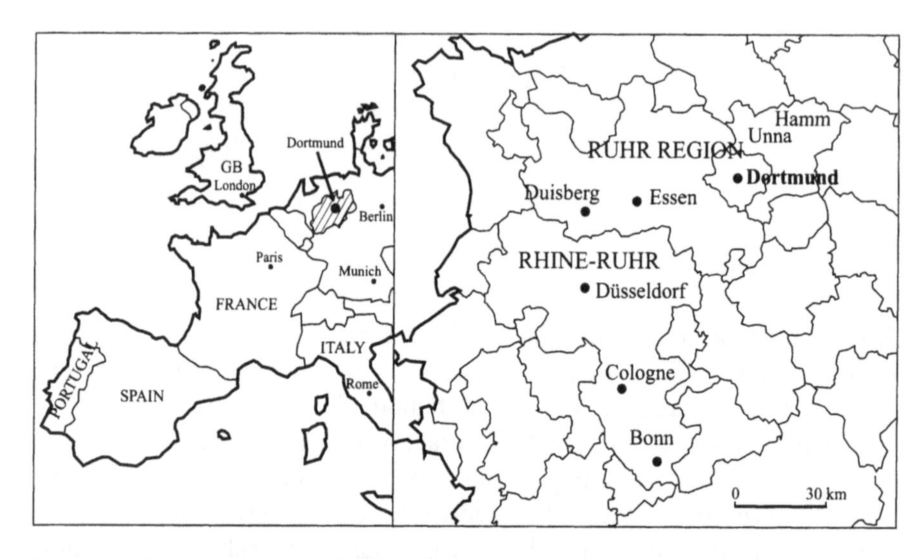

Figure 3.1 Dortmund in Europe and in the Ruhr region

Source: www.dortmund-project.de and Brödner and Rücker (2004).

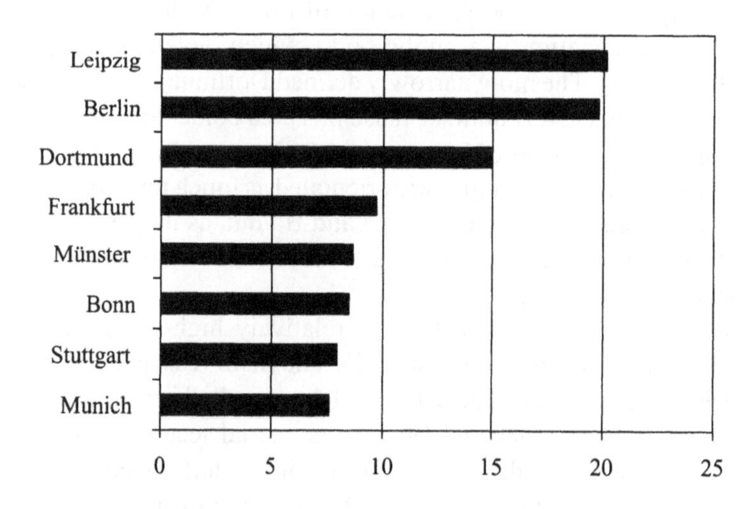

Figure 3.2 Unemployment rates in large German cities, 2003 (%)

Source: Wirtschaftswoche and IW Consult (2004).

3.2 Knowledge Strategy

In the 1960s Dortmund had some 38,500 employees in the mining industry and around 36,600 steel workers. Since then employment in these sectors declined heavily as a result of increasing productivity: in 2003, they employed only 300 and 2,700 workers respectively. Employment in construction companies and beer breweries also declined sharply (Wirtschafts- und Beschäftigungsförderung Dortmund, 1999). In 1998 the steel consortium Thyssen Krupp announced plans to shut down the production of steel in Drotmund completely in 2001. To soften the economic and social impacts of these major declines, the Economic and Employment Promotion Dortmund (Wirtschafts- und Beschäftigungsförderung: WBF), together with Thyssen Krupp, consultancy firm McKinsey, the chamber of commerce and DGB, started an investigation in 1999 to find out how substitute employment could be created. The main research goal was to identify the core competencies of the Dortmund economy, since developing these strengths should lead to significant increases in employment. This research led to the dortmund-project which was accepted by the council of the City of Dortmund in June 2000.

The dortmund-project public-private partnership consists of three innovation spearheads: information technology (IT), micro-electronic mechanical systems (MEMS) and e-logistics. As well as aiming to strengthen these fields, the project should also enhance general economic structures and the image of the city. The dortmund-project is planned to run from 2001 to 2010 and has six goals (dortmund-project, 2000; Küpper presentation, 2004):

1. building new leading economic sectors (IT, MEMS and e-logistics, among others) by stimulating start-up firms and attracting foreign companies;
2. strengthening local companies (this should happen as a result of building new leading economic sectors);
3. enhancing education, qualifications, sciences and research to international standards (to provide new economic sectors with sufficient workers);
4. upgrading Dortmund to a modern economic metropolis by strengthening the quality of life (for example, by enhancing opportunities for leisure activities);
5. creating swift relations for companies by significantly increasing the speed and efficiency of public administration procedures and supporting public private partnerships;
6. significantly increasing employment. When the dortmund-project was started, the goal was to create 70,000 new jobs: 60,000 in new leading

economic sectors (34,000 in IT, 16,000 in MEMS and 10,000 in e-logistics) and 10,000 in already existing branches by investment from the municipality (€65m) and private firms eventually amounting to approximately €500m.

Currently, it is acknowledged by WBF, the municipality and private partners that the last target will not be reached, since the economic boom at the end of the 1990s collapsed at the beginning of the new millennium. However, an official review of results and targets is planned for 2005.

The dortmund-project consists of three fields of action: people and competences; future fields of business; and future locations (see Table 3.1). The fields of action will be discussed in the following paragraphs.

Table 3.1 Dortmund-project – fields of action

People and competences	Future fields of business	Future locations
University city Dortmund	Start 2 Grow: start-ups and company growth	Technologiepark
IT-centre	Locate IT: settlement and founding	City
JOY education campaign	MST Factory: micro- and nano-technology	Stadtkrone Ost
LEAD: centre for personnel management and firm development	TechnologyCenterDortmund and Electronic Commerce Centre	Old airport Dortmund-Brackel Westphalia blast furnaces
Learning alliances	E-port Dortmund (logistics)	Phoenix

Source: Küpper presentation (2004).

The actors in the dortmund-project can be divided in to five pillars (Küpper presentation, 2004):

1 the *city council and committees* decide on main project themes. The dortmund-project reports regularly to the WBF and main city committee;
2 the *mayor and city management* control and accompany the dortmund-project;

3 the *project board* gives advisory support to the *project management*. The chair of the board is the WBF director. Sessions take place once a month;
4 the *steering committee* takes care of inter-regional communication of dortmund-project goals and main projects and accompanies project advancement. It also reflects on the development of joint interests and changes in framework conditions. In addition, it supports projects within the dortmund-project that have a regional dimension. The mayor of Dortmund is chair of the steering committee;
5 the *business sector* invests in location development and participates in so-called infrastructure companies (which stimulate start-ups and enhance growth of other firms) and financially supports the Dortmund Foundation ('Dortmund Stiftung') which co-finances the infrastructure companies. Furthermore, private firms participate in the joint network of the dortmund-project.

The city of Dortmund also tried to promote its development as a knowledge economy by participating in a competition for German cities to become the City of Science ('Stadt der Wissenschaft') 2005. Many projects within the bid aimed to increase interaction between the Dortmund citizens and the science and other knowledge institutions within the city by making the population more aware of the activities that are deployed in scientific education and research. The City of Science proposal also intended to increase linkages between science institutions (Stadt Dortmund, 2003a). Dortmund did not win the competition, but it received good reviews and finished in rank 4–8 among 37 participants.

3.2.1 State of North-Rhine Westphalia

In the Federal Republic of Germany, the regions (Länder) are mainly responsible for R&D and education policy. The regions have a large influence on some sectors like media and health/biotechnology. The region of North-Rhine Westphalia (NRW) is implementing the Technology and Innovation Programme (TIP). This is the follow-up to the Technologieprogramm Wirtschaft. TIP stimulates the development of future technologies and promotes technology transfer in the economy. TIP aims to support SMEs, among others: this provides good opportunities for Dortmund, since many firms in this region are SMEs. The 'Modellversuch Inkubator', for example, aims to stimulate top researchers to start up firms particularly in the field of biotechnology (Forschungszentrum Jülich GmbH, 2003a and b).

Another initiative financially supported by NRW is the Ruhr-Networker: a network of some 200 ICT companies from the entire Ruhr region. Ruhr-Networker organises meetings and events (e.g. trade fairs) to bring local ICT companies and customers within the region together. Ruhr-Networker is financially supported by Projekt Ruhr, an NRW financed agency that stimulates structural change in the region (Brödner and Rücker, 2004).

3.2.2 EU Support

For many years Dortmund has received financial help from the European Union to structurally renew its industrial economy. The city is an Objective-2 area and will receive money at least until 2006. The European Commission approved a programme for urban regeneration in Dortmund under the URBAN II Community Initiative Programme. This will provide for €9.9m over the period 2000–2006. The EU funding has attracted some €18m in further investment from the public sector and €0.8m from private companies, bringing the total amount to €28.7m. The goals of the project include strengthening the local economy in the Nordstadt neighbourhood (European Union, 2004). (An elaboration of the Objective-2 programmes in Dortmund can be found on the website www.wbf-do.de in the downloads section: in a lecture at the European Structure Funds Conference in Dortmund on 25–26 September 2003, Dr Küpper discussed the positive effects of EU funding in Dortmund projects.) At the time of writing, Dortmund is trying to secure EU Structural funds for the period 2006–2013. The economic goals and projects within the dortmund-project contribute to the possibility of the city will succeeding in obtaining new EU funds, since the dortmund-project contents are in line with the EU conditions and aims in supporting economic development (WBF, 2004b).

The financial support for stimulating economic sectors in Dortmund is considerable. In total the state of NRW and the EU have agreed (in 2003/04) to spend €108m as subsidies for 110 projects; the stimulated projects have a total size of €242m. Some important projects are the development of new business locations such as Phoenix and the expansion of the technology and qualification infrastructure such as:

- B1st-Software Factory (ICT): €1.2m in subsidies, €2m in total turnover;
- BioMedizinZentrum Dortmund: €17.0m in subsidies, €21.2m in total turnover;
- Proteom-Kompetenzzentrum: €15.7m in subsidies, €26.6m in total turnover;

- MST.factory Dortmund (MEMS): €13.7m in subsidies, €23.1m in total turnover;
- LogAgency (Logistics): €1.7m in subsidies, €3.6m in total turnover.

The dortmund-project is financed annually with around €5m in project costs; this excludes the personnel costs and the projects that run under the dortmund-project umbrella.

3.3 Knowledge Foundations

In this section, the knowledge foundations – as discussed in the theoretical chapter – will be assessed for Dortmund. They are the knowledge base, the economic base, the quality of life, the accessibility, the urban diversity, the urban scale, and the social equity in the city.

3.3.1 Knowledge Base

Dortmund has a strong knowledge base. The University of Dortmund has about 25,000 students and the University of Applied Science (polytechnic) around 9,000. Dortmund has the highest number of information science students in Germany (over 6,200), with approximately 3,500 information science and 300 ICT students at the university and some 2,300 informatics and ICT students at the polytechnic. Furthermore, the city houses the IT-Centre, which teaches academic courses for ICT professionals (127 students). Dortmund also has a private International School of Management (dortmund-project, presentation) a polytechnic for public management in NRW and an orchestra institute (a branch of the 'Folkwang Hochschule', Essen). Table 3.2 gives an overview of the university student numbers. Dortmund has an average position within the group of 40 biggest German cities when it comes to the ratio of students per inhabitant: the city ranks twentieth with 58 students per 1,000 inhabitants. Nearby cities Düsseldorf, Cologne, Bonn and Bochum rank higher, but Duisburg, Essen, Mönchengladbach and Wuppertal rank lower (Cash, 2004). About 17 per cent of the Dortmund population aged 25–64 had third level education in 2000; this is rather low compared to other European regions (Office of the UK Deputy Prime Minister, 2004).

As well as the high number of students in Dortmund University and the polytechnic, several cities in the vicinity of Dortmund also have large student numbers: for example, Duisburg (15,400 students at polytechnics and

Table 3.2 Student numbers at Dortmund University, 2002–2003

Study	Number of students
Informatics	3,523
Economics and social sciences	2,791
Educational sciences and sociology	2,573
Linguistic science, literature, journalism and history	2,475
Rehabilitation sciences	1,940
Mechanical engineering	1,490
Music, art, textiles, sport and geography	1,445
Mathematics	1,408
Social sciences, philosophy and theology	1,337
Construction	1,326
Spatial planning	1,163
Electronic technology	1,081
Chemical technology	674
Chemistry	609
Physics	533
Statistics	471

Source: Stadt Dortmund (2003b).

university), Essen (24,137), Bochum (41,389), Hagen (44,729 – most of the students study at the distance learning university (Fern-Universität) and are not present in Hagen) and, a little further away, the University of Münster (some 50 km) with almost 43,000 students (and over 9,000 polytechnic students) and the University of Cologne (80 km) which is the biggest university in Germany with over 61,000 students. Cologne has 85,217 polytechnic and university students in total (Stadt Münster, 2003; Stadt Dortmund, 2003b).

Research is also strong in Dortmund: 4,700 employees are doing research in over 20 scientific institutes (Langemeyer, 2004). About 1,000 of these researchers work in institutes other than the university and polytechnic. The institutes together have an annual budget of almost €50m: they include a branch office of the Fraunhofer Institute for Software and Systems Engineering (ISST), the Fraunhofer Institute for Material Flow Systems and Logistics (IML is a key research institution in Dortmund, employing about 400 people), the Max Planck Institute for Molecular Physiology (which is 50 per cent financed by the national government and 50 per cent by the federal states of Germany), the Gottfried W. Leibniz Institute for Labour Physiology (IfADO), the Gottfried W. Leibniz Institute for Spectrochemistry and Spectroscopy (ISAS), the

Institute for Regional and Urban Development of the Federal State of North Rhine-Westphalia (ILS) and the Agency for Social Research (sfs).

3.3.2 Economic Base

Just like other cities in the Ruhr region, Dortmund has a historic background that is dominated by industrial economic sectors, namely the mining industry and steel construction. In past decades these activities experienced major decreases in employment: see Figure 3.3. Therefore, Dortmund has built up innovative economic activities to create new jobs. Up until the time of writing, the growth of new jobs in the tertiary sector has not been enough to compensate for the loss in secondary sector jobs (see Figure 3.4). Table 3.3 shows the employment by sector in Dortmund in 2002. In the period 1980-2000 employment increased most in services for companies (113 per cent growth) and services for households (48 per cent growth), while it declined sharpest in mining (80 per cent decline) and processing professions (48 per cent decline). From 1992 to 2000, Dortmund's GDP grew by 12 per cent compared to 28 per cent in Düsseldorf, 22 per cent in Cologne, 16 per cent in Duisburg and 7 per cent in Essen (Stadt Dortmund, 2003d).

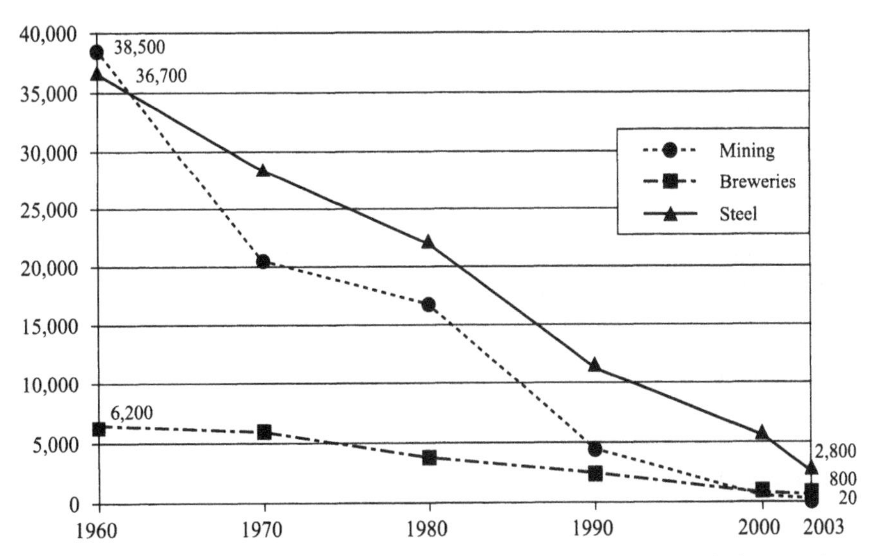

Figure 3.3 Declining employment in Dortmund steel, mining and breweries

Source: WBF-Do (2004).

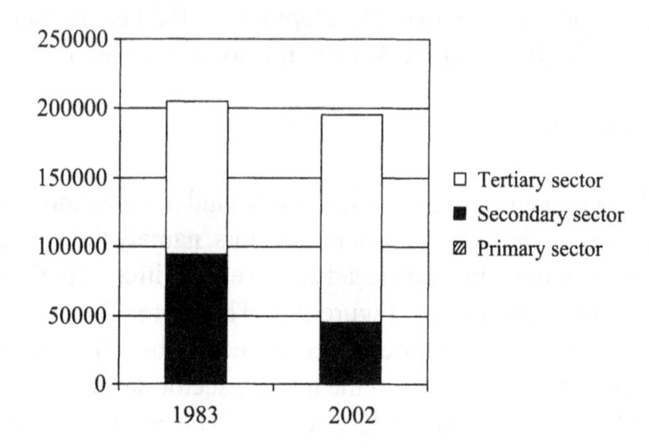

Figure 3.4 Employment development in Dortmund, 1983–2002

Source: Stadt Dortmund (2003b).

Table 3.3 Employment by sector in Dortmund, 2002

Sector	Number of employees	%
Public and private services	46,543	23.9
Real estate, letting, services or firms	31,215	16.0
Trade	30,057	15.4
Processing professions	27,094	13.9
Traffic and media	14,229	7.3
Construction industry	13,554	7.0
Banking and insurances	12,640	6.5
Government	10,833	5.6
Hotels	5,300	2.7
Power and water utilities	2,767	1.4
Mining	379	0.2

Source: Stadt Dortmund (2003b).

Recently the fifty biggest German cities were compared. Regarding the current situation, Dortmund ranked thirty-second in prosperity, forty-first in the labour market situation and twenty-fifth concerning economic structure. The dynamics ranking presented much better results: seventeenth in prosperity growth, twenty-second concerning the labour market improvement and fourth with regard to the development of the economic structure. The Dortmund rise in labour costs, for example, is very low (2 per cent over the period 1998–2002) compared to other German cities: Dortmund ranks their behind Hamm (0.4 per cent) and Bremen (0.8 per cent), while the lowest ranked city is Stuttgart with a 9.8 per cent increase. In contrast, the Dortmund income per inhabitant is growing quite fast, by 1.174 per cent over the period 1998–2003: the city ranks third in this respect. The number of large company headquarters in Dortmund (7) is rather limited: the city ranks fifteenth behind places like Munich (first position, with 94 headquarters), Berlin (ranked fourth with 46 HQs), Cologne (fifth, with 43 HQs), Düsseldorf (sixth, with 37 HQs) and Essen (ninth, with 20 HQs). Cities nearby Dortmund that have about the same number of HQs are Duisburg, Bochum, Bonn and Wuppertal (Wirtschaftwoche, 2004). This research indicates the difficult position of the Dortmund economy and the rather successful improvement of the situation.

In other research, Dortmund ranks fourth out of 40 large German cities with regards to the ratio of newly established companies per firms that stay/close down (1.25). The city ranks high (sixth position) concerning the number of start-ups per 100 already established companies (39.37). Regarding the number of insolvent firms per 100 already established companies (5.82), Dortmund ranks sixth. Overall, Dortmund is ranked twenty-eighth in the list of most attractive German cities concerning real estate (Cash, 2004).

Disposable income in Dortmund is €15,473 per inhabitant: this is higher than Duisburg (about €13,500), but lower than Essen, Cologne (both around €17,000) and Düsseldorf (approximately €19,500). A comparison of the GDP per capita in 2001 in the top 61 European cities puts Dortmund (€26,548) on the thirty-ninth spot just in front of Rotterdam. Frankfurt am Main is ranked first with €74,465, Karlsruhe second with €70,097 and Munich fourth (€61,360). Some other German cities in the ranking are: Düsseldorf (sixth), Cologne (14th), Münster (16th), Bonn (26th), Essen (33rd), Bochum (37th), Duisburg (44th) and Berlin (54th) (Office of the UK Deputy Prime Minister, 2004).

3.3.3 Quality of Life

In contrast to its image as an industrial town, Dortmund is a rather green city:

it ranks seventh among Germany's 40 biggest cities regarding the so-called 'green indicator' (Cash, 2004). Air pollution sharply declined in the period 1994–2002 (Stadt Dortmund, 2003d). The city performs on average regarding criminality (Wirtschaftswoche, 2004). The number of registered crimes (10.1 per 100 inhabitants) is lower than in some nearby cities Cologne and Düsseldorf, but higher than in Duisburg and Essen (Stadt Dortmund, 2003d). There is no serious housing shortage in Dortmund. The municipality tries to attract young people and families to live in the city. Dortmund's image among companies and citizens is that of a sports and shopping city and a Westphalia metropolis. Since previous questionnaires were undertaken, the city has lost some of its image as a beer-drinking and exhibition city: Dortmund's image as a technology city is stable. Citizens are very satisfied with the shopping opportunities, cinemas and concerts; most criticism is aimed at car parking shortages, citizens' participation, the shortage of youth facilities and the bad condition of many schools (however, at the time of writing a good deal of investment is being made in the reconstruction of school buildings). According to Dortmund citizens, their city is lively and extrovert, but not so appealing and safe. The biggest Dortmund attractions are the Westphalia Stadium (where the Bundesliga club Borussia Dortmund plays its soccer matches), Westphalia Park and the Christmas night market. Dortmund's central train station is still seen as a problem by the citizens. The Brückstraße district, with its new concert hall, has seen a sharp increase in its popularity (Stadt Dortmund, 2003c): whereas the neighbourhood was viewed upon as a twilight area in 2000, today it is seen as a colourful mixture of culture, fashion and gastronomy. Halfway into 2004, the financial support from the state of NRW for marketing the Brückstraße district will end. This might prove to be a problem, since the district does not seem to be developing autonomously yet (IHK, 2004). While Dortmund built a new concert hall, the neighbouring cities Bochum and Essen did the same. This is illustrative of the comparatively autonomous positions of the Ruhr region cities in relation to each other: they do not cooperate very intensively and in certain aspects do not tune developments to each other. Besides the attractions mentioned above, Dortmund also houses a casino, a zoo, a large exhibitions and fairs complex (Westfalenhalle) and a theatre/opera house (215,574 visitors in 2002). Large museums in Dortmund are the Museum for Art and Cultural history (35,566 visitors in 2002), the Museum at the Eastern City Wall (27,484 visitors) and the Physics Museum (78,395 visitors) (Stadt Dortmund, 2003b).

In 2001, Dortmund had 590,890 overnight stays; 14.9 per cent of these stays were made by foreigners. This percentage is high compared to the

German average of 1.2 per cent, but a bit lower than the NRW average (15.7 per cent). Table 3.4 compares Dortmund to some cities in its vicinity (Stadt Dortmund, 2003b).

Table 3.4 Overnight stays (excluding camping grounds) in some Ruhr region cities, 2003

City	Number of overnight stays	% of foreign guests
Bochum	422,399	16.7
Dortmund	672,131	19.1
Duisburg	309,999	16.4
Essen	891,503	16.3
Hagen	117,397	19.1
Hamm	116,760	13.8
Unna	258,885	13.6

Note: the numbers for Unna are for 2001, since 2003 numbers were not available.

Source: Informationssystem Tourismus LDS NRW, Düsseldorf 2004; Stadt Dortmund (2003b).

Since 1990, Dortmund has experienced an increasing number of overnight stays: in 1990 the number was about 450,000, whereas in 2003 there were over 670,000 overnight stays (WBF, 2004a; Informationssystem Tourismus LDS NRW, Düsseldorf 2004). To promote the city centre, a company was founded: City-Marketing GmbH (WBF, 1999). The budget is shared in private-public partnership 50:50 by the business sector and the city of Dortmund.

3.3.4 Accessibility

Dortmund's internal accessibility is good: the city has little traffic congestion and a rather extensive public transport network which includes buses and a subway. The number of bus lines increased from 27 in 1983 to 53 in 2002 (Stadt Dortmund, 2003b). Companies in Dortmund view car accessibility as the most important location factor; about 78 per cent of the firms rate car accessibility in Dortmund as (very) good in 2003 (compared to around 82 per cent in 2000). The availability of car parking is also essential for firms: this aspect is rated good as well. Some 63 per cent of the citizens were satisfied with Dortmund's public transport in 2003 (compared with around 55 per cent in 2000); this made public transport one of the best-rated aspects of the city. Public transport is used more often: the share of frequent users increased by 20 per cent in the period 2000–2003. However, only 21 per cent of citizens are satisfied with the state of the streets (compared to 26 per cent in 2000). Satisfaction rates for cycle tracks are even lower (Stadt Dortmund, 2003c).

External accessibility is rather good for Dortmund. One out of five firms use rail transport and the canal harbour, while over 25 per cent of the companies use Dortmund airport (Stadt Dortmund, 2003c). The Dortmund-Wickede Airport is not very big, although it showed strong growth in the period 1983–2002 (see Figure 3.5): it is rated one hundred and twenty-fourth among European airports regarding passenger numbers in 2000.

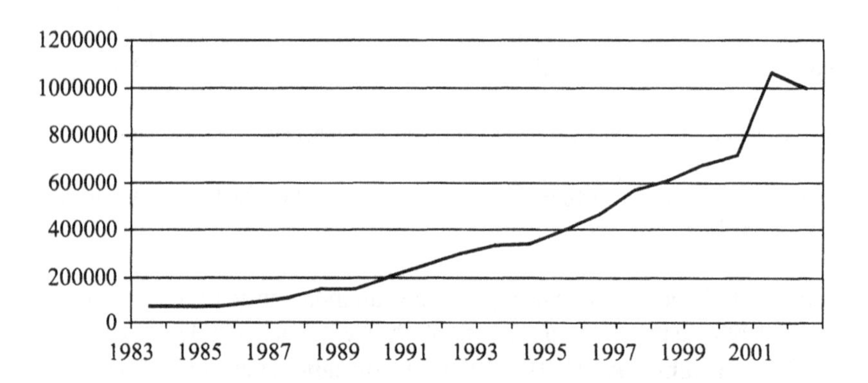

Figure 3.5 Number of passengers travelling through Dortmund airport, 1983–2002

Source: Stadt Dortmund (2003b).

Flights from/to Dortmund are mainly domestic and within Europe. However, several larger airports are nearby (Düsseldorf, some 60 km from Dortmund, ranked 18th in Europe; Cologne, 80 km: ranked 46th) (Office of the UK Deputy Prime Minister, 2004).

The declining heavy industries have resulted in strong decreases in traffic using Dortmund harbour: goods transhipment (which consisted mostly of ore and steel) has declined sharply in the period 1998–2002, from 5,547 to 2,770 tonnes. Within this total figure, the transferred amount of ore declined from 2,025 to zero (Stadt Dortmund, 2003b).

The amount of goods transferred via Dortmund's railways has declined sharply from 52,200 to 16,650 tonnes in the period 1995–2002 (Stadt Dortmund, 2003b). The current railway station connections are rather good: the station is an Intercity trains junction. Furthermore, it has direct international train connections and is part of the ICE high-speed trains network.

One of the challenges for the future is the construction of a new central rail station in the city. Dortmund citizens are no longer satisfied with the current station. In the 1990s plans had already been presented by Deutsche Bahn (the national railways operator) and the Westdeutsche Immobilienbank (a bank specialising in real estate) to build a new, large multipurpose station. The project, called UFO (because of its flying saucer-like shape), would include shops, rental space for service companies, leisure activities, restaurants and a hotel. Construction was planned to start at the end of 1999 and to be finished in 2002. However, the city council demanded that the project proceed on a smaller scale, and subsequently one of the investors dropped out of the project. In 2000, the Dortmund municipality found a new investor to step in. The new goal was to finish the project before the soccer world championship to be organised in Germany in 2006 (Frankfurter Allgemeine Zeitung, 2004a and b). After new organisational and financial arrangements were made, the railway company resumed project participation. The new plan is called '3do' and the railway station itself will be finished in time for the championship.

ICT infrastructure in Dortmund does not seem to be an issue. The liberalisation of the telecommunications market and the huge infrastructure investments by Deutsche Telekom (the German telecom incumbent) and other telecom firms in the 1990s and onwards have made broadband ICT infrastructure quite abundant and affordable in urban regions such as Dortmund.

3.3.5 Urban Diversity

The economic diversity of Dortmund is increasing: while it mainly depended on mining and steel production for many decades, today several sectors are doing well and growing further, e.g., ICT, MEMS, the hotel business, banking and insurance, real estate and government (Stadt Dortmund 2003b; Küpper, 2004). Although the city has a broad range of cultural activities and in that sense is quite diverse, the image of Dortmund to outsiders (especially foreigners) is still very much based on its industrial past. Furthermore, the city does not have an extensive cultural scene consisting of artists and other creative professionals.

A group of people that can contribute to urban diversity (both in a cultural and economic sense) are foreign inhabitants. In 1983 Dortmund had a foreign population of 52,785 people; this number increased to 75,232 in 2002. This implies a rise in the share of foreigners within the total population from 8.9 per cent to 12.8 per cent. Big ethnic minorities within Dortmund are Turks (27,928), Yugoslavs (4,040) and Greeks (3,973). In 1983, 13.0 per cent of unemployed people were of foreign origin; this number rose to 22.2 per cent in 2002 (Stadt Dortmund, 2003b). It appears that foreign Dortmund citizens have a relatively weak position in the labour market.

3.3.6 Urban Scale

Dortmund has a population of 589,212 people. It is the sixth largest city in Germany and is part of the Ruhr region ('Ruhrgebiet': 5.4 million inhabitants). Within the wider metropolitan Rhine-Ruhr agglomeration (11.5 inhabitants) Dortmund is the second largest city (after Cologne, with 967,493 inhabitants). Other large cities in the Rhine-Ruhr region are Bochum (387,588), Bonn (311,047), Essen (588,490), Düsseldorf (573,092) and Duisburg (507,176). The region is a typical polycentric area with no apparent hierarchy of cities. Based on the scale of the region, the potential urban scale for Dortmund is extensive: within a one hour drive by car about 9 million people can be reached. However, in reality the cities in the Ruhr region and beyond do not cooperate extensively: city administrations and universities operate quite independently from their counterparts in other cities. Companies are more willing to interact with partners outside their home city. Several interviewees indicated that interurban cooperation should increase and communication should be arranged especially on the level of the Ruhr region (Brödner and Rücker, 2004; Stadt Dortmund, 2003b; Landesamt für Statistik und Wahlen NRW, 2004).

The more narrowly-defined Dortmund economic region (the Eastern Ruhr region) is made up of Dortmund, Unna and Hamm. This area coincides with the district of the Dortmund Chamber of Industry and Commerce. Most of Dortmund's cross-border economic activities are orientated to these smaller towns in the more rural east, e.g. the district of Unna (429,309 inhabitants) and the city of Hamm (184,559), instead of to the cities of Bochum (389,022) and Hagen (201,532), which are also close to Dortmund (Brödner and Rücker, 2004; Stadt Dortmund, 2003b). The central position of Dortmund can be illustrated by looking at the commuter numbers. The city has about 75,000 incoming commuters, while 55,000 commute outward. Most inward commuters come from the districts of Unna and Recklinghausen, whereas most outward commuters travel to Bochum and Hagen (Dortmund, 2003b).

3.3.7 Social Equity

Dortmund's unemployment rate is quite high compared to many other German cities (see the introduction of this case study); the rate (13 per cent in 2001) is above the German (9.9 per cent) and NRW (9.3 per cent) average (Stadt Dortmund, 2003b). Unemployment among foreign inhabitants is even higher; the problem is most visible among the large Turkish-origin population. The main cause for this high level of unemployment is the low degree of education. During the sharp decline in industrial jobs, these people were hit particularly hard. The problem of relatively high unemployment among foreign inhabitants is apparent throughout the state of NRW (Landesregierung Nordrhein-Westfalen, 2004).

Dortmund contains 33,320 people who receive social benefits (57 out of every 1,000 inhabitants). The rate is higher than in nearby cities like Bochum, Hagen, and Unna, but lower than in Duisburg, Essen and Cologne; it is significantly higher than the NRW rate of 37 (Stadt Dortmund, 2003b). The relatively high unemployment and large number of citizens on social welfare is reflected in a ranking of Germany's 50 biggest cities: Dortmund's social structure level is only ranked thirty-fourth, but its social structure progress is performing better, at twenty-eighth in the ranking. The positive development is also reflected in the large decrease (1.3 per cent) of people receiving social benefits: Dortmund is ranked third (Wirtschaftswoche, 2004). Although Dortmund has some neighbourhoods that are in arrears (e.g. Nordstadt), the city has no major problems concerning crime and no-go areas. Currently, Nordstadt is being restructured with the help of EU money (Urban II programme).

3.4 Knowledge Activities

In the last section, the 'foundations' of Dortmund's knowledge economy were discussed. Now, we turn to knowledge activities and ask how well the city (and the actors in it) manage to: 1) attract and retain talented people; 2) create new knowledge; 3) apply new knowledge and make new combinations; and 4) develop new growth clusters.

3.4.1 Attracting and Retaining Knowledge Workers

How well does the city manage to attract and retain knowledge workers? Dortmund mostly attracts students from the region. Most Dortmund polytechnic students live in NRW (87.8 per cent). Also, 92.2 per cent of university students live in NRW (Stadt Dortmund, 2003b). A weak point is that the university has no lively campus life and the city lacks a student atmosphere; this probably has to do with the absence of a student district. Typically, most of the graduates keep on living in/near Dortmund when they start their working career. Since Dortmund has a large number of informatics students, the city educates more people in this field than it needs for its own economic sectors that depend on ICT. (This situation is quite the reverse in Munich, where the inflow of ICT workers from outside is vast.)

The number of foreign students in Dortmund has risen. In 1983–1984 the University of Applied Science had 462 foreign students (7.4 per cent of the polytechnic students), rising to 1,200 (13.3 per cent) in 2001–2002. In 1985–1986 the university had 1,075 foreign students (4.3 per cent of all university students) rising to 4,415 (12.9 per cent) in 2002–2003 (Stadt Dortmund, 2003d). German cities find it difficult to attract foreign students, since few courses are taught in English: this situation is changing gradually. In Dortmund, the education institutions are quite active in offering English language courses: the university (and its Summer School), the University of Applied Science, the International School of Management and the International Max Plank Research School in Chemical Biology all offer such courses. The Dortmund municipality also tries to increase the international orientation through the PiDo (Pittsburgh–Dortmund) 2010 project. Among other things, it aims to increase the exchange of faculty members and students, and to promote Dortmund to firms from Pittsburgh (dortmund-project, www.pido2010.de). The extensive number of research institutions in Dortmund (including the Max Planck Institute and the Fraunhofer Institute) helps to attract knowledge workers from elsewhere. Most Dortmund-based firms are fairly positive about the availability of employees; they also rate further education possibilities in the city as quite reasonable (WBF, 2004a).

Increasing the number and quality of knowledge workers is an explicit component of the dortmund-project. The reason for this is the expected shortage of knowledge workers in 2010 and beyond, if recruitment and training programmes are not introduced. This project pillar is called 'People and competences'. The IT-Centre offers a two-year educational programme for ICT professionals: the institute currently has 127 students. The JOY campaign for ICT education is meant to increase young people's interest in ICT professions and increase the number of apprenticeships. In the field of micro-electronic mechanical systems (MEMS), a coordination agency is active to strengthen MEMS education potential in Dortmund. Projects concerning ICT and economics are being supported in schools. The Personnel Service Agency tries to attract specialist employees to Dortmund by advertising and recruiting people at the national and international level; the agency also supports firms moving to Dortmund. Furthermore, the LEAD competences agency for personnel management and entrepreneurship tries to improve the skills of employees in ICT, MEMS, biotechnology and logistics. The original goal of the dortmund-project in 2000 was to increase the number of specialist employees in the city to almost 20,000 by the year 2010. Attracting specialist staff via traditional and online recruitment; BA, MA and programmers certificate education in the IT-Centre; easing immigration of knowledge workers through the Green Card taskforce; increasing personnel training and number of apprenticeships; and, increasing ICT and MEMS activity at the university and polytechnic (dortmund-project, 2000).

Table 3.5 Training and attracting specialist employees

Measure	Number of specialists
Attracting specialist staff from outside	7,600
IT-Centre	4,200
Green Card taskforce	3,000
Personnel training and apprenticeships	2,500
Increasing ICT and MEMS activities at polytechnic and universities	2,200
Total projected employment increase	19,500

Source: dortmund-project (2000).

3.4.2 Creating Knowledge

Knowledge creation in Dortmund is rather good. Some 600 professors and 5,000 other scientific workers are active in Dortmund (Stadt Dortmund, 2004b). The University of Dortmund has the biggest Informatics faculty in Germany. In the field of MEMS the city has an advantage, with four research institutes. The Fraunhofer Institute for Material Flow Systems and Logistics contributes heavily to the knowledge creation in logistics and the Max Planck Institute for Molecular Physiology is a key institute within the biotechnology field. To strengthen the links between the scientific institutions in Dortmund, the university, the polytechnic and several other science organisations founded Windo (which stands for 'Wissenschaft (= science) in Dortmund') in 1992. The members regularly meet to discuss matters of general interest, including promoting Dortmund as a city of science (Windo, 2002).

Knowledge creation in Dortmund-based companies is rather high, although the number of large firms in the city is limited compared with other German cities, for instance Munich. In the field of ICT hardware, ELMOS Semiconductor AG contributes significantly to knowledge creation; in software, Materna Information and Communication GmbH is a key player; STEAG microParts GmbH contributes to knowledge formation in the area of biomedicine.

The state of NRW plays an important role in the promotion of knowledge creation, since university and science policies are mostly determined on that level. For instance, the universities of Duisburg and Essen merged because the state of NRW demanded this. Furthermore, NRW contributes around 50 per cent of the Max Planck Institute funds; the other 50 per cent is paid by the national government.

Concerning the number of patent applications, Dortmund (313 patents in 2000) does not belong to the top 32 cities/regions in Germany, which includes several nearby cities such as Düsseldorf (1,901 patents), Cologne (1,090), Bochum/Hagen (689), Duisburg/Essen (654) and Münster (594) (Greif and Schmiedl, 2002). Dortmund's modest position is confirmed by the European Innovation scoreboard 2002: nor Dortmund does belong to the top group of cities/regions (Office of the UK Deputy Prime Minister, 2004). An area of patent application in which Dortmund is a leader concerns mining: in 2002 Dortmund applied for 10.4 per cent of all German patents in this field, followed by Bochum/Hagen (10.0 per cent). While the number of patents in NRW increased by 24 per cent from 6,418 to 7,965 in the period 1995–2000, Dortmund showed a below-average increase of 20 per cent from 260 to 313 patents. However, the number of patent registrations per 100 R&D workers in Dortmund firms is high (21.6) compared to the German average (9.9). The number of patent registrations per 100,000 inhabitants (26.1) and per 100,000 employees (62.6) in Dortmund is low compared to the German average (49.2 and 111.3 respectively) (Greif and Schmiedl, 2002).

Figure 3.6 shows that companies are the main sources of patents, followed by individuals. The research institutes (universities plus public research institutes such as the Max Planck Institute and the Fraunhofer Institute) only have a small slice of the cake. This is in line with the overall German pattern. The number of patents requested by universities and research institutes is low, because in most cases the researchers apply for the patents on an individual basis: this is possible because German universities are lenient towards such activities. German universities have acknowledged that relatively few ideas are being patented. National government has recognised this and therefore changed the law concerning inventions by employees. Polytechnics and universities are also trying to improve the incentives to register patents.The Patent Offensive Westphalia and Ruhr area (POWeR) has been active since 2003: the universities of Münster, Dortmund, Paderborn and Bielefeld try to stimulate a vivid patent consciousness with university employees by organising information and training events.

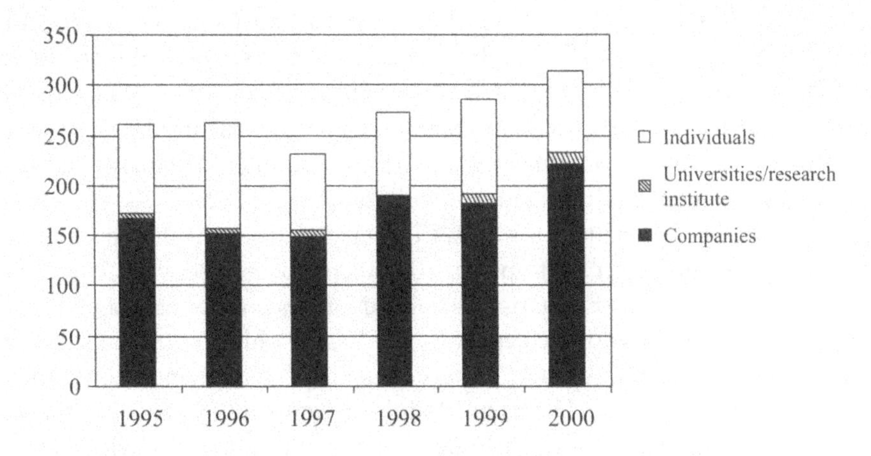

Figure 3.6 Patent registrations in Dortmund, 1995–2000

Source: Greif and Schmiedl (2002).

The patent data have to be interpreted with caution. Firstly, patent data refer to the location where patents have been registered; this is not necessarily the location where inventions took place. This may underestimate Dortmund's innovativeness, since relatively few company headquarters are located there (7) compared to Düsseldorf (37), Cologne (43), Bochum/Hagen (7/3) and Duisburg/Essen (8/20). Additionally, innovations in services are usually not patented, although they make a huge contribution to productivity and quality improvements.

The facts and figures described above position Dortmund not very high in the field of knowledge creation. However, most of the information is based on developments that were measured in the end of the 1990s. Since the dortmund-project only started in 1999 (and most projects within the initiative in 2001), the majority of results will become visible only after a period of some years. Therefore, many of the actions formulated within the dortmund-project may very well be valuable contributions to a knowledge economy (including the creation of knowledge: see the initiatives in the 'People and competences' column in Table 3.1), but have not yet been measured in the data period covered in this paragraph.

3.4.3 Applying Knowledge/Making New Combinations

Dortmund has deployed various initiatives to boost the application of knowledge and to create new combinations. Creating networks of innovation is a key ingredient in this process. Dortmund has a high number of new firms compared to the number of companies shutting down or leaving (1.25): it ranks fourth among 40 large German cities. Also, the number of newly founded companies is high in Dortmund: the city ranks sixth (Cash, 2004). In the period 1999–2002 Dortmund scored well compared to other Ruhr region cities regarding the number of start-ups per 10,000 employees. Between 1991 and 2002 this relative number of start-ups declined for western Germany as a whole and only grew slightly for the nearby city of Essen, while Dortmund showed significant growth. Furthermore, a survey among large German companies ranked the University of Dortmund fourth in Germany (Steemann, 2004). Below, the Dortmund initiatives to boost the application of knowledge and to make new combinations are discussed.

Technology Centre GmbH Dortmund and the Technology Park Already in 1984, the city of Dortmund had become more active in applying knowledge. The Wirtschafts- und Beschäftigungsförderung, together with the chamber of commerce (IHK), the bank Stadtsparkasse Dortmund and the Fraunhofer Institute, developed the idea to build a technology centre and park in the vicinity of the university. In 1985, the Technology Centre Dortmund GmbH and the park were opened. The centre (an incubator) aims to foster the growth of start-up companies, while the park is meant for firms that have already expanded. Goals of the technology centre and park are: stimulating company start-ups and growth; offering needed infrastructure; stimulating cooperation between science and economy (technology transfer); building up regional, national and international networks; and initiating technology and cooperation projects. The Technology Centre Dortmund was one of the first initiatives of its kind in Germany (after Berlin): it was modelled on similar projects in the US and UK. In those days it was very unusual for public and private partners worked together as they did in the Technology Centre Dortmund. Associates in the centre included the Dortmund municipality, the Chamber of Commerce, and several German banks. At the time of writing the Technology Centre has about €1.5m subscribed capital; the shareholders are the municipality (46.5 per cent), several Dortmund credit institutes (25 per cent), the Chamber of Commerce (16 per cent) and the polytechnic and university (12.5 per cent). The centre does not aim to make profits: all money earned is re-invested in improving facilities and infrastructure. The Technology Centre provides a network of qualified specialists to support firms on topics such as business plan development, business promotion, financing, technology transfer, transfer of information and know-how and market penetration (www.tzdo.de).

The timing of the Technology Centre set-up was good, since the state government of NRW and the German government were busy stimulating new economic activities in the Ruhr region, among other places. Within the NRW intitiative 'Zukunftstechnologien in Nordrhein-Westfalen' ('Future Technologies in NRW'), the state government put considerable amounts of money at the disposal of new economic activities, especially microelectronics, measurement technologies, manufacturing technologies, ICT, new materials and environmental technologies. In 1985, all 4,800 m^2 of business space in the centre and park were rented out to 35 firms. The Technology Park Dortmund is now one of the largest of its kind in Germany: some 200 companies are established in the park and centre. Most firms are active in software, services, telecommunication and multimedia. Furthermore, companies in the park and centre are active in MEMS, environmental technology, new materials, robotics, quality protection and control systems. Up to the time of writing, 229 firms had taken up residence in the Technology Park (including 89 firms in the Technology Centre); only 5.8 per cent of the firms that have resided in the centre have gone bankrupt. The centre housed some 50 companies in 2000. The centre and park offer employment to about 8,500 people (TZDO, 2000a; Küpper, 2004; interviews).

The Technology Centre received €1.5m initial capital, mainly contributed by the Dortmund municipality. Currently the city has a share of some 30 per cent in the Technology Centre. Other parties that contribute financially are the state of NRW and the European Union. Furthermore, the Technology Centre earns money by participating in technology centres supported by the EU elsewhere in Europe. Currently, the Dortmund municipality is earning back some money through the Technology Centre.

Start2grow As part of the dortmund-project, the Start2grow competition was introduced. Goals include improving the start-up climate, attracting new companies and strengthening the core economic sectors (MEMS, IT, and logistics). The competition consists of two parts: the first one is a contest for start-up firms and the second is a growth initiative for SMEs that have already been established and employ up to 100 people (see Figure 3.7).

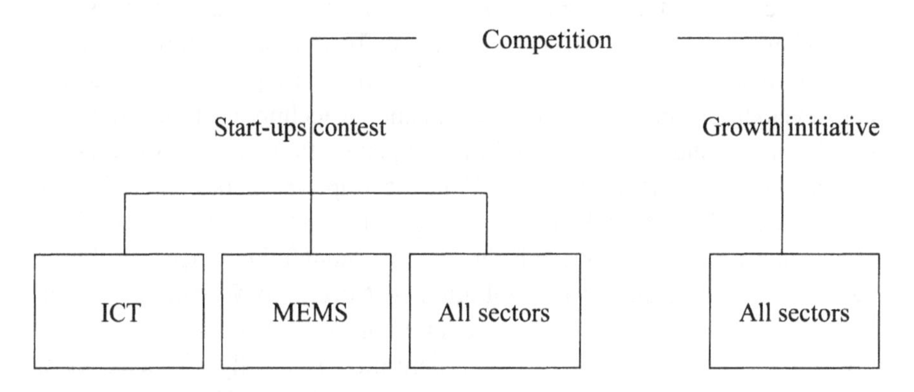

Figure 3.7 Start2grow overview

Source: Start2grow (2004).

The competition includes making business plans and involves supporting activities such as coaching evenings, lectures about specific topics, exchange of start-up experiences and seminars. A network of 500 coaches is available for free advisory support of the start-ups; specialisations include R&D experts, financial and juridical consultants, venture capital providers, and marketing and management experts. The start-up firms that win in the Start2grow competition receive half the prize money right away and the other half only if the company is founded in Dortmund. The prizes are: first prize, €50,000; second prize, €40,000; third prize, €30,000; and fourth to tenth prizes, €15,000 each. The annual contest takes six months. Up to the time of writing a total of 119 firms have been started, including 58 in Dortmund. In ICT, 59 firms were founded, of which 29 were in Dortmund; five MEMS companies, of which three were Dortmund-based, and 55 companies in other sectors, including 26 in Dortmund. The Start2grow contest is financially supported by the state of

NRW (via the Go! Gründungsnetzwerk) and the European Union (Start2grow, 2004).

Infrastructure companies Besides the dortmund-project, the Dortmund Foundation ('Dortmund Stiftung') is involved in the so-called 'infrastructure' companies: organisations that enable economic development (e.g., through education, increasing the application of innovative solutions in firms and supporting start-ups). The Dortmund Foundation has some 100 financial private donors and has over €2m founding capital. The foundation owns all the shares in the DoPro GmbH investment company which holds shares in the infrastructure companies. A maximum of 50 per cent of the foundation's capital is invested in the 'infrastructure' companies. These companies should strengthen the networks in the key economic sectors (IT, MEMS, logistics). Figure 3.8 presents the organisation overview.

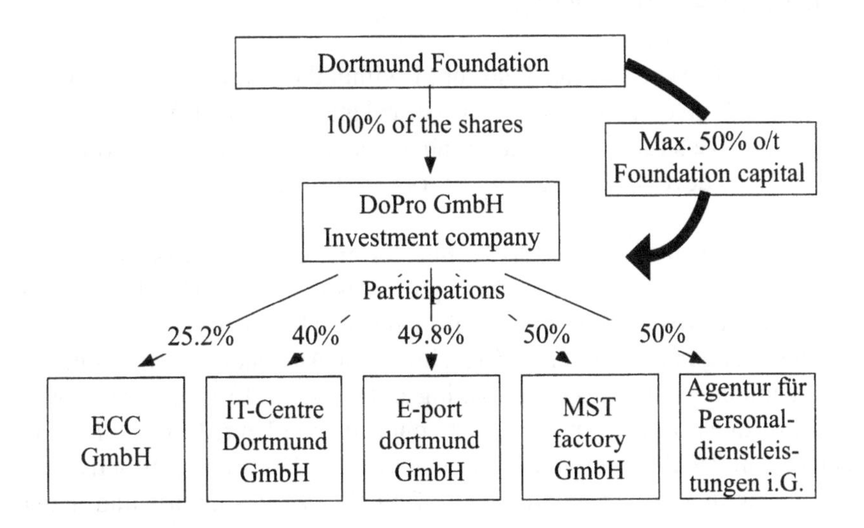

Figure 3.8 Overview of 'infrastructure' companies

Source: Küpper (2004).

The *E-Commerce Centre* (ECC) is located in the city district Stadtkrone-Ost. It contains 30 ICT firms in the field of electronic commerce (via the Internet) with 1,200 employees. ECC aims to increase employment to 5,000 people. The *IT-Centre* is a private education institute for ICT professionals who want to obtain an academic ICT education within two years; it currently has 129 students. The *e-port* is a start-up and competence centre of 5,000 m^2 that aims to foster the use of ICT (e-commerce) by logistic firms; it is a public-private partnership (involving the Technology Centre Dortmund and logistic companies) located in the harbour area. Examples of e-port activities are the third Wednesday meetings (where start-ups and funding capital investors meet) and the DiaLog project (which tries to get firms to cooperate in supply chain collaboration). The MST Factory will try to stimulate start-up developments in the MEMS field; it will be located at the Phoenix West technology park which will be developed an old blast furnace site. And finally, the *Agentur für Personaldienstleistungen* (Personnel Service Agency; currently being founded as LEAD GmbH in the e-port building: location, excellence, acceptance, development) will assist companies in ICT, MEMS, biotechnology and logistics to find appropriate employees by attracting them from other places (nationally and internationally) (dortmund-project, 2000; Küpper, 2004; TZDO, 2004).

Locate IT The competence field for IT within the dortmund-project is called Locate IT. It involves attracting German and international IT firms and building up supportive structures like an e-lab and a venture capital network. Locate IT will also develop facilitating services for IT firms.

IVAM In the field of MEMS, there is an active international interest group of private companies which promotes the use of MEMS technology and the interests of firms in the sector. The group is called IVAM ('Interessengemeinschaft zur Verbreitung von Anwendungen der Mikrostruk turtechniken'): it is non-profit-making and has some 130 member companies and institutes from eight European countries, the United States, Korea and Japan. The IVAM office for NRW is located in Dortmund; its activities are chiefly funded by the federal state of NRW. IVAM supports and manages several federal and national government projects and EU initiatives (e.g., EMINENT: an EU-funded platform for SMEs to accelerate business contacts between suppliers and clients in MEMS) (www.ivamnrw.de). In 2000, the IVAM NRW office founded a venture capital fund for MEMS with a starting capital of around €350,000; it manages some €5m in total (TZDO, 2000).

Stimulating start-ups in polytechnic and university: GDUR and PINC The Technology Centre Dortmund, together with the dortmund-project, the polytechnic, the university and five technology centres in the region (Tec5Plus) will promote start-ups firms at the polytechnic and university through the project GDUR. In the regional Pre Incubator Centre (PINC) potential start-ups can use office space, infrastructure, services and consultancy for 3–4 months for cost based tariffs to find out whether their business idea has 'real world' potential (TZDO, 2004).

Biotechnology Although biotechnology is not one of the three core fields of the dortmund-project, there are some activities in this field to be found in Dortmund (e.g., the Max Planck Institute (MPI) for Molecular Physiology and the Institute of Spectrochemistry and Applied Spectroscopy (ISAS)). Currently, the Bio Medicine Centre ('BioMedizinZentrum') is being built within the Technology Centre Dortmund: it is a competence centre in the fields of biomedicine, bioinformatics, proteomics and bioMEMS. The centre offers laboratory and office space, services and consultancy to start-ups in biotechnology. The Bio Medicine Centre is supported by the state of NRW and the European Union. At the Ruhr region level, the state of NRW – together with municipalities, research institutes and companies (e.g., STEAG microParts and Bartels Mikrotechnik in Dortmund) in the region – is promoting the area as a network for biomedical technologies, called Life Technologies Ruhr. This network is managed by BioMedTec Ruhr and BioIndustry. BioMedTec Ruhr was established in 2001; within this network the universities of Bochum, Essen and Witten/Herdecke take part. BioMedTec Ruhr services include providing information about research institutes in the region, initiating scientific cooperation and mediation of private firm partners, consultancy for start-up companies, venture capital fund acquisition, protection of intellectual property rights, marketing and sales, trade show appearances, and the initiation of further training and education (BioIndustry, 2003). Also supportive of applying knowledge and creating new combinations is BioIndustry, a competence network for the Ruhr region in bioprocessing, bioMEMS and proteomics. The network headquarters are located in Bergkamen. BioIndustry offers start-ups and SMEs know-how and individual contacts in areas such as development, analytics, production and quality management; partners can use laboratories for research. BioIndustry consists of companies (e.g., Protegen AG and Schering AG), research (MPI and ISAS) and education institutes (the polytechnics of Dortmund, Gelsenkirschen and the universities of Dortmund and Witten), technology centres, biotech service firms and (municipal) economic promotion offices (Bio-Gen-Tec-NRW, Arnsberg district government, the Dortmund chamber of commerce, the municipalities of Bergkamen, Ennepe, Witten and Unna) (BioIndustry, brochure). Up to the time of writing, cooperation between actors in the Rhine and Ruhr regions is sparse. Increasing such cooperation is a necessity, but first the relationships between parties on the Dortmund city and region level need to be strengthened: BioIndustry aims to contribute to this goal.

3.4.4 Developing New Growth Clusters

The city of Dortmund has identified three core growth clusters (IT, MEMS, Logistics) within the dortmund-project. Furthermore, the biotechnology activities can also be considered as a new growth cluster. Figure 3.9 shows the growth locations.

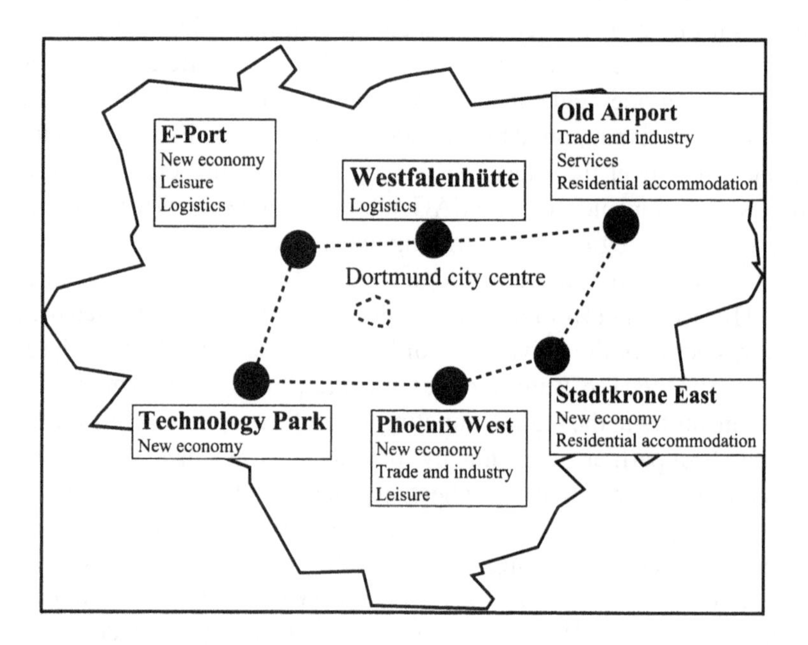

Figure 3.9 Geographic overview of new growth cluster locations

Source: dortmund-project (2000).

E-port is an 'incubator' located in the Dortmund harbour district. The area is developing into a high-grade location for companies in logistics, e-business and leisure activities. E-port aims to stimulate the foundation of e-logistics companies. In the north, former industrial sites were developed into new logistics areas (Hardenberghafen and Ellinghausen), with companies like IKEA logistic centre and DHL logistic centre. The *Technology Park* (15 ha) lies next to the Dortmund University; it contains firms in ICT, MEMS and biomedicine. *Stadtkrone Ost* is an old military barracks ground, where today the E-Commerce Centre and over 30 ICT companies are located. *Westfalenhütte* is an old industrial ground where blast furnaces were situated. This area of over 500 ha is being developed and will contain logistic firms in over an area of 60–100 ha; other professions such as services will also start in the Westfalenhütte area. The neighbourhood will also contain much green space (like parks and forests). The *old airport* Dortmund-Brackel (100 ha) will contain an extensive number of houses besides (service) firms. It will also feature a golf course and several other sport and leisure activities, including the Borussia Dortmund Soccer Park. *Phoenix* is an old industrial area of over 200 ha: steel production ended here at the turn of the new millennium. The revitalisation will consist of two areas: Phoenix West (110 ha) and Phoenix East (98 ha). The area will become a high-grade location for living and working in new economic sectors. Phoenix West (end construction date: 2008) will be an area for high-tech services (MEMS and software) and leisure activities, while Phoenix East will be mainly for living (1,300 new houses), leisure and relaxation. In Phoenix East a lake (28 ha) will be dug to upgrade the living area, improve the environment and to function as a reservoir for high tide water. Phoenix West will contain two incubator facilities: the MST-factory (6,500 m^2; MEMS) and the 'Softwarehalle'. The Phoenix area aims to create 15,000 new jobs and extend the city district Hörde (dortmund-project, presentation II, 2004).

The ICT cluster As mentioned earlier, Dortmund has a huge number of ICT students: they form a powerful reservoir for the ICT firms in the region. Also, the Technology Centre Dortmund and the Technology Park have played a key role in developing ICT activities in the city. At the time of writing the park houses 225 ICT firms with 8,500 employees in the fields of software, telecommunication, multimedia and electronics. Figure 3.10 shows the number of companies and the employment development in the ICT sector. In the Rhine-Ruhr region over 10,000 ICT firms are active; they employ more than 100,000 people (dortmund-project, 2004).

Important institutes for the ICT cluster are: the polytechnic and university, the IT Center; the Fraunhofer Institute for Software and Systems Engineering (ISST); the Fraunhofer Institute for Material Flow Systems and Logistics; and the Forschungsinstitut für Telekommunikation (FTK: research institute for telecommunication). A key networking organisation in the Dortmund ICT cluster is mybird.de: it represents about 850 ICT firms (e.g., Materna, Brockhaus, EDS) from the region. The participants (mostly SMEs) present

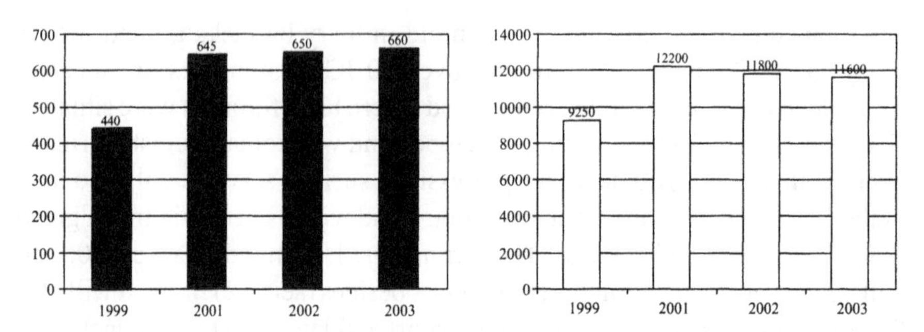

Figure 3.10 Number of ICT firms and ICT employment in Dortmund

Note: no data were available for 2000.

Source: WBF-Do (2004).

themselves via a joint Internet portal that facilitates the search by customers. Mybird also functions as a quality standard (only companies that deliver good quality can be members) and gathers information about customer demands. Furthermore, members tune their business activities to each other. Mybird is financed by member contributions. The state of NRW does not financially support the network; NRW already supports the ICT network of the Ruhr region (WBF, 2004a; dortmund-project, 2004). Both networks cooperate intensely. Plans are in progress to combine the networks.

ICT firms in Dortmund are mostly regionally oriented: they earn the main part of their turnover in a 50 km radius. Most ICT customers are found in the sectors of machinery/equipment manufacturing, ICT, administration/management, energy and waste treatment, automotive, logistics and medicine/health. (WBF, 2004a). Their relationship with the university and polytechnic is rather good: student apprenticeships regularly take place, but the number of joint research projects is limited. A problem for Dortmund's ICT sector is their lack of a clear image: while other German cities are clearly recognised as ICT in, for instance, banking or SAP systems, Dortmund lacks such a profile. The city is now trying to match its ICT sector with activities in (e-)logistics, biotechnology and healthcare: this is not easy, because these sectors have a different way of thinking. The foundation of ICT start-ups in Dortmund is firmly supported both in financial and promotional terms. The city should cooperate more with neighbouring towns.

The MEMS cluster Micro-electronic mechanical systems (MEMS) activities have developed steadily in Dortmund in recent years. The city has four research and incubator institutes in this field (MST-factory, BioMedizinZentrum, the centre for construction and connection technology and the centre for microstructure technology in the Technology Centre). Furthermore, the IVAM NRW headquarters are located in Dortmund. Figure 3.11 shows the number of companies and the employment development in the MEMS sector (dortmund-project, 2004).

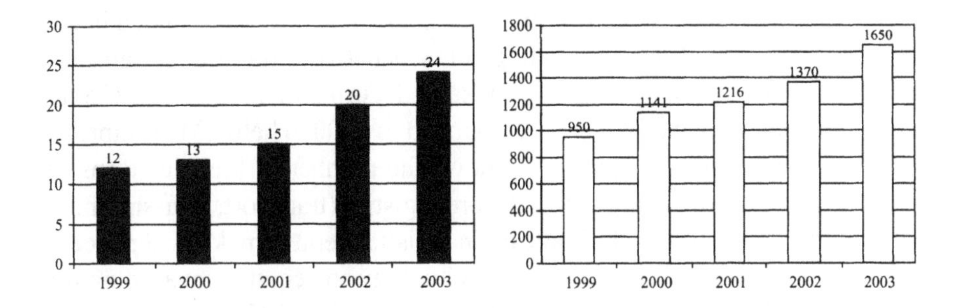

Figure 3.11 Number of MEMS firms and MEMS employment in Dortmund

Source: WBF-Do, 2004.

The total turnover in Dortmund-based MEMS companies increased by 16 per cent in 2002–2003. In 2003, total sector turnover was €248m. Most turnover came from measurement technology, microelectronics and sensors. Customers are mainly the automotive sector, machinery/equipment manufacturing, construction technology and medical technology. Important MEMS firms in the city are: Elmos Semiconductor AG, STEAG microparts GmbH (vaporising technology, micro optics), LIMO GmbH (micro optics, Diode laser) and GfG Gesellschaft für Gerätebau (gas and water measurement technology). The Dortmund-based MEMS firms deliver to customers throughout the world: about 20 per cent of the companies earn considerable turnover abroad. Over 40 per cent mostly have their customer base in German regions other than the Dortmund area and some 20 per cent have a considerable part of their customer base in Dortmund (WBF, 2004a).

The logistics cluster During the formation process of the dortmund-project not all core participants were convinced that logistics should be a core growth cluster. However, in the end it was decided to include logistics in the project.

Turnover in the German logistics sector has risen gradually in recent years. Reasons for this are the increasing importance of e-commerce (where business is done via the Internet and the goods are subsequently delivered through an efficient logistic chain) and the trend towards outsourcing (many firms outsource their logistic operations). In the area of e-logistics (the combination of logistics with ICT to increase efficiency and effectiveness in the logistic process) in particular, large growth opportunities are forecasted. Dortmund has acknowledged these chances and started active supporting policy for the logistics sector. A start-up and competence centre (e-port) was founded in the harbour area and DiaLog (stimulating contacts between traditional transport firms and new e-logistic companies) was started. Furthermore, the eLog-Centre (supported by the state of NRW) was founded in 2000 to help SMEs improve their use of e-logistics. Figure 3.12 shows the number of logistics firms in Dortmund. Most of them are active in core logistics (transport, transshipment and storage: 79 per cent); wholesale trade logistics centres make up 8 per cent of the business and services, suppliers, technology and equipment account for 13 per cent (WBF, 2004a). Two institutions (Dortmund University and the Fraunhofer Institute for Material Flow Systems and Logistics) do research in logistics; the extent of logistics education is rather limited.

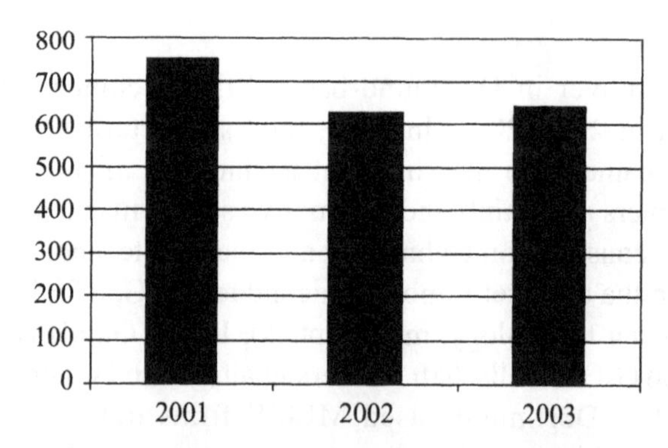

Figure 3.12 Number of logistics firms in Dortmund

Source: WBF-Do (2004.

In core logistics employment declined by 3 per cent in 2002–2003. This decrease is higher than in NRW and Germany, while in the period 1998–2003 core logistic activities grew 7 per cent in Dortmund compared to about 4 per cent in NRW and Germany. Employment in the wholesale logistics centres declined and the number of jobs in services, technology suppliers and equipment increased (see Figure 3.13). A high share of logistic services in the Ruhr region is located in Dortmund (WBF, 2004a).

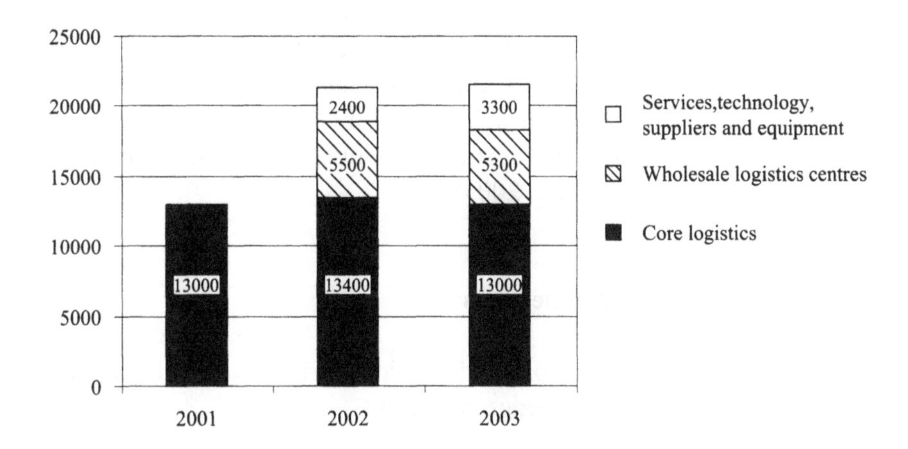

Figure 3.13 Employment in Dortmund logistics

Note: Logistic employment in services and wholesale was not measured until 2002.

Source: WBF (2004).

The importance of the logistics sector for ICT firms is increasing. Currently, about 1,200 ICT employees in Dortmund are active in logistics. Over 8 per cent of all Dortmund ICT firms think the importance of the logistics sector for ICT business will increase in the near future.

Most customers of the Dortmund logistics sector are logistics companies (31.6 per cent): many Dortmund logistics firms work for larger logistic companies. Other important customers are private persons (postal services, courier and express services: 17.5 per cent) and the metal (17.5 per cent) and measurement technology (16.7 per cent) sectors. Most customers are located in Dortmund (75 per cent); only 20 per cent of Dortmund logistics firms have customers from other European countries. Key logistics companies in the city include Deutsche Post AG, Kaufland trade and logistic centre, IKEA storehouse and the Dortmund airport (WBF, 2004a).

The biotechnology cluster Although it is not one of the core clusters within the dortmund-project, the biotechnology sector in Dortmund is of significant size and developing gradually. The Dortmund region houses 23 biotechnology companies; 13 of these firms (they cooperated in a questionnaire research) together have some 450 employees. The biotech firms are active in pharmaceuticals/medicine, food and environment. A large biotech company is Schering AG with 280 employees. Key research institutes in Dortmund biotech are the Max Planck Institute for Molecular Physiology and the new BioMedizinZenter. The university and polytechnic are also important. The biotech sector in Dortmund expects a shortage of well-educated employees in the coming years, if the number of students in biotech fields is not increased (BioIndustry and RAG Bildung, 2003).

3.5 Conclusions and Perspectives

This section contains conclusions and unfolds perspectives for Dortmund's position in the emerging knowledge economy.

3.5.1 Foundations of the Knowledge Economy

The foundations of Dortmund's knowledge economy are moderate to good. While for many decades the city was a strong industrial location, employment in the steel and mining industry has declined sharply since the 1960s. Dortmund has been able to strongly develop new economic activities to create new jobs. However, the new sectors were not able to totally offset the declining industries. Today, heavy industry has been closed down almost completely and development in new economic sectors looks quite positive.

An indication of the value of the knowledge economy foundations in Dortmund is given in Table 3.6. It can be concluded that the city has a rather positive overall score on the foundation indicators. Its *knowledge base* is quite good: the city has many ICT students and various research institutes, among others in ICT, logistics, MEMS and biotech. However, the share of people with third level education (17 per cent of inhabitants aged 25–64) is rather low. Dortmund's *economic base* is also rather good: the city has been quite successful in transforming from an industrial town to a knowledge and services oriented economy. The current economic level of the city is somewhat below average, but the ongoing developments (growth figures) are positive. The *quality of life* is moderate to good: Dortmund is a rather green city and pollution has decreased markedly throughout the years when heavy industries declined. The city has a complete package of leisure facilities and people who were raised in Dortmund tend to stay there after they have grown up. However, the image of Dortmund needs improvement: outside the city (especially abroad) Dortmund is still mostly associated with heavy industry. *Accessibility* is good: Dortmund has little traffic congestion and extensive public transport. Furthermore, the city is a junction in the intercity train and ICE (high-speed train) network. Also, the number of flights and destinations to and from Dortmund's airport is limited. On the other hand, several larger airports are nearby. ICT infrastructure (broadband) seems well developed in Dortmund. *Urban diversity* is moderate: economic diversity is rather broad, but Dortmund does not have an extensive range of artists and other creative workers. Almost 13 per cent of the city's population consists of foreigners.

Table 3.6 An indication of the value of knowledge economy foundations in Dortmund

	Foundations	Score in Dortmund
1	Knowledge base	+
2	Economic base	+
3	Quality of life	+
4	Accessibility	+
5	Urban diversity	□
6	Urban scale	+
7	Social equity	□

-- = very weak; – = weak; □ = moderate; + = good; ++ = very good.

Dortmund's *urban scale* is good. The city is already large enough both to support a specialised labour market adequately and to maintain a range of amenities. The potential for increasing its urban scale is vast: today, the city cooperates relatively little with other large cities nearby, but if such linkages were strengthened the urban scale could be greatly expanded and its attraction for different sorts of people and activities could be improved. *Social equity* in Dortmund is moderate: unemployment is relatively high compared to the NRW and German average. Unemployment among foreign inhabitants is especially high. Dortmund's social structure is below average among German cities, but its position is steadily improving: the percentage of people on social welfare is declining rapidly.

3.5.2 Activities of the Knowledge Economy

Overall, Dortmund scores well on knowledge economy activities. Regarding *attracting and retaining knowledge workers*, the score is positive. The city has a more than sufficient number of ICT students; the size of the foreign student population is quite large and has increased steadily; firms are quite positive about the availability of employees; and, the city has deployed several projects to ensure that sufficient knowledge workers are available in Dortmund in the near future. In *creating knowledge*, Dortmund is performing quite well. The number of patents registered in the city is not very high. However, this might have to do with the limited number of company headquarters located in Dortmund. The city does have various research institutions and in recent years several activities (among others, the dortmund-project) have been initiated to stimulate knowledge creation. *Applying knowledge/making new combinations* in Dortmund is rated good. The number of start-ups is quite high and the city has an extensive number of activities that enhance knowledge application (e.g., start-up competitions and the Technology Park and Centre) and making new combinations (for instance: combining ICT with logistics and e-logistics). The application of MEMS knowledge in private companies is also very well developed. The *development of new growth clusters* in Dortmund is also rated good. At the end of the 1990s the municipality, together with several other parties, chose to focus on a limited number of clusters (MEMS, ICT and e-logistics). The development of these clusters seems fairly successful. The biotechnology sector – symbolised by the newly built 'BioMedizinZentrum' including the transnational 'Proteom-KompetenzZentrum' – is also developing quite swiftly in Dortmund.

Knowledge economy activities in Dortmund have benefited greatly from financial support from the federal state of NRW, the EU and the industrial conglomerate ThyssenKrupp. The state of NRW has invested a good deal in science and education and has actively promoted knowledge transfer and start-up firms (for instance, in biotechnology). The EU has invested considerably in Dortmund because it is a former industrial area that needed economic revitalisation. ThyssenKrupp acknowledged its responsibility as a former major industrial employer which was downsizing its Dortmund activities. The company helped to find new economic sectors that could create new jobs in the city by financially supporting the research conducted by the city of Dortmund and McKinsey which laid the foundations for the dortmund-project.

Table 3.7 An indication of the value of knowledge economy activities in Dortmund

	Activities	Score in Dortmund
1	Attracting knowledge workers	+
2	Creating knowledge	+
3	Applying knowledge/making new combinations	+
4	Developing (new) growth clusters	+

-- = very weak; – = weak; □ = moderate; + = good; ++ = very good

3.5.3 Organising Capacity; Public-private Partnerships

The strong leadership by the city of Dortmund has proved to be a key ingredient for setting up and successfully running knowledge economy initiatives (such as the dortmund-project). The formation of a vision (modernising the Dortmund economy to create new jobs) was soundly elaborated in a strategy that was based upon the thorough analysis of the economic situation in Dortmund by McKinsey, ThyssenKrupp and the municipality. An earlier study financed by the city of Dortmund had laid the foundation for the growth cluster strategy (Rehfeld, Wompel 'Standort mit Zukunftsprofil-Innovationsschwerpunkte in Dortmund' Institut für Arbeit und Technik 1997/99). The various networks that have been formed (through the 'infrastructure' companies discussed in section 9.4.3, among others) have contributed to involving important private companies and research/education institutes in the region in the various projects that often are based upon public-private partnerships. The willingness of the companies and knowledge institutions to cooperate in the projects is quite high, because the advantages of the initiatives are clear and well communicated (increased productivity/competitiveness and a larger highly skilled labour pool for companies; a growing number of students and research assignments for knowledge institutes).

3.5.4 Perspectives

Considering the knowledge economy foundations and activities in Dortmund, some conclusions can be drawn and final remarks can be made.

- Dortmund seems to be on the right track towards modernising its economy. It has had a hard time adjusting to the sharply declining steel and mining industries, but nowadays new growth sectors (ICT, MEMS, logistics and biotechnology) are developing steadily and the city's economic position is improving.

- The cities in the Ruhr region (including Dortmund) have a multitude of strong (economic) assets and opportunities. However, these factors are only partially beneficial since the cities do not yet seem to cooperate optimally. This does not only apply to the municipalities themselves, but also to several kinds of institutions such as universities and polytechnics. By cooperating more intensively, the cities could spend their money more efficiently and effectively, if they assigned the different areas to those cities/institutions best equipped to deal with them.
- Dortmund is doing a good job of further improving its knowledge force by enhancing education and research. With regard to attracting students, this could be further stimulated if the city developed a livelier campus and/or student district. Increasing the number of courses taught in English could contribute to attracting foreign students.
- The extent of new business location developments in Dortmund is impressive. The city should, however, keep in mind that careful distribution of new locations that become available is needed over time. Otherwise the real estate market will become flooded and will stagnate. This could endanger some of the location developments that it would be worthwhile to realise. Furthermore, the construction of a new central railway station seems key: not only will this improve external accessibility, it will also contribute to the city image and revitalisation of the station area.
- The city of Dortmund took a wise decision to focus on a limited number of economic sectors, because the resources available are limited. Still, the sectors should not be supported without critically assessing their development. Eventually, some might prove to offer less promising possiblities for growth while others that are currently not included in the dortmund-project might start to show considerable development potential. This calls for continuous evaluation of the economic developments. The city of Dortmund is very well aware of this need for evaluation: a review of the development of Dortmund's economic sectors was published and an evaluation of the strategy and targets is planned for 2005.
- A field that might need improvement in Dortmund is its cultural sector including all kinds of creative professions (ranging from art to multimedia). This could also help improve the image of the city outside its direct surroundings (especially abroad) if such artistic/creative developments are

marketed the right way. (Please note: this case study did not specifically investigate the cultural activities in Dortmund, so our view of this field is based on a quite limited amount of information.)

- An asset at present not actively utilised is the large minority of foreign inhabitants (mostly Turkish people). Explicitly initiating economic policies aimed at this group could have beneficial effects in two ways: 1) it would help decrease the relatively high unemployment level among this group of citizens; and 2) such an explicit policy might reveal 'new economic combinations' (e.g., in cultural activities or specific economic activities aimed at people from Turkish origin in places outside Dortmund, like strengthening economic linkages with companies in Turkey).

Note

1 The region is neither politically nor statistically defined. The concept of a European Metropolitan Rhein-Ruhr Region was first suggested in the 1995 North Rhine-Westphalia Regional Development Plan (Grier, 2002).

Discussion Partners

Mr Becker, Rector, Dortmund University.
Ms Martina Blank, Project Manager, Technology Centre Dortmund.
Mr Klaus Brenscheidt, Vice Director, Economic and Technology Policy, Chamber of Commerce Dortmund (Industrie- und Handelskammer).
Mr Dirk Brockhaus, Chair, mybird.de (Brand In Region Dortmund).
Mr Frank Gutzmerow, Project Manager, Economic and Employment Promotion Dortmund (Wirtschafts- und Beschäftigungsförderung Dortmund).
Mr Konrad Hachmeyer-Isphording, Project Manager, dortmund-project.
Ms Mechthild Heikenfeld, Universities and Research Institutes, Department for Mayoral and City Council's Affairs, City of Dortmund.
Ms Cornelia Irle, Regional Affairs, Department for Mayoral and City Council's Affairs, City of Dortmund.
Mr Rolf Kinne, Director, Max Planck Institute Dortmund.
Mr Hans-Jürgen Kottmann, director, IT-Centre Dortmund GmbH.
Mr Roland Krumm, Special Projects, Elmos Semiconductor AG.
Mr Utz Ingo Küpper, CEO, Economic and Employment Promotion Dortmund (Wirtschafts- und Beschäftigungsförderung Dortmund).
Ms Heike Mertins, Project Manager, dortmund-project.
Mr Stefan Röllinghoff, Research and Planning Officer, Economic and Employment Promotion Dortmund (Wirtschafts- und Beschäftigungsförderung Dortmund).
Mr Stefan Schreiber, Director, Economic and Technology Policy, Chamber of Commerce Dortmund (Industrie- und Handelskammer zu Dortmund).

Mr Dieter Steemann, Business Unit Manager, Economic and Employment Promotion Dortmund (Wirtschafts- und Beschäftigungsförderung Dortmund).

References

BioIndustry (brochure), *BioIndustry Kompetenz-Netzwerk für das Ruhrgebiet in Bioverfahrenstechnik, BioMEMS und Proteomics*, www.bioindustry.de.

BioIndustry (2003), 'LifeTechnologiesRuhr – Knowledge, Business, Venture', www.life-tec-ruhr.org.

BioIndustry and RAG Bildung (2003), *Biotechnologie in der Region Dortmund – Beschäftigung, Qualifizierung, Personalentwicklung.*

Brödner, B.H. and A. Rücker (2004) 'ICT Cluster Study Dortmund', in A. van der Meer, W. van Winden and P. Woets (eds), *Urban ICT clusters and E-governance policies in France, Germany, Italy and Spain*, Rotterdam: European Institute for Comparative Urban Research, Erasmus University, www.Euricur.nl.

Cash (2004), *Das große deutsche Standort-Ranking* [*The Large German Cities Ranking*], Nos 1–2.

dortmund-project (presentation), *Menschen und Kompetenzen.*

dortmund-project (presentation II), *Zukunftsstandort Phoenix.*

dortmund-project (2000), *Das Zunkunftspaket – Dortmunds Sprung in die Informations- und Wissensgesellschaft*, www.dortmund-project.de.

dortmund-project (2004), *Das neue Dortmund.*

European Union (2004), http://www.europa.eu.int/comm/regional_policy/country/prordn/details.cfm?gv_PAY=DE&gv_reg=ALL&gv_PGM=2000DE160PC103&LAN=5.

Forschungszentrum Jülich GmbH (2003a), Geschäftsbericht 2002.

Forschungszentrum Jülich GmbH (2003b), 'Programme der Bundesländer – Nordrhein-Westfalen', www.ihk-nordwestfalen.de/industrie_in_muenster/.

Frankfurter Allgemeine Zeitung (2004a), *Die Zeit läuft: Das Prestigeprojekt '3do' hat noch keinen Mieter*, P. V10, 13 February.

Frankfurter Allgemeine Zeitung (2004b), *Vom 'Ufo' zum '3do' – die Bahn hat immer noch Zweifel*, P. V11, 13 February.

Geschäftsricht der Stadt Dortmund.

Greif, S. and D. Schmiedl (2002), *Patentatlas Deutschland – Ausgabe 2002*, München: Deutches Patent- und Markenamt.

Grier, C.H. (2002), *Comparative Analysis of the Rhine-Ruhr Metropolitan Region*, Bezirksregierung Düsseldorf (Düsseldorf Regional Government).

Industrie- und Handelskammer zu Dortmund (IHK) (Chamber of Commerce) (2004), 'Brückstraßenviertel erlebt Renaissance', *Ruhrwirtschaft* 2/04.

IW Consult (2004), 'Deutsche Großstädte im Vergleich', Cologne, www.iwconsult.de.

Küpper, U.I. (2004), 'Das dortmund-project – Kräfte bündeln für den Strukturwandel', presentation for Wirtschaftstagung Leverkusen, 22 April.

Landesamt für Statistik und Wahlen NRW (2004), *Bevölkerung neueste Ergebnisse 2004.*

Landesregierung Nordrhein-Westfalen (2004), *Zuwanderung und Integration in Nordrhein-Westfalen*, www.nrw.de.

Langemeyer, G. (2004), *Vorwort Internet, Bildung und Forschung*, www.dortmund.de.

Office of the UK Deputy Prime Minister (2004), *Competitive European Cities: Where do the core cities stand?*, London, www.odpm.gov.uk.

Stadt Dortmund (2003a), *Wettbewerb 'Stadt der Wissenschaft 2005' des Stifterverbandes für die Deutsche Wissenschaft.*

Stadt Dortmund (2003b), *Dortmunder Statistik – Jahrbuch 2003*, Statistik und Wahlen.

Stadt Dortmund (2003c), *Dortmunder Statistik Nr. 165 – Bürgerschaft und Verwaltung.*

Stadt Dortmund (2003d), *Dortmunder Statistik Nr. 163 – Lebensraum Dortmund.*

Stadt Dortmund (2004a), *Dortmunder Statistik Nr. 168 – Bevölkerung, Jahresbericht 2004*, www.dortmund.de/statistik-wahlen.

Stadt Dortmund (2004b), 'Hochschul- und Wissenschaftsstandort Dortmund', presentation sheets.

Stadt Münster (2003), *Multitalent Münster – Wege der Stadt der Wissenschaft 2005 – Die Details.*

Start2grow (2004), 'Gründungswettbewerbe und Wachstumsinitiativen', Powerpoint presentation.

Steemann, D.K. (2004), Powerpoint presentation WBF-Dortmund.

TechnologieZentrumDortmund (TZDO) (2000), 'Die richtige Idee', *TZDO News*, Issue 1.

TechnologieZentrumDortmund (TZDO) (2004), 'LEAD führt Unternehmen zum Erfolg', *TZDO News*. Issue 1.

Windo (2002), *Wissenschaft in Dortmund.*

Wirtschafts- und Beschäftigungsförderung Dortmund (WBF) (1999), *Zukunftsprofile des Wirtschaftsstandortes Dortmund: Technologiestadt und Metropole – Werkstattbericht.* www.wbf-do.de.

Wirtschafts- und Beschäftigungsförderung Dortmund (WBF) (2004a), *Dortmunder Branchenbericht 2004.*

Wirtschafts- und Beschäftigungsförderung Dortmund (WBF) (2004b), http://www.wbf-do.de/home/news_detail.jsp?cid=2300.

Wirtschaftswoche (2004), *Lichtjahre entfernt – Städtetest*, 15 April.

Chapter 4

Eindhoven

4.1 Introduction

At the beginning of the twentieth century, the city of Eindhoven was no more than a small agricultural town, with some 5,000 inhabitants. The foundation of Philips Gloeilampen NV in 1891 marked the beginning of the rapid development of the city (Adang and van Oorschot, 1996). Nowadays, Eindhoven is the fifth city of The Netherlands, with 206,000 inhabitants (2003). It forms the centre of the region of Southeast Brabant, which is often referred to as the Greater Eindhoven area, or the 'Eindhoven region'. Like the city of Eindhoven, the population of this region has grown fast in the last century. Today, it counts some 670,000 inhabitants.

Figure 4.1 Eindhoven in The Netherlands

Geographically, the region of Eindhoven is situated at some 100 km east of the Randstad, the cultural and economic 'gravity centre' of The Netherlands. Nevertheless, its location is by no means peripheral, particularly in a European perspective. The region is situated within the rectangle formed by the Randstad, central Germany, the Ruhr area and the Belgian cities of Brussels and Antwerp.

In Section 4.2, we will explore the knowledge economy strategies of the local authorities. Then, in Section 4.3, we will analyse the foundations of the knowledge economy for Eindhoven. In Section 4.4, we will go into Eindhoven's four knowledge economy activities. Finally, Section 4.5 presents some conclusions and perspectives.

4.2 The Knowledge Economy Strategy of the Eindhoven Region

The knowledge economy strategy of the municipal government, which can be found in, amongst other places, the administrative vision for the 2002–2006 period, can be summarised as follows:

- concentrate on technology development, which is already a regional strength (Eindhoven: leading in technology);
- apply technology in a larger number and variety of economic sectors, for instance, within the medical sector (Eindhoven: leading through technology);
- be more aware of the regional strengths and communicate it to the world (in and outside the region of Eindhoven);
- create better linkages between small and medium-sized enterprises (SMEs) and education institutes.

These four elements will be discussed in this case study.

Besides this strategy, there are two economic action programmes which are relevant for the transition towards the knowledge economy in Eindhoven. These programmes, Stimulus and Horizon, are described below.

4.2.1 The Stimulus Programme

Stimulus is an economic stimulation scheme for the region of Southeast Brabant. The Stimulus programme was set up in 1993 to fight the economic crisis in the region stemming from the heavy restructuring of Philips, the largest employer in the region, and the bankruptcy of DAF Trucks. The plan contained measures to strengthen the regional socioeconomic structure and to reduce unemployment. The total project turnover was about €1bn over ten years, including subsidies (see Box 4.1). At the time of writing, the programme is in its final phase. For Eindhoven it will end in 2006, because it is no longer an Objective-2 region.

One important role of Stimulus was to stimulate cooperation between different actors in the region (firms, institutes, educational institutes) by financing cooperation projects. Stimulus financed 25 per cent of an innovative project if two or more firms were involved; if one of these firms was of small or medium size, the contribution rose to 35 per cent. If a public research institute was also involved, the percentage increased to 45 per cent of the project. More than 100 companies were involved within Stimulus. An important result of the Stimulus programme was attracting TNO Industry (a technology research institute) to Eindhoven. About 400 knowledge workers are employed in this organisation. In addition, Stimulus contributed to the development of high-grade public transport (HOV-line) and supported ASML, a producer of wafer steppers (for making computer chips), to develop new products.

Another innovative example of the Stimulus programme is the KIC-project (*kennis intensief cluster*; knowledge-intensive cluster) of Océ, the copier producer. This project was aimed at stimulating the joint development of new products and technologies of Océ and its more than 100 regional suppliers, for the purpose of upgrading the skills level and the quality of co-production and co-development of the suppliers, and keeping more work in the region.

Related to Stimulus is the Stimulus Venture Capital Fund (SVCF). This fund is meant for SMEs (fewer than 250 employees) in the manufacturing sector of Southeast Brabant. Promising companies which cannot make use, or make insufficient use, of the regular capital market can receive risk capital from the SVCF for a period of 5–7 years. With this fund it is hoped to stimulate starting companies or further stimulate relatively young companies.

Box 4.1 Stimulus facts and figures 1995–1999

- 400 Projects.
- 1,950 Companies involved in subvention scheme.
- 60 Cluster projects with 200 participants.
- 132,500 M^2 restored business floors.
- 723 Ha. Revitalised business parks.
- 412 Ha. Newly developed business parks.
- 33,500 Courses for (un)employed.

- 4,000 New jobs (end of 1999).
- Investment incentive: €620m.

4.2.2 *The Horizon Programme*

At the end of 2001, the former mayor Welschen, together with the Eindhoven region (SRE: 22 municipalities), came up with an initiative after Stimulus: a commission led by the president of the Technical University of Eindhoven (TU/e) created a strategic action plan. This plan aimed to concentrate on opportunities instead of problems. The essence was to catalyse developments that contributed to the reinforcement of the economic structure over the mid-term for the whole region of Southeast-Netherlands, in which Eindhoven is located. The programme aims to:

1 improve the labour supply. Reducing the structural labour shortage by, among other things, solving a mismatch between the national education system and the regional labour market needs (within the Eindhoven region there is a larger demand for technical educated workers than the national average);
2 increase commercialisation of technology by improving innovation and/or market competences;
3 decrease cyclical sensitiveness by diversification; apply technology to a wider range of economic sectors.
4 internationalise. The international profile of the region is considered to be too low; both the product 'Eindhoven region' and communication about it need attention.

Twenty-five projects have been identified within Horizon. Although the programme is still relatively new, it is already considered quite successful because of the willingness of all relevant actors to participate. This has to do with the region's culture of mutual solidarity. All participants appear to endorse the importance of the new programme for the economic development of the Eindhoven region.

Each project has its own leader. These leaders are people with no direct interests but with considerable knowledge and reputation in the region. The leaders have the task to create sufficient support for the projects. The pattern of Horizon is to build commitment. A steering group, which includes all key players, meets twice a year: the key players are not allowed to send a substitute.

The daily board meets four times per year. A monitoring commission informs the SRE about Horizon developments.

Horizon is not a subsidy programme. The SRE funds Horizon with €770,000; the province and firms also contribute financially. This implies €35,000 per project. Until now the programme did not receive EU funding. The programme aims to set in motion a multiplier effect; more funds have to be generated by coming up with appropriate projects. The funds available should thus work as lubricant for obtaining financial means through other organisations.

Examples of projects within Horizon are:

- encouraging the flow of (technology) students from lower via intermediary to higher education (MBO-HBO-WO) within a dual system (learning and working). Through this project, it is hoped to improve the knowledge level of the region and to raise the number of students in the higher education institutes (the Dutch name of the project is 'Competentie OntwikkelingsPunt' (COP));
- metal house (Metaalhuis) is an initiative in which new knowledge on metal manufacturing is disseminated among SMEs. The idea behind this project is that, by disseminating knowledge, more SMEs can be kept in and attracted to the region. Within this initiative, evenings are organised around concrete themes, such as how to deal with customer management, cheap labour from Eastern Europe, and how to organise exporting activities. At these meetings, entrepreneurs tell their colleagues practical stories about the problems they faced and how they solved them. The meetings are informal and there are no barriers to join the meetings. The number of attendants is high. The Stimulus project has contributed financially to the Metaalhuis. Now, some small sponsors have to be found to cover the entire budget;
- realising new techno start-ups (Incubator 3+). This includes new housing for start- ups with growth possibilities at R&D locations, such as the High Tech Campus Eindhoven and the TU/e campus. Buildings and facilities for incubators have been present in Eindhoven for some time, but now the incubator facilities are better incorporated into the Incubator 3+ programme. A foundation, covering all relevant parties in the region, operates as steering committee. '3+' stands for more, faster and better techno start-ups. The most important asset of the Incubator 3+ project is the availability of pre-seed and seed capital. Also, there are facilities for firms that have come to the end of the start-up phase: these people are supported in finding laboratory facilities and housing. Over the pilot period

from April 2003–April 2004, 70 companies made use of the facilities of Incubator 3+. Ten of them obtained pre-seed financing to further elaborate their plans;[1]

- development of the Philips High Tech Campus to an international R&D site: High Tech Campus Eindhoven. External technology companies may also locate on the campus to encourage new synergies and cooperation on R&D. This should contribute to the concentration of knowledge-intensive high-tech businesses and support the 'open innovation' model that the region seeks. Research and development is no longer about top researchers working in splendid isolation. To create effective and efficient R&D and to speed up innovation processes, leading companies adopt 'open innovation' – working closely together with suppliers, universities, technological institutes, customers and sometimes even competitors. Proximity becomes more important, as various clusters around the world prove (for example, Silicon Valley). Royal Philips, one of the world's leading high-tech companies, has decided to open up its state-of-the-art high-tech campus in its hometown of Eindhoven for other companies. A number of new buildings have been built to help fulfil this aim. At the time of writing, as well as Philips, EMEA Recruitment Services, Dalsa corporation, Fluxxion, and the Foundation for Fundamental Research on Matter are located on the campus. At the high-tech campus there are specific technical facilities and service facilities, like restaurants, where knowledge workers can easily meet each other.

4.3 Knowledge Foundations

In this section, the knowledge foundations – as discussed in the theoretical chapter – will be assessed for Eindhoven. They include the knowledge base, the economic base, the quality of life, the accessibility, the urban diversity, the urban scale, and the social equity in the city.

4.3.1 Knowledge Base

The knowledge base of Eindhoven is predominantly technically oriented. The most important institutions in this respect are the Philips high-tech campus, TNO-industry, the Technical University of Eindhoven (TU/e) and the polytechnics. The first two will be discussed below, the TU/e in Section 4.4.2 and the polytechnics in Section 4.4.1.

Private R&D in the Eindhoven region Compared with other important industrial regions in The Netherlands, Southeast Netherlands scores very highly for innovation: 45 per cent of total Dutch R&D expenditure is spent within the region. Within the province of North Brabant, the R&D level exceeds the Lisbon challenge; with 3.2 per cent of gross domestic expenditure on R&D. With these figures, Eindhoven ranks among the top three of most knowledge-intensive regions in Europe (see Box 4.2), together with Stockholm in Sweden and Uusima in Finland (Horizon Programme Agency, 2003). The region's high knowledge intensity is generated by large firms such as Philips, ASML, Océ, and PDE Automotive, which spend large amounts of money on R&D. Tegional knowledge institutes such as the TU/e, institutes for higher education and private research firms also contribute to the innovativeness of industry.

In 2000, the R&D costs of Philips were 7.3 per cent of sales (€2,766m) (van den Biesen, 2001). Worldwide, there were 22,000 people working for Philips in R&D: 19,000 in product divisions and the Centre for Industrial Technology and 3,000 in Philips Research. Of the latter group, 1,650 R&D employees are working at the Natlab (physics laboratory) in Eindhoven, 400 in Aachen (Germany) and a group in Leuven (Belgium). The Natlab is the largest private R&D centre of The Netherlands and is located at the Philips high-tech campus.

Box 4.2 Regional Innovation Scoreboard 2003: Indicators

Human resources

Population with tertiary education (% of 25–64 years age class).
Participation in life-long learning (% of 25–64 years age class).
Employment in medium-high and high-tech manufacturing (% of total workforce).
Employment in high-tech services (% of total workforce).

Knowledge creation

Public R&D expenditures (GERD – BERD) (% of GDP).
Business expenditures on R&D (BERD) (% of GDP).
EPO high-tech patent applications (per million population).
EPO patent applications (per million population).

Transmission and diffusion of knowledge

Share of innovative enterprises (% of all manufacturing enterprises).
Share of innovative enterprises (% of all services enterprises).
Innovation expenditures (% of all turnover in manufacturing).
Innovation expenditures (% of all turnover in services).

Innovation finance, output and markets
Sales of 'new to the firm but not new to the market' products (% of all turnover in manufacturing).

The Noord-Brabant Province, with the Eindhoven region as a high-tech nucleus, ranked third (2002) and fourth (2003) on the European Regional Innovation Scoreboard.

Research: TNO-industry A recent new element in Eindhoven's knowledge infrastructure is the research institute TNO-Industry. This institute carries out contract research for firms, very often in cooperation with technical universities. It aims at developing solutions in the field of product development, production technology and materials technology. Most of the 400 employees are high-level scientists. In 1997, the Institute decided to move to Eindhoven, to the campus of the TU/e. The reasons for this were twofold. Firstly, the majority of TNO's clients are situated in or very near to the Eindhoven region. Secondly, it was felt that TNO could realise synergy from its location so close to the University of Technology, which may help to offer better solutions for its clients. Cooperation will take the form of knowledge exchange, staff exchange, and joint projects in the top-institutes metals and polymer. Furthermore, research cooperation is planned in the field of materials technology. The Stimulus subsidy scheme has contributed €7m to attract TNO.

Table 4.1 shows the number of students in Southeast Brabant. What is striking in this table is the large number of technical students: almost half of all students in Southeast Brabant (more than 21,000). In addition, all students at the university are technical students.

4.3.2 Economic Base

The Eindhoven region ranks among the more prosperous in The Netherlands. Incomes per capita are well above the Dutch average. The region of Eindhoven is an important centre of employment. It counts some 307,000 jobs on a population of 670,000.

During the last 25 years, the economic fortunes of the regions have changed several times. Up to the mid-1980s, the region did relatively well, with growth rates higher than the national average. From then on, however, regional employment growth has lagged behind national averages. Between 1986 and 1991, the yearly employment growth rates dropped from 5.3 to

Table 4.1 Number of students in Southeast Brabant (per subject, in 2000/01)

	Agri-culture	Economy	Care and well-being	Technical	Various	Total
Basic level (VBO)	185	679	1,486	2,170	49	3,922
Intermediary level (MBO)	–	5,529	5,654	8,360	–	19,543
Polytechnic	–	5,528	4,268	4,671	3,922	18,119
University	–	–	–	6,076	–	6,076
Total	185	11,466	11,408	21,277	3,971	48,307

Source: CBS, in NV Rede (2002).

0.9 per cent. From 1991 onwards, things got worse. On balance, the region started to lose jobs. Up until 1994 the regional employment figures continued to drop. The main factors behind the economic problems of the region were a reflection of severe difficulties of the two leading industrial firms in the region: Philips (electronics) and DAF (lorry construction). In the early 1980s, Philips employed 35,000 people in the region. This figure had dropped to 16,000 by 2000. This decrease was also due to outsourcing of services in the region. Philips has moved many manufacturing activities to 'low-wages countries' (for instance, to China). This process is expected to continue. In the mid-1990s, Philips moved its headquarters to Amsterdam. Important reasons for this decision were the vicinity of the international airport Schiphol and the presence of a more international business community. Although moving the Philips headquarters to Amsterdam concerned only some 100 of the 16,000 Philips employees in the Eindhoven region, it was considered to be negative for the Eindhoven region, since the fact that the decision-makers are located elsewhere might have some consequences for any feelings of solidarity in the company towards the Eindhoven region. In 1993, the DAF company collapsed, involving the loss of another 2,500 jobs. Additionally, the large network of external suppliers in the region that worked for DAF was also severely hit. After these difficult years, the region recovered strongly in the mid-1990s.

Along with a general recovery of the economy in the second half of the 1990s in The Netherlands, the severe restructuring of Philips and DAF in the beginning of the 1990s has generated positive spin-off effects in the form of newly started businesses and the outsourcing of activities. Nowadays, although the economic structure of the Eindhoven region is still dominated by industry, predominantly manufacturing companies, employment in trade, business services and health and welfare services has grown substantially. The share of manufacturing industry in the regional economic structure (in terms of both employment and added value) is one of the highest in The Netherlands. The region hosts some fast growing high-tech companies such as ASMLithogaphy, Simac and Neways (see Table 4.2 for the most important employers). The industry is very modern and knowledge intensive. Many companies are related to Philips.

Table 4.2 The most important employers in the Eindhoven region

Firm	Industry	Employees in the Eindhoven region
Philips Electronics	Electronics	20,000
Rabobank	Financial	4,000
Daf Trucks	Automotive	3,750
Stork	Manufacturing	3,160
ASML	Manufacturing	2,130
ATOS Origin	ICT	1,600
VDL	Automotive	1,500

Source: NV Rede (2002).

Currently, the economic downturn has quite a strong impact on the region: this has to do with a domino effect. Companies such as ASML obtain 90 per cent of their input from firms in the region. Thus an economic downturn has negative accelerating effects within the region. In this respect, the regional-economic structure of the Eindhoven region is still considered to be too homogeneous and, consequently, too vulnerable to market changes. A more diversified structure is desirable. Supply industries should not be dependent on just a few large clients but on a larger portfolio of clients.

4.3.3 Quality of Life

The appearance of the city of Eindhoven is determined by urban design of the twentieth century. In the city centre, buildings and housing estates offer a cross-section of twentieth century (housing) architecture. The city's fast growth and Eindhoven's preoccupation with technology have also left their mark on the city's appearance. Although Eindhoven is the fifth city of The Netherlands, up until the 1980s the urban environment did not live up to that status. The town centre looked chaotic, and apart from the Evoluon, the exhibition centre for new technology, there were no special attractions.

Since the 1980s Eindhoven has put a great deal of effort into city renewal. A high quality shopping centre (Heuvelgalerie) and a concert hall (Muziekcentrum Frits Philips) have been opened, as well as a multifunctional building (the Witte Dame, a restructured large Philips building). Moreover, the Van Abbe Museum of Modern Art has been extended. Eindhoven is thus trying hard to bring its urban environment into line with its economic position. In addition, one of the advantages of Eindhoven is its attractive surroundings. In the immediate vicinity there is a good deal of open space, relatively speaking, with, amongst other things, woods and heathlands.

One of the drawbacks, however, is a lack of attractive facilities for (young) knowledge workers. There are insufficient attractive central facilities (restaurants and pubs) where people can meet each other. There is no vibrant metropolitan atmosphere in the city centre for this. Some argue that the City of Eindhoven should change its policies in this field. Facilities like hotels, restaurants and leisure should be concentrated in certain areas to improve the attractiveness of the city centre.

Eindhoven does not have a metropolitan image, and it will be quite difficult to acquire such an image. Some argue that Eindhoven must not promote itself as a metropolitan area but, rather, stress the village-like living style in the city. Thus, instead of trying to be a metropolis, Eindhoven could designate itself 'the nicest provincial city in Europe'.

4.3.4 Accessibility

During the last decade, the number of passengers using the regional airport of Eindhoven has grown dramatically (210,000 in 1995 and 420,000 in 2003). In particular, the advent of the low-cost carrier Ryanair stimulated this growth. There are scheduled flights to Amsterdam, Birmingham, Gerona (Barcelona), Hamburg, London (Heathrow and Stansted), Paris (CdG) and Rotterdam. The regional airport is considered to be an important location factor for knowledge-intensive companies. Inhabitants and workers of the Eindhoven region can make use of three other airports: Amsterdam Schiphol, Düsseldorf and Brussels. The rail connections to these airports could, however, be improved. The rail connection to Schiphol will be better as from 2005. In addition, Eindhoven has no good connections to the European high-speed-train (HST) network. There are no high-grade rail connections to HST stations, such as Antwerp and Cologne (van den Berg, de Langen and Pol, 1997).

One of the major points of attention for Eindhoven's accessibility is the completion of the highway ring. On the east side of Eindhoven there is no highway link. This is one of the causes of regular congestion on the existing highways west and south of Eindhoven. Several alternatives for a new link at the east side of the urban region are now being investigated. The realisation of this link is, however, being opposed by inhabitants of villages east of Eindhoven, who fear the negative effects of the new infrastructure. All our discussion partners agreed, however, on the crucial importance of this road connection for the economic development of the urban region.

Within the Eindhoven region an innovative project, *Kenniswijk* (knowledge district), is running, aimed at speeding up the development of the information society. Within a designated area (parts of Eindhoven, Helmond and Nuenen, in total almost 84,000 inhabitants), homes will be connected to a fibre-optic broadband infrastructure. Companies, local government and other organisations will develop all kinds of services for the population. The project is considered a test bed for the information society of the future. It was initiated by the Dutch Ministry of Economic Affairs, and carried out by a public-private partnership of 27 shareholders. Yearly research will be undertaken to study the effects of the project. *Kenniswijk* is considered to be part of the government's KE activities. For this project the local government will construct the ICT-infrastructure itself, but arranges matters, for instance through demand bundling.

4.3.5 Urban Diversity

Within the Eindhoven region there is an increasing number of foreign inhabitants, with 20.1 per cent of the working population of the region (262,000 in 2000)[2] being of foreign descent. The majority comes from Western countries.[3] Research shows that foreign inhabitants have on average €1,000 per year less to spend than the Dutch inhabitants (see Table 4.3). There are, however, large differences between Western and non-Western inhabitants. The former earn €2,700 more than the latter. It can be assumed that most of the Western foreigners are knowledge workers. In 1996, when 7 per cent of the working population was unemployed, 14 per cent of that percentage were foreigners while 6 per cent were Dutch. Unemployment amongst foreigners was thus relatively high. In 2002, with 3 per cent unemployment, there was no longer much difference between foreigners and Dutch inhabitants (4 per cent and 3 per cent respectively).[4]

Despite the substantial percentage of foreign workers, there is no real international community in Eindhoven. There are two international schools, a primary school with 425 pupils and a secondary school with 300 students

Table 4.3 Disposable income of Dutch and foreign inhabitants in Eindhoven

	Total	Dutch	Western foreigners	Non-Western foreigners
Number of persons with a full-time job	135,500	106,000	15,400	13,000
Percentage		79	11	10
Average disposable income	€16,100	€16,300	€16,600	€13,900

Source: Gemeente Eindhoven (2003).

(from 38 different nationalities). At the time of writing the schools are in the process of merging in an attempt to strengthen the role of this school. The international school contributes to attracting more foreign well-educated employees to Eindhoven (who want to send their children to an international school). In addition, the cultural facilities within Eindhoven do not have an (extensive) supply of products for an international community; most of the products are in the Dutch language. This is considered to be a barrier to attracting (more) foreign knowledge workers.

4.3.6 Urban Scale

Eindhoven is the fifth city of The Netherlands. With 206,000 inhabitants in the City of Eindhoven and 670,000 inhabitants in the region, it can be considered a relatively small city. This relatively small size is considered to be an advantage, as key people all know each other. However, there are also drawbacks, such as insufficient (national and international) political power, and insufficient attractiveness of metropolitan facilities.

This is also being acknowledged by the local decision-makers. One of the instruments to increase the political force is the Horizon Programme. The cooperation in the 'triple helix' (government – knowledge institutions – business world) fuels the drive to excel as a top technology region. The Horizon Programme functions as a catalyst for economic development and also demands proper spatial conditions for the further improvement of the business climate. The Eindhoven Region's important contribution to the Dutch economy has recently been acknowledged by the Dutch government. The Eindhoven Region has been appointed as a 'Brainport' in the national Spatial Planning Memorandum.

The City of Eindhoven also wants to set up a network with the cities of Aachen (Germany) and Leuven (Belgium). The three cities are all cities with a relatively high degree of R&D in technology. They all house Philips' research facilities: among other things, this company initiated cooperation between the urban regions. The three cities established their cooperation in a declaration of intent to become a top-tech region by 2010. Some compare the city triangle of Aachen, Leuven and Eindhoven with Silicon Valley and want to call it the 'Intelligent Systems Valley'. The scale is about the same and both regions contain medium-sized urban centres with high-tech activities. One of the ideas for stimulating cooperation is to construct up-to-date glass fibre (ICT) infrastructure between the three urban regions.

4.3.7 Social Equity

In Eindhoven there is a relatively large group of long-term unemployed. In 1993, when many people lost their jobs at Philips and DAF, unemployment rose from 30,000 to 40,000. In 1996, 7 per cent of the working population of the Eindhoven region was jobless; in 2002, this share was 3 per cent.[5] In 1996, the unemployed were fairly consistently distributed among the poorly educated (9 per cent), average educated (7 per cent) and well educated (6 per cent). In 2002, predominantly low educated people were without a job (6 per cent); 2 per cent of the average educated and 3 per cent of the well educated were jobless. In 2003, unemployment rose within one year from 18,000 to 25,000.

At the time of writing, some well-educated technicians are also unemployed. In Eindhoven there are specific programmes to educate the temporarily unemployed. People who lost their jobs because of the economic downturn can obtain dedicated education in order to strengthen their knowledge, capabilities and job opportunities in anticipation of better economic circumstances.

About 12 per cent of the total number of households lives around the poverty border (social minimum; the Dutch average is 13 per cent). In total, they number about 20,000 inhabitants. Within this group, there is a large share of poorly educated people and a relatively strong presence of ethnic minorities.[6]

4.4 Knowledge Activities

In the last section, the foundations of Eindhoven's knowledge economy were discussed. Now we turn to the knowledge activities and ask how well the city (and the actors in it) manage to: 1) attract and retain talented people; 2) create new knowledge; 3) apply new knowledge and make new combinations; and 4) develop new growth clusters.

4.4.1 Attracting and Retaining Knowledge Workers

Eindhoven has quite a gravitational pull for knowledge workers. This has to do with the good opportunities to find a job. Even so, it is hard to keep Eindhoven's attraction rate for knowledge workers at a high level. Possible causes for that are the relatively weak international image of Eindhoven: the city is not sufficiently attractive to students and Eindhoven does not have a sufficiently vibrant metropolitan atmosphere.

Technical education in Eindhoven In the region of Eindhoven, technical education is present at all levels, from secondary education to the university level (see Table 4.1 and Figure 4.2). Below, two institutes will be described: the Fontys polytechnic and the Design Academy. The Technical University of Eindhoven (TU/e) will be dealt with in the next section.

Fontys has establishments throughout The Netherlands, but its centre is Eindhoven, with 16,000–17,000 students. Fontys has three faculties and 25 disciplines. It focuses on education in technical (HTS) and economic (HEAO) studies. However, the polytechnic also offers education in other fields. Fontys offers arts studies in Tilburg, since this is a field that does not belong in Eindhoven. The three Fontys Eindhoven establishments are all centred around the TU/e. One of the points for attention is to make the polytechnics more attractive for teachers and to upgrade the quality of the teachers. Attracting and retaining sufficient qualified staff appears to be difficult; this will, however, be crucial considering the competition with other polytechnics and the university.

Design and creative industry and design educationn are both important elements in Eindhoven's knowledge infrastructure. The City of Eindhoven houses the Design Academy (evolved from the School for Industrial Design), and attached to this, the European Design Centre, both of which educate students in the designing of industrial products. At the Eindhoven University of Technology the Master's Industrial Design programme started in September 2004 when the first students of the Bachelor's Industrial Design programme (started in September 2001) have received their Bachelor's diplomas. As the importance of industrial design in new products is increasing, these institutes are considered to be an important asset for the Eindhoven region in the transition process towards the knowledge economy.

Vocational training The VMBO is a secondary school for handicrafts students, amongst others. Six years ago this school system was dramatically restructured nationwide. Several school levels were merged into one new system. Many people admit that these changes were not an improvement.[7] One disadvantage is the combining of pupils with learning backlogs with regular pupils resulting in lower average achievements. Another drawback is that handicraft students have to do abstract subjects such as mathematics, which they are often not good at, resulting in a lower percentage of pupils passing their exams. Within The Netherlands there is the conviction that the school system has to be changed again. This feeling is, however, even stronger in Eindhoven, where technical education is essential to fill the demand for technical employees.

Eindhoven has an important manufacturing industry. During previous decades, many production plants moved to countries where it was possible to pay lower wages. This is certainly true for relatively large plants. There are, however, still economic possibilities for relatively small manufacturing industries with high added value. In such industries there is a need for craftsmanship. That is an important reason to educate sufficient artisans. This component is often missing in the ideas within The Netherlands to stimulate the knowledge economy. All attention is focused on high-grade education and research, but it is easily forgotten that knowledge must be applied in small industries with high added value, for which you need good craftspeople. A better-functioning artisan education is thus also considered essential for the Eindhoven region. The regional strategy on technical education involves all levels of vocational training.

In The Netherlands more investment in education is needed According to our discussion partners, investing in education will be a key factor to improving regional competitiveness. In previous decades the number of students has dramatically increased, which has led to an explosion in educational costs. As a consequence, governments cut financial contributions to educational institutes. In the short run this is good for government budgets, but in the longer term it could badly harm the countries' economic growth. Other countries appear to perform better in this respect. In the Anglo-Saxon countries where private systems are prevalent, increasing investment in education is the trend. The Scandinavian countries are also performing better regarding investments in education. It is argued that if the Dutch government is not willing to invest more in education, it should give schools, polytechnics and universities much more freedom to set their own tuition levels. As a consequence they would have the possibility of obtaining more means to raise the quality of the education.

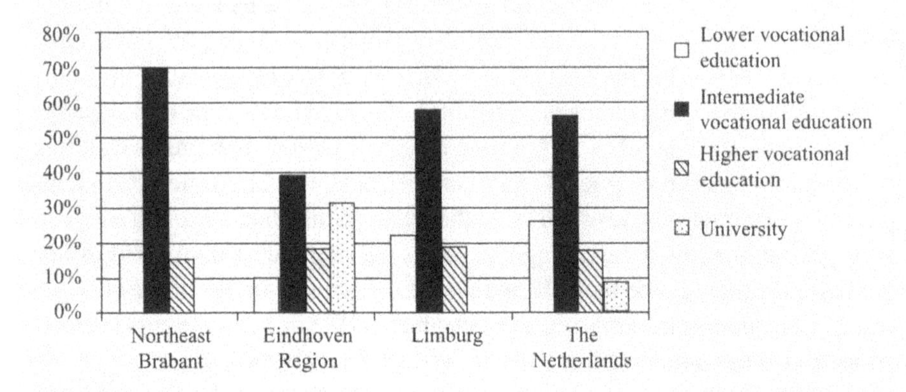

Figure 4.2 Distribution of technical students over education levels (in %) (2001–2002)

Source: CBS, Eindhoven University of Technology, Tilburg University, Maastricht University.

However, this is not happening, which puts the educational institutes in an awkward position.

Technology and the new generation Another worry is the shortage of students in sciences. In a city that wants to strengthen its role as a technology city, it is important to make children enthusiastic about technology and the sciences. However, young people's interest in sciences is still relatively low. This also applies to teenagers with a foreign background. For stakeholders in Eindhoven, it is worrying to see that a large number of non-native Dutch teenagers mainly has an interest in white-collar professions and studies, while there is a growing need for more technically-oriented students.

The people shortage in sciences also has to do with an image problem: being a 'nerd' in The Netherlands is not seen as fashionable. The Netherlands is very much a trading nation. In addition, engineers are not known as enthusiastically career-minded. For Eindhoven, it is considered crucial to turn this tide. Policy measures to be taken could be to offer attractive technology education at the primary school and to create exhibition and learning centres on technology (for instance, in the New Evoluon and in the Eindhoven Experience Centre).

Matching demand and supply The matching of demand for and supply of education could be better. By enhancing the networks between the higher education institutes and the companies this should be improved. Both types of organisation should inform each other regularly of their needs and their education programmes.

4.4.2 Creating Knowledge

Since the 1960s, private knowledge development in The Netherlands has been dominated by six large manufacturing companies: Philips, AkzoNobel, Unilever, Shell, Océ and DSM (Stichting Nederland Kennisland, 2003). Together, they generate about 44 per cent of Dutch R&D investments. Three of these companies (Philips, Océ and DSM) are located in the southeast part of The Netherlands.

At the Philips high-tech campus the number of generated patents is relatively high. The Eindhoven region is European leader, with the largest number of patents (550) per 100,000 active workers (second is Oberbayern with 160 (CBS, Etin Adviseurs)). Most of these patents are generated at the Philips campus. The Philips Natlab is, however, quite a closed institution; there is a good deal of internal knowledge exchange, but not much with external partners.

The Eindhoven University of Technology (TU/e) An important player in the research field is the TU/e. This university was founded in 1956. After the war, Dutch policy makers were convinced of the leading role of the industry in the reconstruction of the country: the number of technically skilled people at the highest level had to be increased. The presence of Philips in the region, in particular the NatLab, was an important argument for founding the new university in Eindhoven.

In The Netherlands as a whole, there are only three dedicated technical universities. The other two are located in Delft, in the western part of The Netherlands, and in Twente, in the east. Recently, the three technical universities decided to increase cooperation. In the field of research, the three universities are going to mutually fit their programmes within the '3TU Institute for Science and Technology'. Abroad, the three technical universities will promote themselves as one organisation.[8]

The TU/e employs 3,100 people within nine departments. These are Biomedical Engineering, Building and Architecture, Electrical Engineering, Chemical Engineering and Chemistry, Applied Physics, Technology Management, Mechanical Engineering, Mathematics and Computer Science and Industrial Design. Every year, some 1,000 students graduate. Seventy per cent of the 6,000 students originate from the surrounding region (Brabant and Limburg). Doctoral students, however, are attracted from more distant places;

between 50 per cent and 60 per cent of the PhD students (500–600) come from abroad. The university offers a complete range of education in many technical disciplines, and carries out a broad range of research, both fundamental and for the industry. The research priorities are clustered into three research areas: biomedical technology, adaptive systems, and new materials. The university stimulates interdepartmental cooperation, because it expects breakthroughs especially in the border regions of scientific fields.

The Technical University of Eindhoven scores very high in Europe regarding scientific publications. On the impact score of publications, it is ranked third after the universities of Cambridge and Oxford (EC, 2003).

4.4.3 Applying Knowledge/Making New Combinations

An excellent knowledge base is not sufficient for a flourishing knowledge economy. It is equally important to apply knowledge through (new) products for which there is a (potential) need in society. Thus to make money from the knowledge created in the region, it should be applied in an appropriate way. One of the sayings in Eindhoven is therefore 'knowledge, skills, cash' (*Kennis, Kunde, Kassa*).

To apply knowledge it is normally important to have good relationships between relevant partners, such as between knowledge institutes and firms. It is argued that within the Eindhoven region there is a good atmosphere of cooperation. This can be explained by a strong solidarity feeling within Eindhoven, amongst other things. Relevant actors often prefer to find partners within the Eindhoven region itself. Moreover, there are a number of relevant networks that help actors to find partners to cooperate with. Examples are the manufacturing circle (*de fabrikantenkring*), the employers' circle (*de BZW-kring)* and the industrial circles.

Good internal networks: however, external networks could be strengthened Within Eindhoven internal networks function very well. Key people know each other, and communication lines are short. What can be improved, however, is the communication with and lobbying of external partners, such as the national and European governments. This communication is sometimes too diffuse. By coming up with one clear regional message for outsiders, better results could be achieved.

Business ambassadors of the region Another key to success is trust and willingness to look beyond the short term. Stakeholders are often willing to help and support promising cultural and economic initiatives. New initiatives contributing to the welfare of the urban region can be expected from both multinational and medium-sized companies. Medium-sized companies are expected to show more corporate community involvement than the large multinationals. But the multinationals also play an important role in supporting regional economic development. Businesses are aware of the growing importance of acting on a regional level. The 'open innovation' model and the transformation of the Philips High Tech Campus into the High Tech Campus Eindhoven prove the involvement and commitment of stakeholders.

Start-ups The Incubator 3+ initiative was mentioned under the Horizon programme. Within the Eindhoven region there have been more start-up initiatives. In particular, NV REDE manages five incubator centres. Over the period January 1998–December 2002, 163 techno start-ups were established in the Eindhoven region: 75.9 per cent of them were profitable and 21.6 per cent were R&D intensive. Fifty-seven per cent of these techno start-ups were in ICT and the Internet, 8 per cent in mechatronics, 2 per cent in medical technology, 2 per cent in embedded software, 1 per cent in biotechnology and 30 per cent in other technologies.[9]

Higher education institutes linkages It is argued that the application of scientific knowledge created within the Eindhoven region could be much better. A considerable barrier in this respect is the fact that scientists often do not have direct incentives to look for applications for created knowledge. Scientists are predominantly judged on the basis of their publications. They participate in worldwide networks and spread created knowledge over the world. However, they could be more involved in high-grade production activities within the region.

The TU/e does have good contacts with the large companies. The university employs 120 full-time professors, compared to 130 professors who also have jobs in private firms. For instance, quite a number of Philips employees also work as professors at the TU/e. The TU/e research programme is being evaluated: in the electro-technology faculty talks with Philips have been conducted to evaluate the degree to which the TU/e activities connect with Philips' business. The results indicate that the connection is quite good. However, Philips micro-electro is situated in Leuven, because that city is a considered to be a better knowledge place for this technology field than Eindhoven.

There is a need for better relations between the polytechnic and the university. New knowledge is particularly being created within the latter institute, while the former concentrates on education. The polytechnic can be considered to be a *circulating system* for knowledge and the university a *creating system* for knowledge. However, within the *circulating system* there is a need to continuously update knowledge, in order to be sufficiently attractive and to better fit to societal needs. The polytechnics therefore want closer links with the universities to obtain this knowledge earlier. A major barrier to realising this is (again) the lack of incentives for scientists to exchange such knowledge.

SMEs and education institutes linkages SMEs should operate more and be positioned more as innovative industries. This is one of the key points of the KE-vision of the Eindhoven administration. Smaller firms, however, have too still few linkages with knowledge institutes. The linkages between universities and multinationals are good, but there is little interaction between universities and SMEs. Smaller firms are often not interested in longer-term projects. Moreover, SMEs find the universities difficult to approach. In addition, the universities are not particularly interested in linkages with SMEs, because they normally have neither research departments nor (large) budgets for research.

In this respect, the polytechnics are considered to be more natural partners for the SMEs than the universities as the polytechnics are more regionally oriented and more focused on practical experience than the universities. The relations between polytechnics and companies are in practice good, mostly because of student apprenticeships.

One of the key questions will be how the interests of SMEs towards long-term R&D activities can be stimulated. Maybe there should be appropriate interest organisations of SMEs to improve the linkages between SMEs and HEIs. The knowledge networks developed by the Horizon Programme (metal, (bio)medical, car, sports, broadband, automotive, emmbedded systems and design) can play an important role in this. In addition, many cooperative efforts with SMEs are troublesome because of tight rules and subsidy restrictions. More freedom for policy makers in firms, government and knowledge economy institutions is therefore desired. Also, organisations such as Syntens (an innovation network for SME entrepreneurs) and Senter (an agency of the Ministry of Economic Affairs, responsible for the carrying out of programmes in technology, energy, environment, export and international cooperation) play a supporting role as intermediaries between knowledge economy institutions and firms. Syntens, however, is directed from a national level; perhaps it would be better to steer this initiative on a regional level, since at this level, there would be a better knowledge of the specific regional circumstances.

New combinations Much is expected from the application of technology in some new fields such as health care and sports and one of the spearheads of the knowledge economy strategy is the combination of technology and design. By combining manufacturing with design market, opportunities can be improved (for example, Philips Senseo, Cappellini furniture with Philips electronic devices).[10] People from the Design Academy are considered to be good observers of trend. They could thus help Eindhoven's companies with relevant market information. A new field in this respect is discovering trends in food and nutrition. The contacts between the Design Academy and the other knowledge institutes such as the TU/e, Fontys, ROC appear to be quite good. However, it seems likely that there will be more new market opportunities when more knowledge and experiences are being exchanged.

The design activities could become of increasing importance for Eindhoven. Manufacturing companies such as Brabantia could move their production activities to other countries, but could still keep its design department in the region.

Combining new technology with healthcare can lead to new products and services. With the increasing number of elderly people and increasing attention on health, this is seen as an important growth market. Much is also expected from combining popular sports with technology: for instance, Eindhoven intends to build a new national swimming-pool stadium with high quality measuring systems (like the sports science institute in Hamburg).[11] Such initiatives could lead to new expert products. This could be a reason for the local government to invest more in some innovative elements, such as an innovative swimming pool, but also, for instance, an innovative library.

Embedded systems A promising field of knowledge-intensive activities is 'embedded systems' – electronic systems (hardware and software) that reside within devices (such as machines, instruments, household appliances and vehicles) and control their functioning. Embedded systems will play a crucial role in all advanced industrial products; these systems are related to mechatronics. By enhancing mechatronics with software control you get embedded systems (or embedded mechatronics) (TU/e, 2004). Embedded systems form part of the adaptive systems, one of the three research priorities of the TU/e. Within the Eindhoven region Eindhoven Embedded Systems Institute (ESI at the TU/e) is active in this field. One of the Horizon projects is to further develop the dissemination of this institute's expert knowledge in technology-focused SMEs. In addition, extra attention within this programme will be given to the relationships of the ESI with SMEs.

The Polymer Institute[12] The Polymer Institute is a good example of an organisation in which knowledge creators and producers cooperate. The Dutch Polymer Institute (DPI) is one of the four leading technology institutes established in 1997. DPI is a public-private partnership between the main polymer producing and processing industries in The Netherlands and knowledge institutes (universities and TNO) that have a track record in the research of polymers and polymer processing. DPI focuses academic research on issues that are relevant to polymer industries. The research is characterised by a strong multidisciplinary 'chain-of-knowledge' approach.

DPI was initially established for a four-year period. An international committee evaluated DPI positively in 2001 and consequently the Dutch government (the Ministry of Economic Affairs) agreed to supply funds for a second period of six years (instead of the initially intended four years) up to 2008. This will enable DPI to expand into the rest of Europe, which was one of the original objectives. DPI is funded by industry (25 per cent), universities/TNO (25 per cent) and the Ministry of Economic Affairs (50 per cent). In four years the budget nearly tripled to €11m in 2001.

The participating industries and founding fathers of DPI are AKZO Nobel, Dow Chemical, DSM, General Electric Plastics, Montell (now Basell), Océ, Philips, Shell and TNO. Meanwhile, ATO, Avantium, Avery Denisson, Chemspeed, ECN, Kraton, NTI, Sabic and SEP also became DPI partners.

The main participating knowledge institutes are the Universities of Groningen, Twente, Eindhoven and Delft and TNO. Over the last few years DPI has become increasingly active outside The Netherlands. Universities in Germany, Italy, UK, South Africa and Greece have already joined DPI. It is anticipated that the number of non-Dutch research groups involved in the DPI programme will grow the coming years.

DPI's turnover rose from €3.6m 1998 to €16.1m in 2003; the total number of publications currently stands at 500 and reported inventions have already reached some 35. The number of researchers located in the different participating cities has grown to 220 (130 FTE).

4.4.4 Developing New Growth Clusters

According to NV Rede (2002), there are four spearhead clusters in the region of Eindhoven: medical technology, automotive, ICT and mechatronics. Table 4.4 gives an overview of firm and job development in the four spearhead sectors. The ICT and the mechatronics clusters are the largest regarding the number of generated jobs.

Medical technology Within the Eindhoven region, there is a platform – Medical Technology – for knowledge exchange and development. Almost 300 companies with about 5,800 employees are considered to be part of this cluster. The Technical University of Eindhoven invests about 10 per cent of its research capacity in medical technology, often in cooperation with regional industries and medical institutes.

Automotive The regional automotive cluster consists of some large original equipment manufacturers (OEMs), DAF and Philips, together with some main suppliers and many small supplying firms. This spearhead sector contains around 55 firms with about 5,500 jobs.

ICT Job growth has been very high in the ICT sector (almost 22 per cent). However, after 2001 there has been a decline of jobs in this sector (after the booming period with the transition to the new millennium and to the Euro). Important for the ICT sector is the continuing product and process innovation. It is, however, becoming increasingly difficult to innovate within one company, because of the short life-cycle of products and the substantial costs and risks related to this innovation. Consequently, there is more and more cooperation between ICT companies in product development.

Mechatronics Table 4.4 shows that the mechatronics cluster is the most important for employment in Southeast Brabant. More than 40,000 people work in this sector; there was a growth of more than 11 per cent in jobs in the 1997–2001 period.

The word mechatronics is a combination of mechanics and electronics. It refers to the combination of precision mechanical engineering, electronic

Table 4.4 Developments in the four spearhead clusters of Southeast Brabant

	Number of firms			Number of jobs		
	1997	*2001*	*1997–2001*	*1997*	*2001*	*1997–2001*
Medical technology	236	284	20.3%	5,226	5,791	10.8%
Automotive		56			5,494	
ICT	1,191	1,973	65.7%	18,075	22,003	21.7%
Mechatronics	1,321	1,603	21.3%	36,251	40,350	11.3%

Source: Vestigingenregister Noord-Brabant, ETIN-adviseurs (in NV Rede (2002)).

control and system thinking in the design of products and manufacturing processes. The number of jobs involved in the mechatronics cluster in the Eindhoven region is great: a total of about 39,000 people (Horizon Programme Agency, 2003). There are important differences between types of firms in the cluster. Some, such as Philips, are very large and operate on global markets; others are much smaller and have a more regional focus. To differentiate between firms a distinction between types is depicted in Figure 4.3. Horizontally, a distinction is made between 'makers' and 'thinkers'; vertically between product- and capacity-oriented firms (van den Berg, Braun and van Winden, 2001).

Within this division, companies in the first category are the licence takers. They produce high volumes of products for large markets. This type of firms does not invest heavily in research and development. Rather, they use technology that has already been developed elsewhere, and apply it to their products. These firms are 'makers' more than 'thinkers'. This type of firm is relatively scarce in the Eindhoven region. Some 10 per cent of the firms in the mechatronics cluster are estimated to be active in this sector.

The second group, the licence providers, are the large and smaller industrial firms in the region that make substantial R&D efforts: they are 'thinkers'. Some 30 per cent of the mechatronics firms in the region are estimated to belong to this category. The most important are Philips (the electronics multinational),

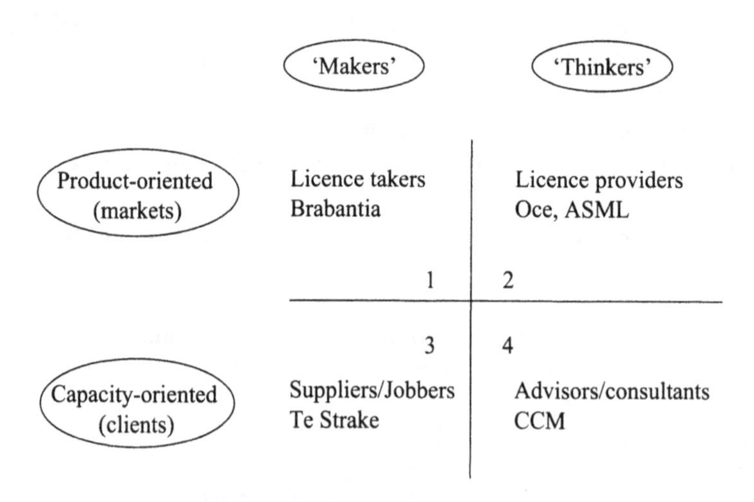

Figure 4.3 Firm typology within the mechatronics cluster

Source: Based on Van Gunsteren (1996).

Océ (copiers producer), Stork (machine building), DAF (lorries), and ASML (wafer steppers).

In most cases, the core activities of these firms are R&D and marketing. Other functions are outsourced. Research and development is executed partly within the confines of the firm, but to a large extent in cooperation with other firms. To that end, firms organise extensive R&D networks with technology partners. The large licence providers (the OEMs) in the Eindhoven region are considered to be fundamental 'engines' in the mechatronics cluster. Research, development and licence provision are not limited to these giants: many smaller firms put in substantial R&D efforts and participate in technology networks as well.

The third type of firm, the suppliers and jobbers, supply modules or parts to other firms, mostly to the licence providers. Within this category there are large variations as to the technological level and sophistication of the firms. Some jobbers are very straightforward executors of standardised products: They receive a set of specifications from the commissioning firms and produce a number of copies. Other supplying firms are much more sophisticated. They do not just produce a single part, but complex modules with a high technology content. Some of these firms are active in R&D as co-developers with the commissioning firm. They are the so-called main-suppliers and form a vital constituent of the mechatronics cluster. The growth of this type of firm has been enormous, mainly through the inclination of large OEMs to concentrate on core activities, outsource complete modules to reliable suppliers, and involve suppliers in product development. An example of a fast growing main supplier is Te Strake. In 1993 this company's turnover was only €12m. By 2002 it had risen to €37m.

The fourth type of firm is the advisors/consultants in the region. The majority within this type are engineering offices and other types of technology advisors. In the Eindhoven region, there are around 100 of these offices. Another characteristic of this type of firm is that they work for a limited number of clients and not for large markets. A successful example within this category is CCM (Centre for Construction and Mechatronics). With about 100 employees, CCM carries out research and design activities for innovation projects of industrial firms. CCM's employees are highly-educated technical specialists (about 40 per cent are academics) in mechanical engineering, physics, electronics and informatics. The approach is clearly mechatronic: technical specialists from different fields work together on integral solutions. In some cases, CCM also exploits its inventions commercially. The advisors/ consultancy firms are 'supportive' to the mechatronic development and production firms described above. It is illustrative in this case that the majority of CCM's clients are located within the Eindhoven region.

It can be assumed that in a Western urban economy focused on knowledge-intensive activities, it is predominantly the activities within the second category that are interesting. They can be considered core activities – engines – within the knowledge economy. Related to them are the activities in the third and fourth categories, which support the engines. The least interesting category is the first one. These are not knowledge-intensive activities and those involved are often inclined to move their activities to lower-wage-paying countries.

4.5 Conclusions and Perspectives

In this case, we have seen that the City of Eindhoven is further strengthening its economic specialisation in technology. This is supported by the national government, which appointed the Eindhoven region a 'Brainport' in the national Spatial Planning Memorandum. At the same time, Eindhoven wants to widen the technology applications in order to reduce the economic vulnerability of the region. With the Stimulus programme, Eindhoven was already active during the 1990s in stimulating new combinations in technology. At the time of writing, the Horizon programme has to strengthen the position of Eindhoven within the international knowledge economy.

4.5.1 Foundations of the Knowledge Economy

The foundations of Eindhoven's knowledge economy are moderate to very good. Table 4.5 gives an indication of the values of these knowledge-economy foundations. The knowledge base is considered to be very good. There are important public and private research institutes within the region, such as the Technical University of Eindhoven and TNO-industry. Moreover, 50 per cent of the total Dutch R&D expenditure is made within the region. The economic base is valued as good, though it is still considered to be too vulnerable, as the regular economic downturns showed. Technology development is seen as a strategic, competitive advantage but it should be applied within a larger number and variety of economic sectors. Quality of life is valued as moderate. On the one hand, Eindhoven does not have a metropolitan and international atmosphere and there is a lack of attractive facilities for (young) knowledge workers. On the other hand, in the last few years many new urban facilities have been added and the city possesses attractive (natural) surroundings.

Table 4.5 An indication of the value of knowledge economy foundations in Eindhoven

	Foundations	Score in Eindhoven
1	Knowledge base	++
2	Economic base	+
3	Quality of life	+
4	Accessibility	□
5	Urban diversity	□
6	Urban scale	□
7	Social equity	+

-- = very weak; – = weak; □ = moderate; + = good; ++ = very good.

The accessibility of the urban region is valued as moderate. The growing regional airport is a positive aspect; however, the rail connections to international hubs (airports and HST-stations) and the frequently occurring congestion on the highway ring are negatives. The urban diversity is valued as moderate. About one fifth of the working population is of foreign descent. More than half of them are from Western countries; most of them are assumed to be knowledge workers. However, Eindhoven still lacks a real international community. Also the urban scale of Eindhoven is valued as moderate. The city is relatively small, which is positive for its organising capacity, though negative for its political power and attractiveness for metropolitan facilities. One of the policy spearheads in this respect is to set up intensive cooperation with two other high-tech cities, Aachen and Leuven. Social equity is considered to be good. Unemployment is relatively low and the number of low-income households is below the Dutch average.

4.5.2 Activities of the Knowledge Economy

The knowledge activities are moderate to very good (see Table 4.6). Attracting and retaining knowledge workers in Eindhoven is considered as neutral. The region has a gravitational pull on (technical) knowledge workers, but there are concerns that this will be insufficient in the near future. This can be explained on the one hand by the (local) lack of metropolitan attractiveness (quality of life), and on the other hand by the (national) shortcomings in education. It was argued that education has to be seen much more as an strategic investment for the position of The Netherlands within the knowledge economy. Consequently, the government should invest more in education, or give the education institutes more freedom to do this themselves. There are major concerns about the attractiveness of technical education for new students. This is a nationwide problem, but actions can be taken on a regional scale; for instance, by promotion activities in the New Evoluon. Education in design is a positive. This will attract a broader variety of students (arts as well as technical students) and can help to change the image of Eindhoven and of technical education.

Creating knowledge in Eindhoven is considered to be very good (as is the knowledge base). There are three large private companies in the wider region which invest relatively highly in R&D (Philips, Océ and DSM). Philips scores very good regarding the number of generated patents. The Technical University of Eindhoven ranks third in the impact score of publications, after Cambridge and Oxford. Cooperation between the three Dutch technical universities might strengthen their international knowledge-creating role.

Applying knowledge and making new combinations in Eindhoven are valued as good. One of the explanations for that is the good atmosphere for cooperation. There is a strong feeling of solidarity towards the region and stakeholders are often willing to support regional initiatives. For the latter, predominantly medium-sized companies are important. Start-up policies have been quite good, but will be improved and concentrated under the Horizon programme. An important field for improvement is the linkages between SMEs and knowledge institutes. Better cooperation between SMEs appears to be important to achieve this. One of the spearheads of the Eindhoven KE-policy is to apply technical knowledge within more economic sectors. This has to contribute to a broadening of the economic base. Much is expected in this respect from combining technology with healthcare, sports and design. Promising new institutes for applying knowledge are the Embedded Systems Institute and the Polymer Institute.

Development of new growth clusters is considered to be good in Eindhoven. There are four spearhead clusters within the region: medical technology, automotive, ICT and mechatronics. The latter two are the largest regarding the number of generated jobs. All four are clearly techno-clusters and medical technology particularly is seen as a promising new application of technology. Growth in the medical technology cluster can contribute to the broadening of the economic base, thereby reducing the vulnerability of the regional economy. In Eindhoven, it is believed that there is a good location climate for particularly small-scale product-oriented 'smart' companies within the four above-mentioned clusters.

Table 4.6 An indication of the value of knowledge economy activities in Eindhoven

	Activities	Score in Eindhoven
1	Attracting knowledge workers	+
2	Creating knowledge	++
3	Applying knowledge/making new combinations	+
4	Developing new growth clusters	+

-- = very weak; – = weak; □ = moderate; + = good; ++ = very good

4.5.3 Perspectives

Be aware of your assets and communicate this One of the strengths of the region of Eindhoven is the feeling of solidarity among important economic actors, particularly SMEs. They are willing to undertake additional efforts in actions that contribute to the local economy and urban attractiveness. One of the drawbacks, however, is an insufficient awareness of the strengths of the region. Eindhoven already has many knowledge-intensive activities but many actors (within as well as outside the region) are insufficiently aware of this. Moreover, regional strengths are insufficiently communicated to other places by the inhabitants and firms of Eindhoven. In this respect, the Dutch attitude of 'Just behave normally, then you're already standing out sufficiently' also applies to Eindhoven. The local authority wants to change this: Eindhoven firms and citizens should be proud of their region and communicate this to the outside world. One of the challenges for Eindhoven, therefore, is to firstly create more awareness regarding the existing knowledge-intensive activities and thus increase self-esteem and, secondly, to communicate the regional strengths nationally as well as internationally.

In this respect it is important that the Dutch ministry of VROM has recently appointed three economic core regions in The Netherlands: Rotterdam with its harbour; Amsterdam with Schiphol Airport; and the southeastern part of the country, including Eindhoven as 'Brainport'. This is very important for the region of Eindhoven as a national acknowledgement of its economic strength.

Reducing economic vulnerability and extending networks to enlarge the region's critical size One of the main concerns has been, and remains, the region's vulnerability to economic downturns. This should be reduced by developing more application fields for technology, such as healthcare, sports and design.

Moreover, by extending regional networks, Eindhoven hopes to enlarge the critical size of the region to make it more attractive for international companies and for locating metropolitan amenities and knowledge facilities. In this respect, much is expected from the recently initiated international cooperation (by Philips, amongst others) between Eindhoven, Leuven and Aachen.

Do not forget the quality of the base of the knowledge pyramid Much is often said regarding the need for improvement at the top level of knowledge institutes. It is, however, important to realise that the entire knowledge pyramid needs quality improvement. More investments should be made in the quality of education, not only at the top level, but over the entire trajectory. One of the major points of attention for the Eindhoven government is to stimulate more knowledge development within SMEs. The strength and attractiveness of regional SMEs could be enhanced by better interaction and cooperation with regional knowledge institutes.

Notes

1. Horizon Nieuwsbrief, Uitgave, 1 November 2003.
2. Source: www.statline.cbs.nl. The data are for the *Stadsgewest Eindhoven.*
3. Western countries are Europe except Turkey, North America, Oceania, Japan and Indonesia.
4. Source: www.statline.cbs.nl. The data are for the *Stadsgewest Eindhoven.*
5. Source: www.statline.cbs.nl. The data are for the *Stadsgewest Eindhoven.*
6. Source: Eindhovense Armoedemonitor 2001.
7. See, for example, NRC Handelsblad, 16 March 2004, p. 2 ('We hebben het vmbo veel te groot gemaakt'. Kamerleden erkennen dat er veel mis is gegaan bij fusie mavo en beroepsonderwijs).
8. See NRC Handelsblad, 16 March 2004, p.2 ('Technische universiteiten federatie' and 'Nederland te klein voor drie TU's').
9. Horizon, *Top Technology: Crossing borders, moving frontiers.*
10. Van der Kwast, Paul, 22 January 2004, 'We zijn slimmer dan we denken ... maar hoe lang nog?', *Intermediair.*
11. NRC 20 January 2004, 'Zwemstadion krijgt eigen "meetstation"'.
12. Source: www.polymers.nl.

Discussion Partners

Mr W. Claassen, Alderman of Social and Economic Affairs and District Gestel.
Mr J.M.H.M. Claessens, Delegate Director, Stimulus.
Mr F. Damen, Manager, South Region, Ministry of Economic Affairs.
Mr B. van Gijzel, Commissioner, KMWE Precision.
Mr J.M.H.M. Hendriks, CEO, Philips Electronics Netherlands.
Ms E.P.J. Lemkes-Straver, Programme Director Horizon and Vice President, NV REDE, Economic Development Corporation, Eindhoven Region.
Mr A.H. Lundqvist, President, Technical University of Eindhoven (TU/e)
Mr H.J. Mengelers, Director, TNO Industry.
Ms M. Mittendorff, Vice-Mayor, Alderman of Education, Sports, Urban Policy, Safety, Finance and District Strijp.
Mr A. Sakkers, Mayor of the City of Eindhoven.
Mr E. van Schagen, CEO, Simac.
Mr J.A.P.M. Smeekens, President, NV REDE, Economic Development Corporation Eindhoven Region.
Mr K. Tetteroo, Chair of the Board, ROC Eindhoven.
Mr J.H.M. van de Vall, Chair, Chamber of Commerce, East Brabant.
Mr N.M. Verbraak, President, Fontys, Polytechnic of Professional Education.

References

Adang, A.J.V.M. and J.M.P. van Oorschot, (1996), *Regio in bedrijf, Hoofdlijnen van industriële ontwikkeling en zakelijke dienstverlening in Zuidoost-Brabant.*
Berg, L. van den, E. Braun and W. van Winden (2001), *Growth Clusters in European Metropolitan Cities*, Aldershot: Euricur Series, Ashgate.
Berg, L. van den, P.W. de Langen, and P.M.J. Pol (1997), *De positie van Eindhoven in het internationale personenspoorvervoer, Een strategische analyse*, Rotterdam: Euricur.
Biesen, Jan van den (2001), 'Research Basic to Philips', OECD Workshop on Basic Research, Oslo, 29–30 October.
Eindhovens Dagblad (2003), 'Philips pleit voor versterken van positie Zuidoost-Brabant', 8 October.
European Commission (2003), *Third European Report in Science and Technology Indicators,* Luxemburg: Office for Official Publications of the European Communities.
Gemeente Eindhoven (2003), 'Inkomen in Eindhoven, Gegevens uit het Regionaal Inkomensonderzoek 2000, Afdeling Bestuursinformatie en Onderzoek van de dienst Algemene en Publiekszaken', BiO-rapportnummer 956.
Horizon Programme Agency (2003), *Top Technology: Crossing borders, moving frontiers*, Eindhoven: Kempen Druk.
NV Rede (2002), *Feiten en Cijfers Regio Eindhoven 2002*, Eindhoven.
Stichting Nederland Kennisland (2003), *Kenniseconomie Monitor, Tijd om te Kiezen*, Heerhugowaard: Hollandia Equipage.
Technical University of Eindhoven (2004), 'Research Profile TU/e, Science and Engineering of Biomedial Technologies, Adaptive Systems and New Materials', draft, January.

Chapter 5

Helsinki

5.1 Introduction

In the last decade, the population of the metropolitan region of Helsinki has grown by 150,000. The capital region of Finland has 1.2m inhabitants. The region is one of the fastest growing urban regions in Europe. The most dynamic element of the local economy was the ICT cluster led by Nokia. From 1993 to 2001, the number of employees working in ICT industries rose from 40,000 to 70,000. One of the strengths of the Helsinki region appears to be the close relationships between the local universities and industrial firms. It is argued that this can be explained by the Finnish way of life with short social distances between people making it easy to network (Holstila, 2003).

In Section 5.2, we will explore the knowledge economy strategies of the national government of Finland and of the local authorities. Then in Section 5.3, we will analyse the foundations of the knowledge economy for Helsinki. In Section 5.4, we will go into the four knowledge economy activities for Helsinki. Finally, Section 5.5 presents some conclusions and perspectives.

5.2 Knowledge Economy Strategy of the Urban Region

5.2.1 Impact of the Economic Crises

The start of the knowledge economy in Finland can be dated back to the early 1980s during the economic crisis. That crisis made it clear that dramatic measures were necessary for an economic recovery in Finland. This necessity became even clearer with the collapse of Soviet Union in the early 1990s. This collapse implied the fall of Finland's most important trade partner and consequently a rapid increase of the unemployment (about 20 per cent of the Finnish workforce). One result of this realisation was Helsinki's first internationalisation strategy, initially formulated in 1995 and exemplifying new attitudes regarding cooperation and development. The city's long-term success was perceived to depend on education, science and research. In addition, cultural and environmental factors were seen as important. In the mid-1990s, there was already an economic recovery in Finland, thanks to the success of high technology companies, with Nokia playing the star role.

5.2.2 The Role of National Organisations

In Finland, national actors have played a crucial role in stimulating the knowledge economy. Many efforts have been made to improve the quality of education and research and development and to enhance cooperation between education and research institutes, business companies and public organisations.

In 1985, the Science and Technology Policy Council (STPC) was set up to develop a vision of and to elaborate a knowledge economy strategy for Finland. The STPC is chaired by the Finnish prime minister. The other members are the ministers of Education, Economic Affairs and Finances and no more than four other ministers. Moreover, six members of the STPC are representatives of business companies, universities, education, employers and employees. Four seats are reserved for independent experts. The STPC meets four times a year to discuss specific themes and/or to decide about specific funds.

National institutes such as Tekes, Sitra and the Academy of Finland play an important role in funding research and development and stimulating cooperation between relevant actors. The role of these organisations is described below.

Tekes, the National Technology Agency founded in 1983, finances R&D projects of companies and universities in Finland. The funds are awarded from the state budget via the Ministry of Trade and Industry. Tekes has 300 employees and a yearly budget of €400m (Nauta et al., 2003). Tekes stimulates cooperation between research and business organisations. In 2001, Tekes funded 2,261 research and development projects to a total value of €387m (Tukiainen, 2003). Two-thirds of this funding was in the form of grants and loans to company research and development projects and one-third to university and research institute projects (www.tekes.fi.). The Uusimaa province, in which Helsinki is located, receives about 40 per cent of the Tekes funding. The regional distribution of Tekes funding roughly follows the regional distribution of companies R&D inputs. Tekes is by far the largest funding provider in the region. The central idea in funding is to give companies and research institutes incentives to cooperate. Another objective is to promote new entrepreneurship.

Sitra is the Finnish National Fund for Research and Development. It is a public foundation under the supervision of the Finnish Parliament. The Fund was set up in conjunction with the Bank of Finland in 1967 in honour of the fiftieth anniversary of Finnish independence. The Fund was transferred to the Finnish Parliament in 1991. Sitra has 95 employees and a yearly budget of €70m (www.sitra.fi). The Fund aims to promote Finland's economic prosperity by encouraging research, backing innovative projects, organising training programmes and providing venture capital. Sitra aims to further economic prosperity in Finland by developing new and successful business operations, by financing the commercial exploitation of expertise and by promoting international competitiveness and cooperation. The focus of Sitra's corporate funding is directed towards enterprises at the start-up stage. Sitra's corporate funding activities are divided into four sections: technology, life sciences, regional funds and pre-seed funding. In addition, there is the commercialisation of technology. Sitra cooperates with public sector bodies such as the Finnish National Technology Agency (Tekes), the Finnish Industry Investment Ltd., Finnvera, the Academy of Finland, and Employment and Economic Development Centres (TE-centres). In addition, Sitra tries to coordinate its funding with that of other providers of capital.

Though Sitra and Tekes are both aiming at financing start-ups, in practice they particularly fund established organisations.

The *Academy of Finland* distributes research money among the universities and monitors the scientific quality of the Finnish research. In 2003 the Academy of Finland had a budget of €185m and a staff of more than 100 employees (Nauta et al., 2003). The independence of the Academy is important to improve fundamental research at the Finnish universities. One of the aims is to stimulate cooperation between the Finnish researchers and others in the EU.

The idea behind the national *Centre of Expertise Programme*, launched in 1994, is to focus local, regional and national resources on the development of internationally competitive fields of know-how. The programme pays special attention to SMEs to develop selected internationally-competitive fields of expertise, and stimulate technology transfer from universities to firms. The programme covers the whole country and it is carried out in regional Centres of Expertise, appointed by the Council of State, that work closely with universities and companies in their respective sectors. The Ministry of the Interior has an important role in the national coordination of the Centre of Expertise programme.

Equity or efficiency aims In the Helsinki region, there is a fear that, increasingly, funds for applying knowledge from Tekes and Sitra will go to regions outside Helsinki. Among politicians there is a discussion whether an efficiency or an equity policy is desirable; thus the increase of support for lagging regions or for leading regions. During our discussions, it was argued that some politicians in national government do not want to support the 'unhealthy' growth of Helsinki. They believe the city already receives enough benefits. Regional actors argue, however, that the wealth created in Helsinki has to be invested inside the region to maintain growth.

5.2.3 Knowledge Economy Strategies within the Helsinki Region

Local authorities in the Helsinki region played a specifically *accommodating* role regarding the knowledge economy. They supplied the necessary public products, such as business space and (transport) infrastructure, to facilitate the economic developments. However, more recently the city has become more proactive in stimulating knowledge-intensive economic activities. Helsinki has a strategy for enhancing knowledge and business development. The latter strategy was started in 1999 and was revised twice. The city does not have a specific (codified) knowledge economy strategy. During our discussions, it was stated that the activities of Culminatum (see Section 5.4.4) come closest to a knowledge economy strategy.

Urban policy programme for the Helsinki Metropolitan Area Within the Helsinki Metropolitan Area, a specific common urban policy programme has been developed to stimulate the competences within the region, on the one hand, and on the other, the internal socioeconomic cohesion. This local programme covered the period 2002–2004. It can be considered as the regional adjunct of the national policy to stimulate the knowledge economy and socioeconomic equity. The programme was started in October 2000 by the mayors of the cities of Helsinki, Espoo, Vantaa and Kauniainen and by the executive directors of the Helsinki Metropolitan Area Council and the Uusimaa Regional Council. The main goal of the programme was to strengthen competitiveness, knowledge and citizen participation in the Helsinki Metropolitan region. Three challenges were formulated in the programme (Holstila, 2003):

1 economic competitiveness based on knowledge and skills;
2 maintaining the social coherence of the region;
3 the need caused by recent rapid population growth to build new housing and neighbourhoods.

In addition to the organisations mentioned above, the Ministry of the Interior, Culminatum and the Association of Finnish Local and Regional Authorities were also represented in the project. The mayors gave Culminatum, in cooperation with its shareholders, the mandate to put the first challenge mentioned above into practice. This involves the following projects (Holstila, 2003):

1 to establish an International University of Helsinki (this initiative is already stopped, see Section 5.4.1);
2 to set up Helsinki-Tallinn as 'Twin Cities of Science and Knowledge' (see also Section 5.4.1);
3 to improve business intelligence in science parks and campus areas;
4 to prepare a regional land use strategy for knowledge-based activities;
5 to set up an R&D programme for the Helsinki Region Innovation System;
6 to develop a knowledge-intensive business services (KIBS) sector for the regional economy.

5.3 Knowledge Foundations

In this section, the knowledge foundations – as discussed in the theoretical chapter – will be assessed for Helsinki. They include the knowledge base, the economic base, the quality of life, the accessibility, the urban diversity, the urban scale, and the social equity in the city.

5.3.1 Knowledge Base

The Helsinki economy is distinctively knowledge intensive. Research intensity increased in Finland throughout the 1990s. The region of Helsinki is by far the most important region for R&D activity in Finland: over 50 per cent of Finnish R&D expenditure is spent in the Helsinki region. Private-sector involvement is high, accounting for some 62 per cent of R&D investments (van den Berg and Russo, 2003). According to Eurostat, the proportion of employees working in the field of R&D is 3.9 per cent in the Helsinki region, which is the highest figure among city regions in the EU, the European average being 1.4 per cent (Holstila, 2003).

The knowledge intensity of the Helsinki economy is also illustrated by the high educational level of its population: 40 per cent of the 25–64-year-old population in the Helsinki region has completed at least 13 years of education. The aim is to achieve 70 per cent. According to our discussion partners, this is an ambitious, though realistic, aim. Education is almost free. Not having income barriers to studying at a university or polytechnic is seen as important for raising the average education level in Finland, because more people are able to study. On the other hand, there are considerable quality incentives for students. It is, for instance, difficult to enter a higher education institute because of a *numerus-fixus* policy for students: consequently, they have to pass specific entrance exams.

Some 50 per cent of Finland's academics live in the Helsinki region, and the share of inhabitants with a higher education qualification reaches 40 per cent in Espoo and more than 50 per cent in the small city of Kaunianen (City of Helsinki Urban Facts, 2002a). In a comparative analysis of the 37 Finnish urban regions, the FUN project (Urban Facts, 2001) ranks Helsinki as first both for diversity and for strategic specialisations; on this account it is the region with the best preconditions for development, together with Tampere, Oulu and Turku.

Altogether Finland has 20 universities throughout the country, all state-run and engaging in both education and research. The Helsinki region hosts eight universities with some 60,000 students, consisting of one multidisciplinary institution, three specialist institutions and four art academies. The University of Helsinki, with its more than 37,000 students in 2001, is the largest. The region's share of all university students in Finland is two-fifths, while its share of postgraduate degrees is even larger (Susiluoto, 2002).

During the 1990s, a new educational system – polytechnic institutes – emerged in Finland. Studies for a degree at a polytechnic take some four years to complete and there is a more vocational orientation than in a university degree. The educational level in the Helsinki region is expected to rise as a result of the strong growth in the number of students completing qualifications at polytechnics. Helsinki's polytechnic, Stadia, is the second largest in the country, with some 8,000 students and 30 educational programmes (see also Section 5.4.1).

Table 5.1 provides a breakdown of registered students, graduates and foreign students in the main institutes. In the last ten years the number of university students has increased by about 40 per cent. However, entry into universities is very competitive, and universities organise various types of entrance examinations.

The supply of higher education facilities in the Helsinki metropolitan area is rich, diverse and qualitatively strong (van den Berg and Russo, 2003). The

Table 5.1 Higher education students in the Helsinki metropolitan area, 1999–2000

	Attendees	Degrees taken	Foreign students
City of Helsinki			
University of Helsinki	37,244	4,034	1,190
Helsinki School of Economics and Business Administration	2,757	402	–
Swedish School of Economics and Business Administration	2,200	285	–
Sibelius Academy	1,700	147	–
University of Art and Design	1,600	244	240
Theatre Academy	–	56	–
Academy of Fine Arts	250	62	–
Espoo			
Helsinki University of Technology	14,264	1,053	790
Helsinki metropolitan total	60,015	6,283	1,865
Other HEI: Polytechnic Stadia	7,500	1,275	106
Other polytechnic schools	16,343	1,137	–

Source: van den Berg and Russo (2003).

University of Helsinki is the country's leading institute of higher education. Business studies can be pursued at the Helsinki School of Economics and Business Administration and the Swedish-language Swedish School of Economics and Business Administration. The Helsinki University of Technology is the most important institution for higher technical education and research. The Academy of Fine Arts, the Sibelius Academy, the University of Art and Design of Helsinki and the Theatre Academy of Finland provide higher education excellence in the arts. Other educational opportunities are provided by the Helsinki Summer University, the Open University in Finland and the Third Age University.

5.3.2 Economic Base

As is typical of large cities, the industrial structure of the Helsinki region is strongly service dominated. Four out of five of those employed work in some part of the service sector. Economic growth has been fastest in financing and business services. The Helsinki region has a strong profile of private offices, expert work and management. The most important central government decisions are made in Helsinki. Information sector jobs are concentrated in the Helsinki region, particularly in the Helsinki metropolitan area, which contains almost half of this sector's jobs in Finland. The information sector has been an important driving force behind the region's growth since the mid-1990s. This sector, together with production based on high-level technology, keeps up a diversified service structure and a high level of economic activity (Susiluoto, 2002). In the 2001–2003 period the number of jobs in business services and construction has increased. Decreasing numbers of jobs were found in transport and logistics, financing, health and social services and ordinary industries.

Nokia is a very important firm for both the Finnish national economy as a whole and for the Helsinki region. It has been the leading engine in the growth of the ICT sector in Finland, transforming the country's economy and modernising its export structure. Also, the direct and indirect growth effects of Nokia (and more generally the ICT sector) are great in the Helsinki region; for example, a notable part of all new office space construction in the area has lately been Nokia- and ICT-connected (Susiluoto, 2002). Nokia has all kinds of activities with universities, such as lectures and teaching, research funding, sponsoring, hiring trainees and directing theses (Tukiainen, 2003).

According to the OECD (2003) the specialisation of the Helsinki ICT cluster has introduced considerable vulnerability, as it is dependent on one single sector. 'Demand for mobile technologies, products and services has already demonstrated susceptibility to global economic slowdowns. In the long term, market growth may not prove as rapid as it was during the late 1990s due to the transition from original demand to replacement demand' (OECD, 2003, p. 14). Another study is more positive about Helsinki's economic future. It states that the sector's dependency on Nokia is not a risk in the long run as the knowledge capital remains in Finland regardless of decisions of the globalised firm. Critical mass has been achieved and the cluster can develop on its own (Tukiainen, 2003).

5.3.3 Quality of Life

According to our discussion partners, quality of life is considered to be very good in the Helsinki region. Finland ranks sixth in the latest United Nations survey of quality of life. Indeed, the quality of life in its capital city can be judged very high. The region offers metropolitan amenities, but the countryside is not far away. These characteristics, among others, make Helsinki a pleasant place to live. On the other hand, the climate – cold, rainy middle seasons and long, dark winters – partly offsets these advantages. In general, the Finnish capital attracts many young people from other parts of the country who prefer the urban way of living in Helsinki (van den Berg and Russo, 2003). During our discussions it was stated that when foreign investors, entrepreneurs and visitors make their way to Helsinki they are usually very pleased, but it is difficult to attract them for the first visit.

Recently, the opinions of Helsinki inhabitants on public services were studied. It turned out that Helsinki citizens were very content with public transport, library services, markets, various cultural services and safety of residential areas in 2001 (Susiluoto, 2002). The results for the most important basic services such as child day care, schools and health centres were also good. In 2001, citizens were most discontented with such factors as availability of housing and care for the elderly. Altogether opinions have remained fairly stable over the years, even though it seems that the overall economic situation is reflected in the way people use and value public services, as people tend to be more critical in good times.

In the light of international comparisons Helsinki can be characterised as a safe city (Susiluoto, 2002). Unlike many other cities, street safety is at a high level: there is, for instance, no tradition of heavy street violence. Largely due to the fact that the differences between social classes are moderate, Helsinki is a safer city than is usually perceived by its own inhabitants.

5.3.4 Accessibility

The peripheral location of Helsinki in the European perspective is considered to be a weak point. Indeed, distances to important European economic centres such as London, Paris, Brussels and Frankfurt are relatively great. Accessibility by air is, however, considered to be relatively good. The Helsinki-Vantaa Airport is Finland's only international TEN-linked airport with regular direct flights to Finnish destinations and to numerous cities in Europe, North America, Russia, the Baltic countries and the Far East. The development of the airport and its related services forms part of the gateway-strategy of the Helsinki Metropolitan Area. The intention is to make the region a more attractive location for foreign companies willing to exploit its position as a gateway between eastern and western Europe. The Aviapolis in the vicinity of the airport is one of the largest construction projects in the Helsinki region. In addition, several investments are planned in railway lines for high-speed trains, including a new line to St Petersburg. This fits the target to strengthen the position of the region as a gateway between the European Union and Russia.

Internal accessibility within the Helsinki region is considered to be quite good, though infrastructure for several transport relations can be improved. One example regards the problems concerning the construction of a metro line to Espoo: the firms and Otaniemi Business Park want this line to be built, but the city of Espoo is resisting it. The city is afraid that the metro line will result in high density housing; something that Espoo does not want.

The ongoing suburbanisation of higher education places additional demands on public services like transport. The Helsinki region is one of the most sparsely populated conurbations in Europe, according to an overview by the OECD (2003). The campuses of universities and polytechnics lie relatively far apart from each other. Since innovations often come about when different sciences meet, the interaction between campuses should be increased (Holstila, 2003). Though plans are under way for the realisation of high-quality transport links between the main 'knowledge districts' of the Helsinki region (Figure 5.1), the budget for this work is far from readily available.

5.3.5 Urban Diversity

As stated, economic diversity in the Helsinki region is considered to be relatively low. The economy is dominated by the high-tech sector and by the Nokia company. This can make the economy relatively vulnerable to structural economic changes.

The city of Helsinki is, however, becoming increasingly culturally diverse. Although the city does not have the image of being so, there is considerable attention on a large variety of cultural initiatives such as the Arabianranta cultural cluster, the University of Art and Design, the Art Academy and the Cable Factory. Since Helsinki was a cultural capital of Europe in 2000, culture has been considered instrumental in stimulating the economy. Representatives of cultural organisations are even involved in initiatives to strengthen the regional economy (see the text on the Club of Helsinki in Section 5.4.3).

Urban diversity is to be found particularly in the City of Helsinki. Conversely, within the City of Espoo, where Nokia and the HUT are located, there appears to be hardly any urban diversity. Apparently, for activities focused on applying high-tech knowledge, a highly diversified urban environment is not a necessity.

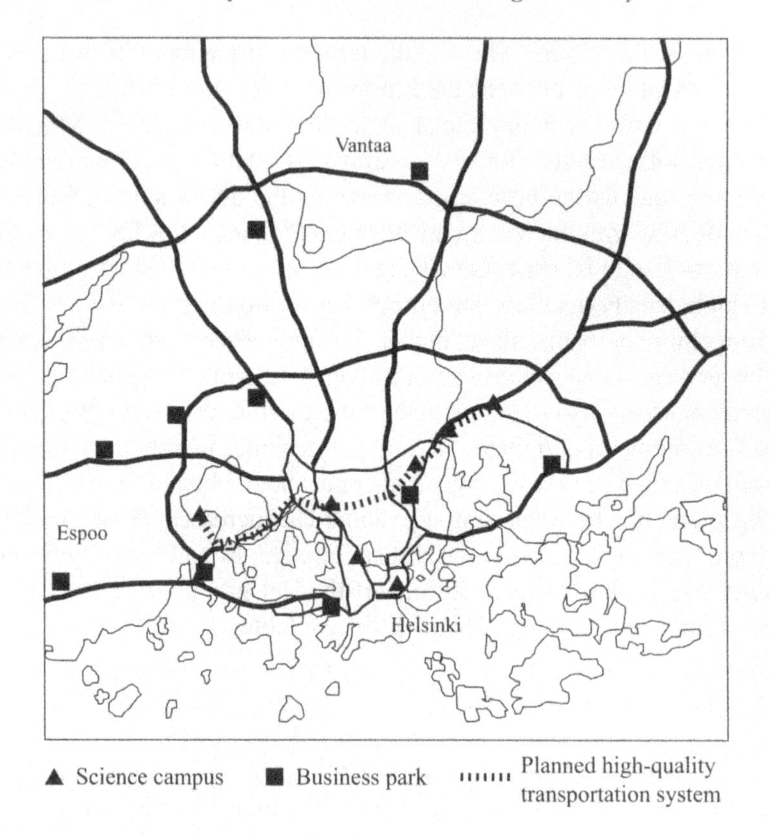

Figure 5.1 The new 'Science Corridor'

5.3.6 Urban Scale

The Helsinki region is considered to be a large one within Finland. With its 1.2m inhabitants, more than 20 per cent of the Finnish people live in the capital region. Internationally, however, Helsinki urban region is a secondary one, compared with regions such as Amsterdam, Prague and Zurich. For Helsinki this might be a reason to look for closer relationships with other cities. An example is the cooperation with Tallinn. By such strategic cooperations, the city can enhance its relevant economic region, and consequently its market for metropolitan amenities and infrastructure.

5.3.7 Social Equity

In the early 1990s, Helsinki appeared to be a socially balanced city with small differences in socioeconomic level among its inhabitants and its districts. The economic growth of the city seemed endless until that period, and that allowed the extension of a welfare model with good provision of social services and high-income redistribution. However, in the same period a serious economic recession hit Helsinki. Massive unemployment brought new, until then unfamiliar social problems with it, related to education, age, gender, and housing. In 1994, the unemployment rate reached 18 per cent. The problem was aggravated by a tendency to centre on certain urban areas. Moreover, in the last ten years the share of foreign inhabitants has increased from 1 per cent to 5 per cent of the population, which also has a considerable impact on the socioeconomic development of Helsinki. The majority of immigrants come from Russia, Estonia, Somalia and Vietnam. However, the immigration figures are not comparable to those of some larger cities in other European countries (UK, Germany, France, Benelux) with numbers up to 30 per cent or 40 per cent of ethnic minority people (van den Berg, van der Meer and Pol, 2003).

In Helsinki, the main driving force for new economic growth is the (high-tech) information sector, which emphasises the role of education as a labour market resource. As a result, the growth affects different areas at a different pace, depending mainly on the educational standard of the population. The less educated and more working-class areas, concentrated particularly in the eastern part of the urban region, are lagging behind while the western areas, with a better-educated population, are leading the upswing. Consequently, the already-existing educational divide of the city is gradually breeding both unemployment and income differences (Susiluoto, 2002). One of the options to stimulate the local economy in the eastern part of the urban region is that of building business parks. In fact, there is a good deal of empty land in the eastern part of the city. One drawback is the lack of universities nearby, which might be a major argument for knowledge-intensive businesses not to locate there. A business park was built in the eastern part, but not many businesses have been attracted yet.

In the last few years it seems that the acute problems that hit the city have been decreasing. Unemployment in particular has shown an important decrease at the same time as Helsinki's economy has recovered. In 2002, the unemployment rate was 6.2 per cent (City of Helsinki Urban Facts, 2003). Although fears expressed during the recession years that social exclusion would happen on a larger scale have not materialised, there is the perception that action should be taken to prevent the worsening of existing problems. Therefore prevention rather than cure of social exclusion has become a policy objective of the Helsinki government.

5.4 Knowledge Activities

In the last section, the foundations of Helsinki's knowledge economy have been discussed. Now, we turn to knowledge activities and ask how well the city (and the actors in it) manage to: 1) attract and retain talented people; 2) create new knowledge; 3) apply new knowledge and make new combinations; and 4) develop new growth clusters.

5.4.1 Attracting and Retaining Knowledge Workers

Natural attraction within Finland Within Finland, the Helsinki region functions as a magnet for symbolic analysts. Knowledge workers are almost naturally attracted by the capital city. According to Susiluoto (2003), the most important reasons for locating in the Helsinki region are human resources and the high quality of local education. Its knowledge and economic base is much broader than those of the other Finnish regions. Organisations like the Helsinki University of Technology in Espoo and the Nokia company are contributing in a substantial way to the region's attractiveness. For the Finnish inhabitants, the location climate appears to be sufficiently attractive. Quality of life is valued as high and there is a broad variety of cultural amenities. The city has a cultural policy that is explicitly considered to be instrumental in creating an attractive atmosphere.

One of the drawbacks, however, is the supply of accommodation, which is considered to be too limited and the average size of houses (particularly in Helsinki) too small. Projects such as the construction of new housing in the old port area have to contribute to improve the quality of the environment for knowledge workers.

In the Helsinki region, there are relatively many knowledge workers and students in science and engineering (S&E).[1] Data on S&E graduates were unfortunately only available at country level. However, because of the fact that Helsinki plays quite a dominant role in Finland, national figures can give a fair indication of the importance of S&E. Amongst other things, Figure 5.2 shows that the share of S&E graduates in Finland is 31.9 per cent, higher than, for instance, the UK (25.4 per cent) and much higher than The Netherlands (15.2 per cent). The only European country in which this share is higher is Germany (32.5 per cent). This is considered to be one of the strategic strengths of the Finnish region. This large share can be explained by the high quality of the education system, but also by the economic successes of the ICT sector and particularly the Nokia company. Consequently, S&E studies have a good image and S&E workers have a high status and good career opportunities.

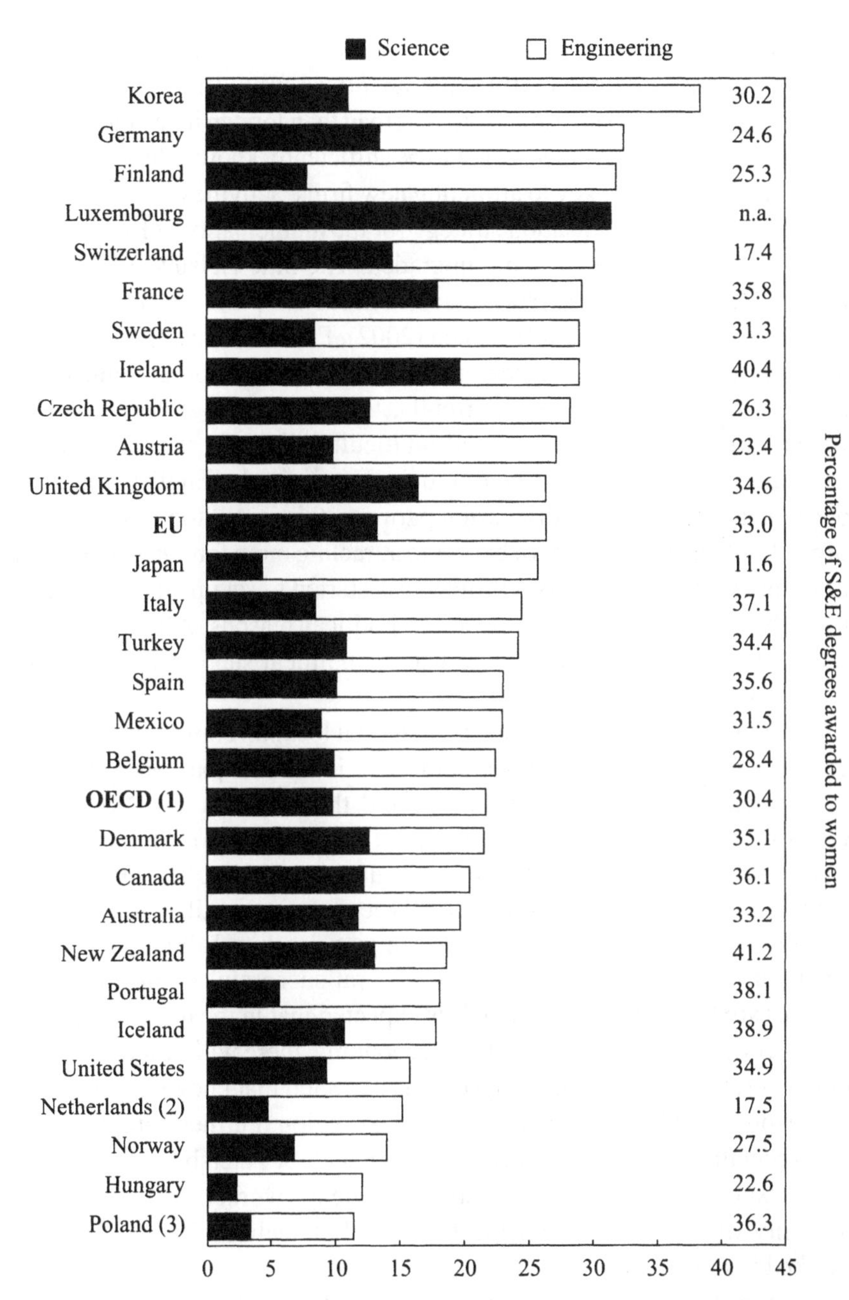

Figure 5.2 Science and engineering degrees as percentage of total new degrees, 2000

Source: http://www1.oecd.org/publications/.

International relationships During our discussion, it was stated that the Helsinki region does not really have an *international* location climate. Currently, more Finnish investment goes abroad than foreign investment comes into Finland. The country appears not to be sufficiently known abroad. For this reason, *marketing* is the key to attracting new firms, activities and knowledge workers. An internationalisation strategy for the municipality of Helsinki was drafted in 1998 and is now being upgraded. The nine Helsinki universities work together on marketing Helsinki as an attractive place for students.

According to Castells and Himanen (2002), Finland's success story in the late 1990s was based on the universities and the education system, which were very nationally oriented. Currently, the key challenge for Finland is a process of internationalisation and multiculturalism. At the nine Finnish universities lectures are increasingly being given in the English language. However, according to our discussion partners, this is still too low. This low share of English lectures forms a barrier to attracting more foreign students and knowledge workers. Recently a project was started to set up an international university. This project was stopped for political reasons. All education in Finland has been free to date: there was no political support for providing paid international education.

In order to further strengthen its international location climate, the Helsinki region is aiming for further cooperation with its natural partners within the region, such as Russia (St Petersburg) and the Baltic States. Nowadays, there is more and more cooperation with Tallinn (Estonia). Moreover, this cooperation is being supported by an (regional) urban policy project (see Section 5.2). At the moment, however, this cooperation is still predominantly a uni-directional relationship: Helsinki can give much to Tallinn, but Tallinn cannot yet give much back. Currently, several successful Finnish business strategies are being used in Tallinn. It is expected that in the near future, with the further development of the Estonian economy, there will be a much more reciprocal relationship. In this respect, it is expected that knowledge workers from Estonia will be attracted to the Helsinki region. Furthermore, there are high expectations for improved relationships with St Petersburg.

Two examples of institutes attracting knowledge workers are the Helsinki University of Technology and the Helsinki Polytechnic Stadia. They are described below.

In technological and scientific research, the *Helsinki University of Technology* (HUT) plays a significant national role; in many fields of technology the HUT is the only place in Finland where university-level teaching and research are carried out. It is the oldest and largest university of technology in Finland, dating back to the nineteenth century. There are about 100 laboratories in 12 departments at the HUT, and four separate research institutes. The HUT is located in the Otaniemi Park in Espoo. The main language of instruction is Finnish: some basic courses can be taken in Swedish. The HUT also offers courses in English. In addition to the international nature and high quality of research, freedom, innovation and the ethical nature of research have been fostered. In research, the HUT is committed to preserving the balance between basic and applied research, and efforts have been made to ensure that research findings can benefit both business and society at large. The Department of Architecture at the HUT was nominated a national Centre of Excellence in Artistic Activities for 2001–2003 by the Ministry of Education. The *Otaniemi International Innovation Centre* offers its services to HUT staff and students in applying for research funding, commercialising the results, and managing property rights. The HUT also participates in extensive cooperation networks with Europe, the United States, and Japan (van den Berg and Russo, 2003).

Helsinki Polytechnic Stadia is a multidisciplinary institution of higher education in applied science. It was established in 1996 by merging eight existing municipal higher education institutes. Stadia is a fully-fledged department of the Helsinki City Council. At Stadia, the open-learning environment produces professional experts, of whom 88 per cent become employed immediately after graduation. The number of full-time students is 7,500. Close to 30 different degrees are available. Stadia has three faculties: Health Care and Social Services, Culture and Services, Technology (hosting respectively 46 per cent, 11 per cent, and 43 per cent of the students). For every five applicants, approximately one student is chosen to study at Stadia. Applications for the faculties of Culture and Health Care by far exceed the available places: it has been noted that technology studies attract a decreasing number of students. The Polytechnic's activities take place on about 16 campuses across the city and the faculties run their programmes quite independently within the common frames, aims and regulations (van den Berg and Russo, 2003).

Until recently, polytechnics in Finland did not carry out R&D activities. Currently, R&D activities make up for 3–7 per cent of the polytechnics' activity volume (measured in Euros). The Minister of Education wants to increase this to 10 per cent. In future years, birth rates will decline, resulting in fewer students at (among other institutes) polytechnics. Carrying out R&D, therefore, means not only diversifying their activities but it can also help to attract good students (from within Finland as well as from abroad).

5.4.2 Creating Knowledge

Figure 5.3 shows that in Finland there is relatively high investment in knowledge: 6.2 per cent of the GDP in 2000 (3.4 per cent in R&D, 1.7 per cent in software and 1.1 per cent in higher education[2]). Only in Sweden and the United States were investments in knowledge higher (7.1 per cent and 6.8 per cent respectively). The average annual growth rate of the investments in knowledge is also high in Finland at 8.8 per cent; only in Sweden and in Ireland has this growth rate been higher (respectively 9.7 per cent and 10.8 per cent). As Helsinki is the most important knowledge economy region of Finland, it can be expected that investment in knowledge as a percentage of the regional GDP is even higher.

Number of patents Table 5.2 shows the number of patent applications by year per million inhabitants.[3] It teaches us that, regarding patent application per inhabitant, Finland ranks second in Europe, after Switzerland. Moreover, Finland also shows a relatively high growth (218 per cent in eight years). It is assumed that a large share of these patents was applied in the Helsinki region.

Table 5.2 European patent applications per million inhabitants

	1991	1995	1999	% increase 1991–99
Finland	83.1	136.6	264.6	218.0
France	84.9	86.1	116.9	38.0
Germany	141.1	158.6	248.5	76.0
Netherlands	95.5	111.5	181.7	90.0
Spain	8.3	9.9	18.0	117.0
Sweden	107.1	171.5	239.2	123.0
Switzerland	234.4	238.5	339.2	45.0
UK	60.1	65.0	93.8	56.0
European Union	73.0	82.4	125.0	71.0

Source: www1.oecd.org/publications/.

Important players in the knowledge production in the Helsinki region are the University of Helsinki (UH) and the Technical Research Centre of Finland (VTT). They are described below, as is the Centre for Knowledge and Innovation Research (CKIR).

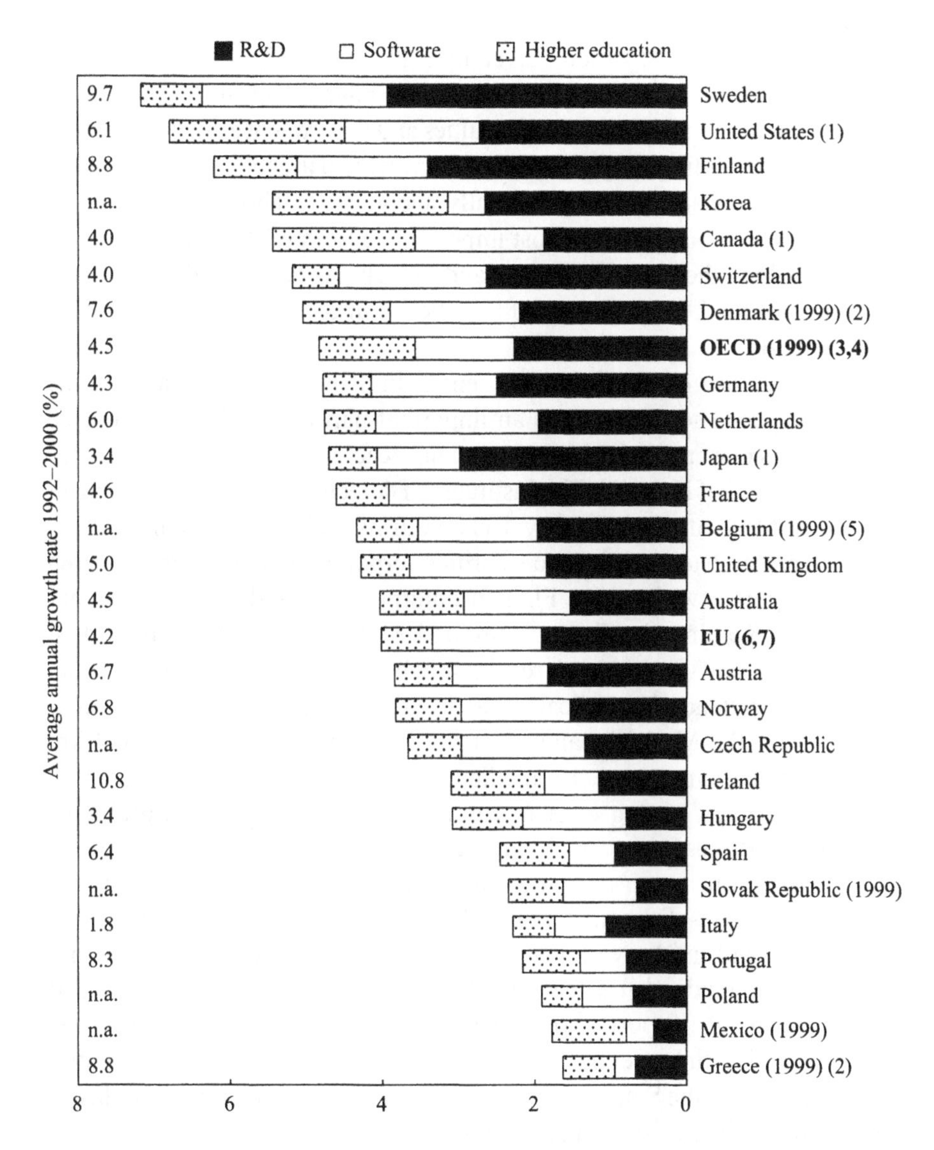

Figure 5.3 Investment in knowledge as a percentage of GDP, 2000

Source: OECD Science, Technology and Industry Scoreboard 2003, 'Towards a Knowledge-based Economy', www1.oecd.org/publications/.

The largest and most varied university is the *University of Helsinki* (UH). The UH is the oldest and largest university in Finland. High-level research is carried out at the departments of the faculties and department affiliated research stations. Basic funding for research comes from the national budget; however, the volume of outside funding currently accounts for about 50 per cent of the total research budget. The most important sources of outside funding for research include the Academy of Finland, which has designated the 11 units of research of the UH as Centres of Excellence in 2000–2005 (van den Berg and Russo, 2003).

The University of Helsinki's Biocentre, built in 1995, is located in the Viikki Campus. The Biocentre is an important part of the Helsinki Science Park, and it is currently a home of four major teaching and research institutes of the University of Helsinki. The Institute of Biotechnology is an independent research institute of the university. Related firms include Alkomohr Biotech Ltd. and a research unit of the Orion-Farmos corporation. Many of the research groups cooperate with major Finnish enterprises including Valio, Medix biochemica, Kemira, the Enso group, and MetsaSerla, with an important objective of identifying research results suitable for commercial exploitation. A centre of excellence for medical research and training, Biomedicum Helsinki is situated on the Meilahti campus of the University of Helsinki, 3–4 km northwest of the city centre. Biomedicum unites investigators and research groups to perform cutting edge research on currently important problems in medical sciences and to facilitate new drug development. It is the aim of Biomedicum to attain an important liaison function between academia and industry. State-of-the-art research facilities are provided to some 1,000 investigators and graduate students in residence, and it offers a high-class environment to medical and dental students for the first two years of Medical School. Financial support is provided by the University of Helsinki, the University of Helsinki Central Hospital, the neighbouring cities of Helsinki, Espoo and Vantaa, the State Real Property Authority, but also private industry sponsors part of the research (Susiluoto, 2002).

The Technical Research Centre of Finland (VTT) VTT is an independent expert organisation that carries out technical and technoeconomic research and development work. VTT is the biggest and the most important research institute in Finland. It is a contract research organisation involved in many international assignments. With more than 3,000 employees of which about 2,200 work in the Helsinki region (Espoo), VTT provides a wide range of technology and applied research services for its clients, private companies, institutions and the public sector. VTT carries out three types of activities: commercial activities, joint projects and self-financed projects. Of its employees, 12 per cent are doctors and 7 per cent licentiates, 48 per cent have university degree, 29 per cent have college level and polytechnic and 4 per cent basic level education. Its turnover is about €200m. VTT serves over 5,000 domestic and foreign customers annually (www.vtt.fi) (Tukiainen, 2003). It is a solid partner for R&D cooperation with firms in the Otaniemi Science and Technology Park. As previously mentioned, the reason for this is the fact that several firms were started by or employ former VTT employees. It appears, however, that within the Innopoli centre (see Section 5.4.5) companies have only few relationships with polytechnics and universities.

The Centre for Knowledge and Innovation Research (CKIR) (http://ckir.hkkk.fi/) was established in 1999 and is part of the Helsinki School of Economics (www.hkkk.fi). CKIR includes elements from the technological side in its knowledge-based issues: it investigates current technologies, their applications and their market perspectives. In principle, CKIR only does research, not education. The institute receives some funding from the Ministry of Education. CKIR has about 30 researchers and is globally oriented and multidisciplinary. CKIR has three knowledge programmes: 1) knowledge-based organisations and their management; 2) mind-based media and communication technologies (in cooperation with Stanford University); and 3) knowledge societies as innovation ecologies within the global knowledge economy.

According to CKIR, the US is much better in many knowledge fields than Europe is. An example mentioned is the biotech sector. In the US there are four biotech clusters competing with each other; in Europe there are 24 clusters. They will not succeed in this competition without cooperation. To excel, you need a certain critical mass. Therefore, in the EU from the competitive point of view there is a waste of resources. Besides critical mass, strong knowledge and economic bases and networks are also needed. The basic competences are the foundations on which networks should be built. A fundamental weak point of Europe is thus that from the point of view of equality each city has to have a university. Therefore, the aims of the Lisbon Agenda will never be achieved unless dramatic choices are made concerning the real European centres of excellence. According to CKIR, the strong role of the public sector gives the EU a competitive advantage in comparison with the US. The public sector should be used as a strategic partner, as is happening in the Helsinki region. This role could be strengthened and made more efficient.

5.4.3 Applying Knowledge/Making New Combinations

The role of national organisations Within Finland there are several national organisations that play a stimulating role in converting knowledge into business. The STPC played a very important role in stimulating all kinds of partnerships between knowledge-intensive organisations: higher education, businesses companies and public organisations. And, as stated, Tekes plays a dominating role in granting funds to knowledge institutes. One of Tekes' preconditions for providing funds is cooperation between research and business institutes. This is considered to be very helpful for setting up a large variety of new partnerships to finding new practical applications for the knowledge acquired.

In Finland, there are already strong relationships between education and business institutes. It is expected, however, that this will be strengthened by a new Finnish law. According to this law, universities have to develop strategic relationships within their direct region: they have to do relevant regional research, stimulate apprenticeships, etc. In this respect, the Minister of Education has asked all polytechnics and universities to develop a regional strategy.

Moreover, there are six technology-transfer companies in university cities in Finland. They form a close-knit circle cooperating in the task of commercialising research results. They help companies and people thinking of going into business to identify, assess, protect and commercialise different technologies. Sitra owns about one-third of these technology transfer companies. The only company in the Helsinki region is Licentia Ltd (www.sitra.fi) (Tukiainen, 2003).

There are also specific centres that support small and medium-sized enterprises (SMEs): the Employment and Economic Development Centres (TE-centres). These centres provide a comprehensive range of advisory and development services for businesses, entrepreneurs, and private individuals. Their aim is to support and advise SMEs at the various stages of their life cycles by, for example, promoting technological development in enterprises, assisting in matters associated with export activities and internationalisation, implementing regional labour policies, and planing and organising adult training within the official labour policy framework (Tukiainen, 2003).

An inadequate entrepreneurial attitude The CEO of Nokia recently stated that people in Helsinki have perhaps become too cautious; they should become more entrepreneurial and should invest more in new activities with business perspectives. The lack of an entrepreneurial attitude is considered to be a barrier for new economic developments. There are too few start-ups. In this respect, an aspect that needs improvement in Helsinki, and Finland in general, is the attitude towards bankruptcy. It is still considered shameful when firms go bankrupt. This also limits the number of start-ups. Consequently, Finland depends a great deal on large firms and this poses a danger when it comes to the courage to experiment.

Combination of culture and economy In the Finnish culture, design is considered important. Furthermore, one of the strengths of the Finnish economy is the combination of business activities and design. By integrating special design in industrial products, Finnish companies often do well commercially. There is even a specific programme, Design 2005, to stimulate this aspect: this is a national programme to give industrial design more focus in competition strategies.

Furthermore, to strengthen this, Helsinki tries to invest substantially in culture. Helsinki was, for instance, the cultural capital of Europe in the year 2000. One of the permanent revenues of this event was that, since that year, the *rhetoric* of culture has changed. Before 2000, in the debate about culture this matter was viewed upon as having value only in and of itself. After 2000, culture is also regarded as an instrument in other areas, such as stimulating the business economy.

Recently, the economy has experienced a downturn which has resulted in necessary budget cuts, because, among other measures, the national government has cut the amount of taxes that goes to municipalities. Regarding the municipal budget cuts, culture has been hit just as hard as other fields, neither more nor less. Although the culture budget has decreased, at the same time the national government is putting a considerable amount of money into a new opera building, with Helsinki paying a proportion of the costs. Opera can be regarded as an extreme luxury, but on the other hand it is a source of many other things. In the vision of the city of Helsinki, culture remains very important in making the region attractive to knowledge workers.

Urban Professors Programme The city has stepped up its cooperation with the University of Helsinki. Six professorships of urban research in different fields from economics to urban history were founded at the University. These posts are to form an urban think-tank. They have to strengthen the link between the city and its higher education. The main goal of the programme is to develop the area of urban studies in Finland. Currently, six professors are active in this initiative, but more professors are to be appointed: in the technical university, in urban sociology, urban economics, urban history, urban ecology, urban planning, etc. This programme is financed by Espoo, Vantaa, Helsinki, Lahti, the University of Helsinki and the Ministry of Education.

Club of Helsinki: regional players promoting innovation There is a club of regional players promoting technology development and innovation and cooperation. The first Helsinki Club was founded in the midst of the recession of the 1990s. The second Helsinki club was founded in 2003. The participants in the Helsinki Club are the mayors of the three large cities of the Helsinki region, the general directors of Tekes and Sitra, the Minister of Trade and Industry, the head of Nokia, the director of Culminatum, representatives of the opera, television and science parks, the rector of the University of Helsinki, a professor from bioscience, and a representative of the Chamber of Commerce.

The Helsinki Club tries to foster industries in Helsinki. Subjects on the agenda include: housing policy, attracting and maintaining a qualified workforce, and Helsinki's international position.

5.4.4 Developing New Growth Clusters

The programme of Centres of Expertise plays an important role in developing new growth clusters by stimulating strong regional knowledge centres. The programme leads to a redistribution of knowledge capacity (concentration instead of dispersal of capacity). The idea of the Centre of Expertise Programme is to utilise top-level knowledge and expertise as a resource for business operations, job creation and regional development. Initial implementation of the programme over the period 1994 to 1998 was based on 11 Centres. Based on positive results of this work, the Council of State has extended the programme by nominating new fields of expertise and new Centres of Expertise to implement a second national programme over the years 1999–2006. Fourteen regional Centres and two nationally-networked Centres of Expertise have been appointed for this purpose. The concept of the field of expertise has been broadened from the traditional high-tech sectors to include new fields such as new media, cultural business, the recreational experience industry, design, quality and environmental expertise.

The office for carrying out the Centres of Expertise programme in the Helsinki region is Culminatum. This organisation has to contribute to improving the regional relationships between the relevant partners. The aim was to turn leading expertise into profitable business expertise. This organisation, therefore, operates like a kind of broker, bringing relevant partners together. The vision of Culminatum is to develop Helsinki as an 'Ideopolis', an innovation centre of world class based on creativity in art and science. The strategy of this regional organisation has two pillars (Holstila, 2003):

1 to implement regional cluster programmes (Centre of Expertise Programme, the basic function of Culminatum) together with science parks; and
2 to strengthen the knowledge potential and the local innovation environment.

The Culminatum shareholders are the Helsinki region's universities and polytechnics, the cities and Chambers of Commerce of the Helsinki Metropolitan Area, the Science Park organisations and the Uusimaa Regional Council. The budget of Culminatum is about €2m in year 2003. It currently employs ten academically-educated people.

Culminatum is also developing an innovation strategy for the Helsinki region, financed by Tekes, amongst others. The general directors from Tekes, Sitra, Culminatum and Nokia will be involved. The innovation strategy will focus on the question: what unique aspects should Helsinki utilise? Questions relevant in this context are: what is Helsinki's profile? How should Helsinki be marketed? What's the technology foresight? What programmes are currently being run and what programs should be run? What is Helsinki's position in the global market?

The city councils of the Helsinki Metropolitan Area have assigned responsibility to Culminatum for preparing a science and technology strategy for the region, and a common development strategy for the universities and associated business operations. Culminatum primarily provides services for businesses through six Centres of Expertise:

1 active materials and microsystems (in Otaniemi Science Park);
2 gene technology and molecular biology (in Helsinki Science Park);
3 medical and welfare technologies is considered to be a promising cluster. Within this cluster interdisciplinary research into medicine and ICT technology should lead to new profitable business operations. In the coming decades a relatively large share of Finland's inhabitants will retire. Therefore, care facilities for elder people have to become more efficient. New technological developments can, in the first instance, be used within Finland, but should also be exported to other industrialised countries;

4 logistics – Vantaa, where the airport is located, wants to develop as the principal logistics centre of the Helsinki Metropolitan Area and as a leading European area for logistics services;
5 software product business – the Innopoli Centre aims to further improve the dominating ICT cluster in the region. According to the OECD (2003, p. 22), the challenge for the Helsinki region is to develop ICT activities beyond the current cluster scope, e.g., the use of learning and positive externalities in forestry, biotechnology and new media;
6 digital media, content production and learning service – this Centre of Expertise seeks to improve the profile of the cultural industry and the operating conditions in the cultural sector, and to establish networks of experts and support services to improve the cultural sector business environment. The Centre has been set up in the Arabianranta cultural cluster surrounding the Helsinki University of Art and Design.

5.5 Conclusions and Perspectives

5.5.1 Knowledge Economy Strategy

The success of the current knowledge-intensive activities in the Helsinki region can be explained by the high quality of education and the good inter-relations between research and business. National institutes appear to play an important role in this respect. The most striking institute is the STPC, chaired by the Prime Minister, setting out strategic themes for the knowledge economy. In addition, funding organisations like Tekes and Sitra prescribe cooperation between knowledge institutes and businesses. In the first instance, local authorities in the region of Helsinki did not have an explicit knowledge economy strategy. They played predominantly an accommodating role by, for instance, supplying the necessary infrastructure and business space. In recent years, however, they have developed more explicit policies to enhance regional competitiveness as well as cohesion. However, they still play a marginal and complementary role in relation to the national knowledge economy strategies.

5.5.2 Foundations of the Knowledge Economy

The *knowledge base* of the Helsinki region is considered to be very good; it is seen as a decisive factor for the success of the Helsinki knowledge economy. On the one hand, students receive their education nearly for free; on the other hand, there are considerable quality incentives for them. Moreover, the share of employees working in R&D is the highest in the EU regions. The *economic base* in Helsinki is judged as good. However, dependency on the ICT sector makes the regional economy susceptible to global slowdowns. On the one hand, the *quality of life* is considered to be very good, with a relatively safe and clean environment. On the other hand, the region does not really have the environment of a vibrant knowledge city and during wintertime it might be too cold to attract many foreign symbolic analysts. The external and internal *accessibility* of the Helsinki region is quite good, though some public transport facilities have to be improved. One obstacle to the further broadening of the knowledge economy could be the peripheral European location. The *urban diversity* is considered to be rather low. This is particularly the case in Espoo, the gravitation point of successful ICT activities. Apparently metropolitan diversity is not a necessity for them. The local authorities in Espoo are even concerned about the advent of metropolitan developments stimulated by the eventual extension of the metro line to this city. Some consider the low urban diversity as a barrier to attracting foreign workers. The relatively large investments in culture in the Helsinki region are a positive element. The *urban scale* is relatively small. This might be one of the explanations of the relatively small economic base of the region. Cooperation with other cities in Finland, with other Scandinavian countries and with Baltic cities might help to create a larger critical mass. *Social equity* is judged as relatively high. The Helsinki government is focused on preventing rather than curing social exclusion. In addition, the high social equity (and high average education) is also seen as very helpful in creating a large critical demand for knowledge intensive products.

5.5.3 Activities of the Knowledge Economy

The capacity of the Helsinki region to attract and retain knowledge workers is considered to be good. Within Finland, the Helsinki region is very attractive to symbolic analysts. The metropolitan attractiveness of Helsinki and Espoo, with its technical university and Nokia headquarters, appear to be major attraction factors. There are relatively many science and engineering (S&E) students and workers, because of good career perspectives and image. Cultural policies are considered to be instrumental for an environment that attracts and retains symbolic analysts (also from abroad). One of Helsinki's drawbacks is a shortage of good housing. Moreover, Helsinki lacks an international location climate; the city does not attract many foreign knowledge workers.

Table 5.3 An indication of the value of knowledge economy foundations in Helsinki

	Foundations	Score in Helsinki
1	Knowledge base	++
2	Economic base	+
3	Quality of life	+
4	Accessibility	+
5	Urban diversity	□
6	Urban scale	□
7	Social equity	+

-- = very weak; – = weak; □ = moderate; + = good; ++ = very good

Internationalisation strategies have been set up to change this situation such as, for instance, marketing programmes and university lectures in English. In addition, by setting up strategic relations with cities such as Tallinn and St Petersburg, it is hoped not only to serve those markets better but also to attract workers from these regions. Moreover, in this way, the urban scale and critical mass for new promising knowledge-intensive economic clusters can be extended.

Creating knowledge is one of the strengths of the Helsinki region. Research and education is at a high level and there appear to be sufficient incentives to keep them there, such as the national funding of academic and business research (under quality restrictions) and the *numerus-fixus* policy for students. R&D investments in Finland are at a high level (6.2 per cent of GDP) as is their annual growth (8.8 per cent). In addition, the number of patents per inhabitant is one of the highest in Europe. More European cooperation and specialisation is needed, however, to compete with R&D in the US. Currently, there are too many relatively weak R&D institutes within Europe. One competitive advantage of Europe and particularly of Finland appears to be the strong public sector, which can play an important role in supporting pioneering knowledge economy activities.

The capacity within the Helsinki region to apply knowledge is judged as being very good. In the Helsinki region there are strong relationships between education and business institutes. This is stimulated by national funding programmes, which prescribe cooperation, but the national culture of consensus and trust is also helpful. One of the strengths of the Finnish economy is to combine technology knowledge and design. This is one of the

Table 5.4 An indication of the value of knowledge economy activities in Helsinki

	Activities	Score in Helsinki
1	Attracting knowledge workers	+
2	Creating knowledge	++
3	Applying knowledge/making new combinations	++
4	Developing new growth clusters	+

-- = very weak; – = weak; □ = moderate; + = good; ++ = very good

explanations for the success of the Nokia company. Such combinations are to be further strengthened. One promising new combination in this respect is between welfare activities and technology. There also appears to be a fruitful combination between the local economy and culture. Stimulating start-ups might become more important to widen the economic base. One of the barriers, however, appears to be a lack of entrepreneurial attitude. According to our discussion partners, there are too few start-ups of new promising knowledge-intensive companies. This appears to be one of the side effects of a welfare state: inhabitants become more risk averse. More dedicated policies and education programmes are to be developed to limit this barrier.

The ability to develop new growth clusters in the Helsinki region is good, but could be improved. Because of the relatively small economic base of the Helsinki region, there is an urgent need to develop new promising growth clusters. The centres of expertise programme could contribute to this aim, by supporting the most promising regional R&D activities. Regions in Finland can apply for specific clusters to obtain national funding. The programme of Centres of Expertise can thus play an important role in stimulating strong regional knowledge centres and can lead to a redistribution of knowledge capacity within Finland (a redistribution which will also be necessary at European scale to ensure the competitiveness at global scale according to the Lissabon aims). Within the Helsinki region, there are six such clusters of knowledge intensive activities designated, one of which is medical and welfare technologies. Within the region, a relatively large number of growth perspectives are expected from this combination of welfare and technology. Another cluster is the already dominant ICT cluster, whose scope is to be broadened to other economic activities.

5.5.4 Perspectives

From the case study it was found that there is a fear that other regions in Finland are going to be given more support out of national funds, to the detriment of the Helsinki region. In the viewpoint of a globalising economy, it was questioned to what extent Finland can afford such an equity policy. To be sufficiently competitive in the worldwide economy, the engine of Finland – the Helsinki region – must obtain sufficient 'fuel' to further improve its economic and knowledge base. It could be argued that the other regions will benefit of the spread effects of the national engine; but they could also face decline parallel with an eventual slow-down of the Helsinki economy. Whether an efficiency or equity policy is wise for Finland's economic growth cannot be deduced from this case study, which has a limited scope.

Notwithstanding the economic downturn, the Finnish economy performs relatively well. But there is a vulnerability in the economic structure because of too much dependence on one single economic sector. There appears, however, to be a sufficient sense of urgency that much has to be done to guarantee sufficient economic growth and diversity in the near future.

Notes

1 S&E degrees include the following fields of study according to the 1997 International Standard Classification of Education (ISCED). Science includes: life sciences, physical sciences, mathematics and statistics and computing. Engineering includes: engineering and engineering trades, manufacturing and processing and architecture and building (http://www1.oecd.org/publications). Data on S&E graduates were unfortunately only available at country level, and not at the level of urban regions.
2 European averages of investments in knowledge are 1.9 per cent in R&D, 1.4 per cent in software and 1.1 per cent in higher education.
3 Only European countries with a score in 1999 above 150 were included in this table, plus some countries relevant for this study.

Discussion Partners

Mr Ilkka-Christian Björklund, Deputy Mayor, City of Helsinki.
Mr Timo Cantell, Senior Researcher, Urban Facts Department, City of Helsinki.
Ms Nina Kalliovaara, Project Researcher, Urban Facts Department, City of Helsinki.
Mr Marko Karvinen, Manager, Urban Policy Development Projects, Urban Facts department, City of Helsinki.
Ms Seija Kulkki, Director, Centre for Knowledge and Innovation Research.

Mr Jussi Kulonpalo, PhD student, urban governance in European cities.
Ms Minna Maarttola, Planner, Business Development Unit, City of Helsinki.
Ms Asta Manninen, Director, Urban Facts Department, City of Helsinki.
Ms Marja-Liisa Niinokoski, Director, Culminatum.
Mr Juha Nymann, Helsinki Polytechnics Stadia.
Mr Jorma Nyrhilä, Deputy Director of Helsinki Chamber of Commerce.
Ms Heli Penttinen, Project Manager, Centre for Knowledge and Innovation Research.
Mr Jaakko Pajula, Director of Helsinki Entrepreneur's association.
Mr Heikki Santti, Helsinki Polytechnics Stadia.

References

Berg, L. van den and A.P. Russo (2003), *The Student City: Strategic planning for students' communities in EU cities*, Aldershot: Ashgate.

Berg, L. van den, J. van der Meer and P.M.J. Pol (2003), *Social Challenges and Organising Capacity in Cities, Experiences in Eight European Cities*, Aldershot: Ashgate.

Castells, M. and P. Himanen (2002), *The Information Society and the Welfare State: The Finnish model*, Oxford: Oxford University Press.

City of Helsinki Urban Facts (2002), *Student City*.

City of Helsinki Urban Facts (2003), *Helsinki 2003. Facts about Helsinki*, Helsinki: Lönnberg.

Helsingin kaupunginkanslia (2002), *Portrait Muteis*.

Holstila, Eero (2003), *Building a Creative City. The High Tech Cluster in the Helsinki Metropolitan Region*, Culminatum Ltd, Helsinki Region Centre of Expertise.

Nauta, Frans, Marieke Rietbergen and Joeri van den Steenhoven (2003), *Lessen van Finland, Een overzicht van het Finse innovatiemodel en de beleidsmaatregelen tussen 1980 en 2002*, Stichting Nederland Kennisland.

OECD (2002), 'Territorial Review on Helsinki 2002', OECD Observer, Policy Brief, October.

OECD (2003), *Helsinki, Finland*, OECD Territorial Reviews, Paris: OECD.

Susiluoto, Ilkka (2002), 'Portrait of the Helsinki region: Muteis, macro-economic and urban trends in Europe's information society', European research project.

Tukiainen, Janne (2003) 'The Helsinki Region ICT cluster study', in A. van der Meer, W. van Winden and P. Woets (eds), *ICT Clusters in European Cities during the 1990s – Development Patterns and Policy Lessons*, Rotterdam: Euricur.

Vaattovaara, Mari and Marti Kortteinen (2003), 'Beyond Polarisation versus Professionalisation? A Case Study of the Development of the Helsinki Region, Finland', *Urban Studies*, Vol. 40, No. 11, October, pp. 2127–45.

Chapter 6

Manchester

6.1 Introduction

The city of Manchester is the regional capital and commercial, educational, medical and cultural centre of England's northwest region, the second largest economic region in the United Kingdom (UK) after London. The city has a population of about 440,000 people (2000 mid-year estimate), and the metropolitan sub-region (Greater Manchester) has more than 2.5 million inhabitants (2000 mid-year estimate). The sub-region comprises the cities of Manchester and Salford and the Metropolitan Boroughs of Bolton, Bury, Oldham, Rochdale, Stockport, Tameside, Trafford and Wigan.

Figure 6.1 Manchester in the United Kingdom

Source: http:\\www.manairport.co.uk (edited by the authors).

Manchester is internationally known as the first city of the Industrial Revolution. Like many other industrial cities, Manchester has suffered from the decline of the manufacturing and textile industry, resulting in severe economic and social problems. Between 1972 and 1984, employment in the industrial sector declined by about 207,000 jobs (Phelan, 1997). Nowadays, about 70 per cent of the working population are employed in the service sector, which has experienced an increase in employment over the last two decades. However, the area retains a strong industrial and manufacturing base, employing more than one-tenth (ABI, 2001) of the working population (22,000 in the Manchester area and almost 185,000 in the wider sub-region). About 13,500 people are unemployed in the city (Nomis and MCC Planning Studies, July 2003), which is 7.9 per cent of the economically active population. Across Greater Manchester, just over 44,000 people are unemployed, representing 2.9 per cent of the economically active population (Nomis and MCC Planning Studies, July 2003).

Manchester is part of the *Core Cities* group, together with Birmingham, Liverpool, Nottingham, Bristol, Sheffield, Leeds and Newcastle – the major English regional capitals outside London. They share the challenge of finding their own competitive niche in the British urban system. Manchester uses sports and culture as distinguishing elements (its region has the largest theatre sector in the UK and a wealth of art galleries, museums, libraries, sports and leisure facilities), but the city is also an important commercial, financial and educational centre. There are more than 15,000 people employed in banking and finance and more than 540 financial businesses are present in the city (ABI, 2001).

In Section 6.2, we will explore the knowledge economy strategy of Manchester, laid out in the 'Manchester: Knowledge Capital' Prospectus. Then in Section 6.3, we will analyse the foundations of the knowledge economy for Manchester distinguished in the theoretical chapter. In Section 6.4, we will go into the four knowledge economy activities for Manchester. Finally, Section 6.5 presents some conclusions.

6.2 The Knowledge Economy Strategy of Manchester

In the last decade a number of important vision and strategy documents were published in the Manchester region: in 1994 the 'City Pride Prospectus' (with an update in 1997) and in 2003 'Manchester: Knowledge Capital' (see Figure 6.2). Both documents had to contribute to achieving a more attractive and competitive urban region. In between, the *Core Cities* played an important role, with the publication of 'Cities, Regions and Competitiveness', in getting urban regions higher on the national political agenda.

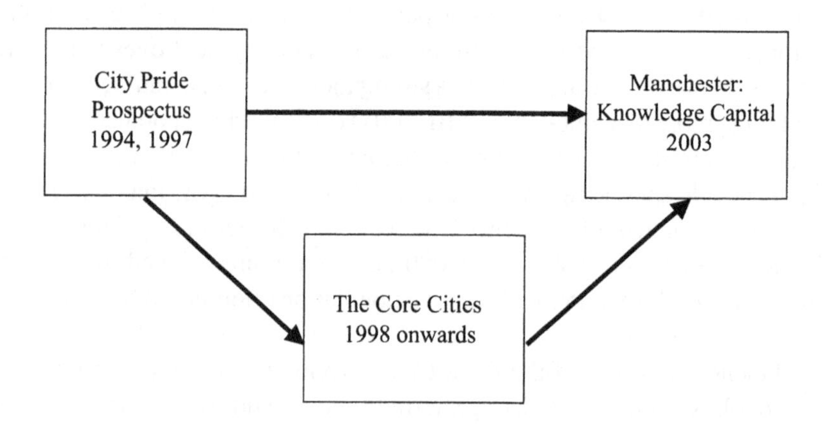

Figure 6.2 Important momentums for the development of the urban vision

The 1994 City Pride Prospectus was written by the Manchester City Pride Partnership (in which the local authorities of Manchester, Salford, Trafford and Tameside participate). That vision saw Manchester developing into a European regional capital and a centre for investment growth; an international city of outstanding commercial, cultural and creative potential and an area distinguished by the quality of life and sense of well-being enjoyed by its residents (City Pride Partnership, 1994). In view of the change of government in 1997 and the New Deals offered across education, health, employment, training and regeneration, the Partnership decided to add a fourth objective in the second City Pride Prospectus: to develop the region into an area where all residents have the opportunity to participate in, and benefit from, the investment and development of their city and therefore live in truly sustainable communities (City Pride Partnership, 1997).

6.2.1 The Core Cities Group: Recognition for the Role of UK Cities

Up until a few years ago there was little recognition for the role and contribution of cities outside London in helping to grow the UK economy. The Core Cities Group was established with its main aim to increase the role and profile of cities in national urban and economic policy. They have, in part, achieved this aim: the important role of cities in the national economy is now acknowledged by many policy makers. Furthermore, the Office of the Deputy Prime Minister started to collaborate with the Core Cities, Regional Development Agencies and other government departments to develop new ideas for enhancing the economic competitiveness of regional cities.

In 2003, this led to a jointly issued report called 'Cities, Regions and Competitiveness', which described itself as a report about strengthening the UK's growth capacity in future global upturns. It used evidence from competitor countries to demonstrate how stronger regional cities could add more cylinders to the UK's economic engine and provide more space for London to further develop its unique global city role. The report thus acknowledged that regional cities can benefit from the spread effects of high growth in London and the southeast, and can thereby support the role of the capital cities, by helping to diminish agglomeration disadvantages. The report contained an action plan for strengthening productivity and the urban renaissance in the major regional cities as the essential foundation for progressive improvements in the performance of all regions.

Following this 'Cities, Regions and Competitiveness' report, the Deputy Prime Minister John Prescott asked the Core Cities to each prepare a prospectus detailing how they would deliver enhanced economic competitiveness and prosperity within their regions. Consequently, Manchester, together with key regional partners, developed the 'Manchester: Knowledge Capital' (MKC) Prospectus. This prospectus can be considered as a further elaboration of the core cities' objectives, but in particular as a further elaboration of the City Pride vision. Major differences between 'Manchester: Knowledge Capital' and City Pride are that there are now more partners involved, the goals have been made more explicit, and it is in particular a unique cross-regional, cross-sectoral partnership that is taking the lead.

Up till now, two actions have been taken: 1) the contextualisation of the knowledge economy within the fabric of Manchester has been researched; and 2) concrete projects have been developed to strengthen Manchester's position in the knowledge economy.

One of the essential points of the MKC vision is to strengthen the links between higher education, business and local communities. Indeed, MKC has been published by a partnership within which such actors are members. The partnership includes Manchester, Salford, Trafford and Tameside local authorities, Northwest Development Agency, Manchester Investment and Development Agency Service (MIDAS), the four Greater Manchester universities, Manchester Enterprises, the health sector and others. The universities play a central role in the stimulation of the knowledge economy, but the city is leading the process.

However, the limited funding and decision-making capabilities of public sector agencies and local authorities is considered as a major handicap to effective operating. In previous years there has been a focus change from the national level towards cities. The Labour government has helped to refocus on major cities but has not given responsibilities back to the urban governments. Most of the strategic decisions (for instance, on transport infrastructure) are still taken at Westminster. According to Westwood and Nathan (2002), if there is to be a true and sustainable urban renaissance, which in turn will feed improvements in national competitiveness, then UK cities must be allowed to have more control over their futures.

The MKC vision is to utilise Manchester's strengths and to try to attract and retain economic assets and investment in the region. The MKC Prospectus 4 mentions as important assets of Manchester (p. 2): the four universities with a combined income of £670m (€944m), 12,500 employees and 88,000 students; a growing financial and professional service sector; Manchester Airport and the Metrolink system (urban rail transport); growing industry clusters, such as biotechnology, advanced engineering, information technology, software, new media and the cultural industries. Critical challenges are also summed up: to address spatial concentrations of deprived residents and unemployed people; to improve quality and choice in parts of the housing market; to increase levels of basic skills and qualification; to increase low economic activity rates; and to increase business survival rates and levels of self-employment.

Critical challenges are identified not only for the Manchester region, but also for the entire UK. An underperformance of the UK against major competitors is observed: employees in France, USA and Germany are 28 per cent, 26 per cent and 16 per cent respectively more productive. Moreover, productivity differences between British regions were recently estimated at 60 per cent (the London region being the most productive). According to the prospectus, the UK's variance in regional economic performance is, arguably, the greatest in the EU.

In the MKC Prospectus, seven key themes are summed up. These are assets of the region, which have to be further strengthened:

1 a dynamic environment: improving the attractiveness and vitality of the urban region, and predominantly the regional centre. *It will include excellent new buildings, green spaces, regenerated communities, better*

housing and signature projects such as waterfront renewal and major new recreational facilities;

2 connectivity matters: improving the regional, national and international, and physical as well as virtual transport infrastructure. *It will include an enhanced West Coast rail link to the southeast (good accessibility to the London region and mainland Europe is considered to be crucial) and mainland Europe, further growth of Manchester Airport and expansion of the light rail system (Metrolink)*;
3 better for business: supporting industrial clusters and attracting and supporting new businesses. *Major investments will be made in additional office developments and science parks. Incubators and workspaces will be created. Venture capital will be increased and there will be fresh strategies for growth in the creative, cultural, digital and biotech industries*;
4 academic acceleration: supporting academic expertise and capital. Manchester's higher education institutes and teaching hospitals have to contribute to global research, social change and wealth generation;
5 genius generation: supporting the next generation of talented, creative individuals. *Brilliant, pioneering individuals will be empowered to embrace better ways of managing knowledge, new ways of working, innovative use of creative technologies and ever-stronger communications.* Richard Florida ranked Manchester as the most creative place in the UK (see www.creativeclass.org);
6 cultural atmosphere: supporting cultural and creative industries and initiatives in order to enhance the economic performance, attractiveness and competitiveness of the Manchester region. It will showcase Manchester as a premier destination for visitors seeking everything from world class sport and leisure events to superb nightlife and the café culture that has become synonymous with the '24-hour city';
7 community and opportunity: targeting disadvantaged communities and contributing towards socially and spatially balanced urban growth. *There will be collaborative skills development partnerships with employers, educators and local communities. Community-based entrepreneurs will be supported and encouraged.*

6.3 Knowledge Foundations

In this section, the knowledge foundations – as described in the theoretical chapter – will be assessed for the Manchester region.

6.3.1 Knowledge Base

Education institutions are crucial actors in the field of the knowledge economy. Manchester is home to a very large student population. Students can choose from three universities in the city and one in neighbouring Salford. More than 87,000 students are enrolled at one of the four universities in the greater Manchester area: the University of Manchester (UM), Manchester Metropolitan University (MMU), Salford University (SU) and University of Manchester Institute of Science and Technology (UMIST) (see Table 6.1). In the period between 1995 and 1998 there has been a large increase in the number of their students. The Manchester-based universities are located close together and adjacent to the city centre, creating one of the largest university quarters in Europe.

The UM is the oldest of the city's universities and offers students a broad range of arts, science and medical degrees. At UM there are important technical disciplines – for the knowledge economy – such as bioscience. MMU is now one of the UK's largest universities and has established extensive links with industry and employers, having a more vocational focus including, for example, specific courses covering regeneration and education in deprived areas. At MMU there are also specialist courses in creative industries; for example, fashion and textile production. This can be seen as building on the inherent industrial strengths within the city (new combinations). UM and MMU work together in linking art and fashion with textile, design and material sciences. The University of Salford has traditionally been a technology-led institution, but is now developing a reputation for being one of the UK's most successful universities in terms of graduates securing employment upon completion of their degrees. The Royal Northern College of Music (RNCM) is one of the leading conservatoires in the UK and attracts highly talented pupils from across the world. Specialised Business Schools form part of both UM and MMU, with respective expertise in research and industry and sector specific courses.

There has been a trend of declining public funds for UK universities. The higher education institutions (HEIs) were therefore stimulated to commercialise to obtain sufficient financial funds. Universities were stimulated to think in a more client-oriented way, to undertake more socially relevant research and to develop activities that can be commercialised. About one-third of the total funds of the universities now comes from private sources.

The further education (FE) sector in Manchester is even larger than the universities. Greater Manchester has the largest number of FE colleges in the

Table 6.1 Students at Manchester institutions: the University of Manchester (UM), Manchester Metropolitan University (MMU), Salford University (SU) and University of Manchester Institute of Science and Technology (UMIST) and the Royal Northern College of Music (RNCM)

	Total students 1995/96	Change in 1995/96–1998/99 %	Total students 1998/99	Change in 1998/99–2001/02 %	Total students 2001/02
UM	22,282	11	24,782	10	27,145
UMIST	6,583	3	6,785	4	7,025
MMU	28,433	8	30,793	3	31,590
RNCM	534	–6	503	11	560
SU	8,935	128	20,395	5	21,490
Total	66,767	25	83,258	5	87,810

Note: The 128 per cent increase in Salford University numbers between 1995/96 and 1998/99 was due to the university merging with University College Salford and the Northern College of Nursing, Midwifery and Health Studies.

Source: Higher Education Statistics Agency (2003) (website).

UK. Within the area there are 25 colleges, with over 200,000 students supported by over £200m of expenditure (Westwood and Nathan, 2002).

6.3.2 Economic Base

In the past Manchester has undergone fundamental economic restructuring. Manchester was not only at the forefront of the Industrial Revolution, it was also a centre of innovation, research and invention; for example UM developed the world's first computer with a stored programme (1948).

The greater Manchester economy has undergone significant changes in past decades. In the 1960s the Manchester economy was dominated by the manufacturing sector, which employed a significant proportion of the working population. Manchester companies were globally competitive in textile industries. Nowadays, however, the manufacturing sector accounts for 7.4 per cent of employment (ABI, 2001).

The fundamental restructuring of the Manchester economy has left a diverse and dynamic commercial base. Economic and industrial strengths across Manchester today include the four universities – together one of the largest employers in the city-region – and financial and professional services – this sector is currently growing faster in Manchester than in any other regional centre. Manchester Airport provides excellent national and international connections with scheduled flights to 180 destinations, greatly enhancing the region's competitiveness and providing significant direct and indirect investment and employment benefits.

In general, however, the economic performance of the Manchester TEC area[1] (measured in gross domestic product growth) remains below the average for the northwest region as well as the UK average (see Table 6.2). Investment levels in the TEC area are well above the national average, mainly because of the major (re)construction projects in the city triggered by the IRA bombing.

Unemployment in the Manchester TEC area is estimated at 6.5 per cent, well above the UK average of 5 per cent. However, unemployment in the TEC area has fallen by one quarter over the past year, following the national trend of lower unemployment rates. Unemployment across the Manchester TEC area varies among the districts, with the city of Manchester accounting for the highest rate of 10.6 per cent.

The economic outlook for the Manchester area is inextricably linked to national and international economic development trends. On the one hand, Manchester holds several trump cards with its airport, educational

Table 6.2 Economic performance of the Manchester TEC area in perspective

	Annual average change in GDP		Annual average change in investment	
	1988–98	**1998–2000**	**1988–98**	**1998–2000**
Manchester TEC	1.2	2.5	3.2	3.5
Northwest region	1.4	2.4	2.6	3.3
United Kingdom	1.8	2.9	0.5	3.2

Source: CE/ER LEFM (1997), in Manchester TEC (1998).

infrastructure and many successful major employers. On the other hand, the levels of unemployment and poverty must be tackled in order to maintain sustainable economic development and growth.

6.3.3 Quality of Life

The quality of life in Manchester is significantly improving as the city tackles the challenges of poor perceptions of crime, safety and attractiveness. The total costs of crime in Manchester were estimated at £989m per year (€1,394m), that is £2,295 per person (€3,234) (Crime and Disorder Team, 2003). Some parts of the city, particularly North and East Manchester, have suffered a collapse of their housing markets, as those residents with the economic means have chosen to move away from the area. In addition, around a decade ago, only a few hundred people resided in the city centre.

However, this has begun to change dramatically. The city centre has become a vibrant place with a growing number of residents, with many modern pubs and restaurants with a high quality of (cultural) life (modern residential buildings and new cultural amenities). In particular the relatively large student population and its reputation for music, sport and creativity contribute to a new image of Manchester. A massive investment programme in housing market renewal is due to start in the most critical areas, with the aim of attracting families back with a much-enhanced quality of life.

The revival of city-centre living is one of Manchester's recent success stories. The image of living in the city centre has changed dramatically. Manchester was a dull, old-fashioned, manufacturing city, but it is transforming into a modern, dynamic, high culture urban centre. Now, an attractive modern built-up area is being created and new leisure infrastructure is being developed.

Many young urban professionals live in the city centre in relatively expensive housing. New developments across the city, most notably in the refashioned Millennium Quarter, at Sport City in East Manchester and around the canals and quaysides in the city centre have added much to the city's growing reputation (Westwood and Nathan, 2002).

An index to measure quality of life has been developed by the Urban Energy Index for the UK. In this index, tracking objective and subjective liveability components, Manchester comes in third place after Cambridge and Edinburgh (Westwood and Nathan, 2002). Elements which were judged as part of the Index include knowledge institutions, cultural and retail amenities.

The city's social, economic and cultural base has also provided the stimulus for cultural and creative innovation. Manchester is known to be at the cutting edge of popular youth culture, more particularly in the fields of music and design. In the late 1970s/early 1980s Manchester bands spearheaded a vibrant music culture in the city, which became renowned across the world. In the second half of the 1980s, Manchester became the centre of 'dance culture' with world-famous nightclubs such as the Hacienda, bringing 'underground culture' to a larger international audience. The band Oasis, one of the most successful in the world, was originally established in Manchester.

As well as the city centre, other local areas in Manchester have been successfully regenerated. Hulme, for example, was in a critical state a decade ago, but now has a much healthier housing market with a much-improved reputation. The Metrolink system contributed successfully by providing clean and safe transport in a regular and sufficient degree.

> A signature of Manchester's renaissance is the amazing array of spectacular, beautifully designed and ingenious new building projects, with international architects such as Daniel Libeskind, Sir Terry Farrell, Will Alsop, Santiago Calatrava, Sir Michael Hopkins, Tadao Ando and Richard Meier making their presence felt right across the cityscape. (MKC, p. 17)

Major events, such as the Commonwealth Games 2002 and the Champions League final in 2003 at the Old Trafford stadium, have contributed to a changing image of Manchester.

Neighbouring local authority areas such as Salford also contribute to and benefit from the improvements in Manchester's profile and attractiveness. The Imperial War Museum North and Lowry Centre are major visitor attractions, with the five-star Lowry Hotel contributing to Manchester's range of quality hotel options. Residential buildings in Salford Quays are in high demand.

6.3.4 Accessibility

Manchester Airport is considered to be very important for the economic development of the region. This publicly-owned international airport offers 180 destinations, guaranteeing swift connections with other economic centres in the global economy, including the city of London. A new two-mile second runway was completed in the year 2000 and by 2015 the airport expects to have doubled its capacity, handling over 40 million passengers a year.

There has been substantial investment in a new urban light rail system: Metrolink. The system carries more than 17 million passengers across the city each year, linking the city centre with the outlying districts of Bury, Altrincham and Eccles. In the near future, the system will be expanded to South Manchester, Ashton, Oldham, Rochdale, and – importantly – Manchester Airport. This extension is part of a major public-private programme costing over £800m (€1,128m) (MKC, p. 19).

The quality of ground transport links (rail and road) to London and other major cities is considered to be very important. In this respect, the north–south links are better than the east–west ones. Major investment is required to improve the West Coast Mainline rail link which connects Scotland, London and Birmingham to Manchester. The recently completed M60 orbital motorway has improved circulation and linkages with the national motorway network but traffic congestion and car parking, as with many major UK cities, are both issues which require long-term, sustainable solutions. Connectivity within the region – ensuring that local communities can easily access job opportunities and facilities located in the regional centre – is also a key challenge for the future.

Fast connectivity to the Internet is considered to be crucial for future economic development. By investing in wireless broadband systems, the region hopes to further its economic attractiveness. Manchester is one of the best-connected cities in the UK after London, in terms of broadband capability and wi-fi.

6.3.5 Urban Diversity

Manchester is considered to have a rich urban diversity, evidenced by its cultural activities, architecture and residents. The heritage of the city's manufacturing and industrial past is redolent in its old redbrick warehouses and other buildings, the canals and the docks; juxtaposed with the modern concrete, steel and glass buildings which reflect the city's new service and knowledge industries.

Characteristically, Manchester also has a rich blend of cultural activities and knowledge-intensive activities, combining music, dance, museums, business parks and university quarters. According to 'Manchester: Knowledge City', Manchester's cultural vitality will, almost instinctively, deliver new commercial opportunities and economic benefits as creativity flourishes and expands the cultural marketplace, securing an ever wider audience (MKC, 2003). The recently published Boho Britain Creativity Index, developed by US regeneration expert Richard Florida, ranked Manchester as the UK's most creative city in terms of its ethnic diversity, proportion of gay residents and number of patent applications.

Related to this is the rich mix of inhabitants in Manchester. On the one hand, there is a problematic part (the unemployed heritage of the manufacturing past), and on the other hand the highly dynamic young urban professionals of the new economy. Furthermore, there is a very dynamic student population and many different nationalities. For the latter, see Table 6.3: an increasing part of Manchester's population is of an ethnic minority. Relatively large population groups are Pakistani (5.9 per cent of the population in 2001), Caribbean (2.3 per cent), African (1.7 per cent) and Indian (1.5 per cent).[2]

6.3.6 Urban Scale

Manchester is the regional capital of the northwest of England, and aims to compete with other secondary cities/regional capitals across Europe and the rest of the world. It thus has to develop competitive advantages in specific niche markets, as is usually the case with secondary cities/regional capitals.

Richard Florida, who ranked Manchester as the most creative UK city, focuses in particular on US cities in his research. Because of the different institutional settings, intra-European comparisons seem, however, to be preferable. Key actors in Manchester tend to compare this English city with, in particular, other European regional capitals/secondary cities, and those which are successful in new economic activities. Examples are Helsinki, Barcelona, Lyon and Frankfurt.

In Manchester key stakeholders are aware of the importance of certain scale economies which are desirable for knowledge-intensive activities. The best illustration is the planned merger of the Manchester universities. This is considered to be necessary for attracting more and better professionals and students, but also to becoming an attractive partner for business activities and to obtaining public funding.

Table 6.3 Population of the City of Manchester and its city centre (1991–2001)

Year	Manchester	City Centre	Source	Ethnic minorities	
				Number	Percentage
1991	404,861	966		51,183	
1998	412,965	4,550	1991 Census		12.6
2001	392,819	5,469	1998 Local Census	74,806	
% change 1991–2001	–3%	466%	2001 Census*	46%	19

Note: Some of these figures are drawn from 2001 Census results, which are currently the subject of a challenge between Manchester City Council and the Office of National Statistics. Quotation of these figures does not imply the Council's acceptance of the Census results.

Source: Office of National Statistics (2002).

6.3.7 Social Equity

Part of the population of Manchester does not benefit from the new opportunities: 7.9 per cent of the economically active population is unemployed (Nomis and MCC Planning Studies, July 2003). Many of them lack the skills and experience to find a job in the new economic activities. Moreover, there is a relatively high illiteracy. According to the Greater Manchester Learning and Skills Council there are 420,000 people in the area with poor basic skills. In East and North Manchester, there is a concentration of people who are unemployed and have poor skill levels.

The drive to regenerate deprived areas forms a central part of the knowledge capital vision. The objective is for all urban areas to benefit from the new economic opportunities. Therefore, explicit support programmes have been set up for (former) deprived areas such as East Manchester and Hulme, encompassing a range of housing, leisure, education, health and skills measures. Currently, East Manchester is the largest of the UK's regeneration sites.

6.4 Knowledge Activities

In the last section, the foundations of Manchester's knowledge economy were discussed. Now, we turn to the knowledge activities and ask how well the city (and the actors in it) manages to: 1) attract and retain talented people; 2) create new knowledge; 3) apply new knowledge and make new combinations; and 4) develop new growth clusters.

6.4.1 Attracting and Retaining Knowledge Workers

Between 1991 and 2001, there has been a spectacular increase of residents in the city centre of 466 per cent. While the 1991 Census identified 966 residents in Manchester City Centre, subsequent local censuses established that this figure had risen to 3,338 in 1996, 4,550 in 1998 and 5,469 in 2001 (see Table 6.3). The large number of residential development schemes which can be seen at the time of writing signifies the anticipation of further growth. Table 6.4 evidences the continued increase in residential development activity: before 1994 there were about 2,300 housing units built, between 1994 and 2002 almost 5,000 units, while at December 2002 about 2,900 housing units were under construction and about 3,900 planned. This would imply a total of more than 14,000 housing units in the city centre becoming available in the near future.

The increase of culture and leisure facilities has contributed in a substantial way to the growing attractiveness of the Manchester region. For example, Manchester attracts particularly people working in media activities, because of the relatively low costs of living compared to the London region.

Manchester hopes to become a knowledge hub for the northeastern part of England: a northern hub of economic, social and cultural activity. For this, the skills base is considered to be quite good, but should be further increased. In order to attract and retain the best students, the City of Manchester tries to increase its attractiveness. A problem in Manchester was that people with good skills tended to move away; often to the London region. Until about five years ago, many Manchester graduates did not want to stay. Although this is changing gradually, one of the problems is still that there are too few high-grade jobs in Manchester. Thus, a proportion of graduates has to leave the city to look for an adequate job elsewhere. Furthermore, there is still a large part of the urban population unemployed, often former manufacturing labourers with low skills and education. Manchester thus still faces considerable problems of social equity, although many policy efforts are focused on reducing these problems.

The MKC prospectus introduces the spatial concept of the 'Arc of Opportunity', stretching from the University of Salford in the west to Piccadilly Station in the east and the HEI Campus in the south. For this area, a coordinated master planning process is intended. 'Land use, building quality,

Table 6.4 Housing in the City Centre of Manchester

	Total student bedspaces	Total private dwellings	Total
Schemes completed:			
Before April 1994	1,003	1,348	2,351
Between May 1994 and February 1998	1,002	671	1,673
Between March 1998 and March 2001	62	1,340	1,402
Between April 2001 and December 2002	146	1,719	1,865
Schemes under construction at December 2002	0	2,881	2,881
Schemes with planning permission at December 2002	0	3,876	3,876
Total	2,213	11,835	14,048

Source: City Centre Housing Schemes, Planning Studies, Manchester City Council. Release date 11 April 2003.

transport, access, exploitation of natural assets including waterways, public spaces, branding and signage will be addressed in unison' (MKC, p. 17). The development of a robust spatial and development framework for this Arc of Opportunity will be essential to ensure that investment opportunities are maximised to realise the total commercial, employment, education, leisure and residential potential of 'Manchester: Knowledge Capital'.

A range of partners, including Manchester City Council and the universities, are considering measures to improve the main Oxford Road corridor, within the Arc of Opportunity, along which many of the city's knowledge institutions are situated. Improving the quality of the street environment, public realm and transport will help to maintain a vibrant, welcoming atmosphere, thereby increasing the area's attractiveness to students and knowledge workers.

The MKC initiative will also address the question of how to attract and retain knowledge workers. Under the heading of 'Genius Generation' it states that knowledge must be captured at every level by encouraging innovative new pathways geared to harvesting the talents of local residents. Specific activities in this respect are incubator facilities and incubator networks. At the Manchester Science Enterprise Centre of the UM, postgraduates are given a desk, phone and computer and exposed to entrepreneurial training and mentoring. MMU's Business School also assists young entrepreneurs with a six-month part-time personal and business development programme. Other actions planned under the heading 'Genius Generation' are encouraging school children to engage with the arts, media and music sectors through workshops and focusing on the development of creative entrepreneurs through foundation and degree level courses.

In the last decade, Manchester has become more successful in attracting and retaining knowledge workers. The city-region attracts, for instance, many workers in media-related business activities. However, because the number of high-grade jobs seems to be still relatively low compared to the number of graduates, knowledge workers leave Manchester to find jobs elsewhere. Measures to improve the number of high-grade jobs (thus improving the economic base, for instance via incubators) seem therefore to be necessary.

6.4.2 Creating Knowledge

At the time of writing, the University of Manchester is in the process of merging with UMIST, with a target date of Autumn 2004 for the merger. The rationale behind the merger is the increase of necessary critical mass. The new institute will have a total of 30,000 students, 9,000 staff and a combined income of £420m (€592m) (MKC, p. 23). Furthermore, it has to stimulate more regional dispersion of top universities in the UK. In the UK there is a trend towards research money going to fewer universities. By merging, UM will increase its chances of obtaining sufficient government funding. The merger between UM and UMIST is thus important for the development of Manchester as a knowledge city. After the merger it will become the largest UK university in turnover (in student numbers, MMU is and will be the largest university of the metropolitan region). In the MKC many advantages are anticipated from the merger of the two universities, which will, at a stroke, create the UK's largest single university, one with a global reputation for research excellence and learning.

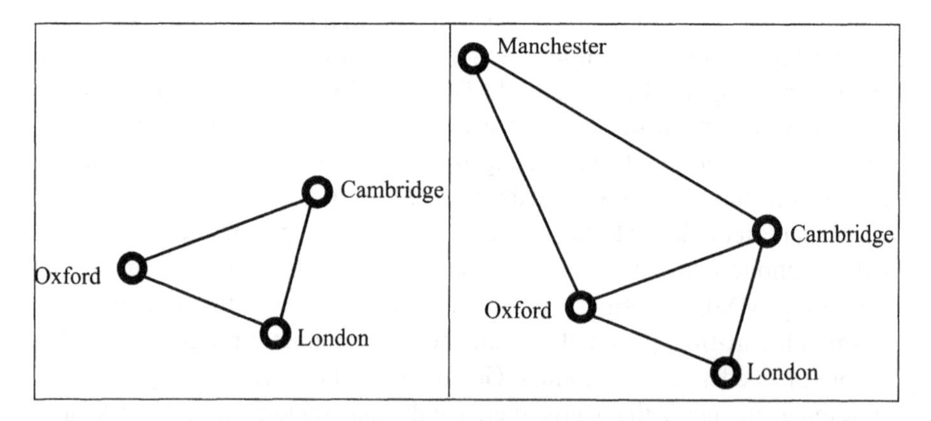

Figure 6.3 A new geopolitical situation: from a 'golden triangle' towards a 'golden quadrangle'

With the new university, the knowledge base of Manchester will obtain more geopolitical weight. A larger size is desirable for attracting international students and for investing in science, contributing towards counterbalancing the knowledge strengths of Oxford, Cambridge and London. It is hoped that a new geopolitical situation will arise. The Manchester universities hope that not all excellent knowledge activities will be concentrated in the 'golden triangle' consisting of Oxford, Cambridge and London, but will include the 'golden quadrangle', also including Manchester (see Figure 6.3). Manchester wants to benefit from crowding-out effects in Cambridge, predominantly regarding bioscience companies. Because of shortage of land and staff, some of these companies have already moved from Cambridge to Manchester.

A merger of all the Greater Manchester universities might be desirable to further enhance Manchester's knowledge profile. At this moment, this is not manageable, but this might change within a decade. Also according to Westwood and Nathan (2002, p. 69):

> a further, future step may be the merger of all four universities and the restructuring/rationalisation of them so they are more outward facing – capable of networking intellectual capital into the business sector. This would involve a significant rethinking of strengths – and weaknesses – of the individual institutions and a subsequent recasting of their specialisms.

The quality of teaching and research at both MMU and Salford University, in the meantime, adds considerable weight to Manchester's total higher education reputation.

One of the spearheads of the MKC is the creation of new, externally-focused centres of academic excellence, including, amongst others, the Nanotechnology Institute, the Dalton School for Nuclear Science and Technology, the Regional Aerospace Centre, and the Institute for Biomedical Informatics.

The MKC also focuses on opportunities for cooperation between the knowledge institutes and the city on urban regeneration, thus also trying to achieve social equity. It anticipates that the Centre for Regeneration Studies involving all four HEIs will provide education and training in regeneration from foundation to Masters levels and that the new East Manchester Partnership will carry out long-term research on the development of sustainable communities.

Manchester is considered to be competitive in creative activities, for which there is a considerable amount of knowledge and skills available. Although there is a vast knowledge base for science development, the region lacked critical weight. The merger of two universities will contribute in this respect to a more competitive position in comparison with the 'golden triangle', for instance, in the field of bioscience.

6.4.3 Applying Knowledge/Making New Combinations

There is explicit recognition in Manchester of the importance of harnessing knowledge for its direct application in the business sphere, thereby maintaining competitive advantage and employment opportunities. Two good examples of organisations supporting the commercial application of knowledge are described below.

MMU MMU has wide-ranging expertise in applied research and vocational programme. Specific projects planned or underway to support the application of knowledge include:

1. the Textile Centre of Expertise: this unit incorporates high-tech with fashion and it links to with the textile industry. The Department of Clothing, Design and Technology has approximately 600 students and is renowned for its teaching of international fashion marketing, clothing and fashion business and fashion design with technology across a range of programmes.[3] A new campus building of MMU will probably be constructed in East Manchester;
2. the Centre of Urban Education: in Manchester a relatively small number of people have an education degree, while there is a high turnover of teachers in the city, particularly in inner-city schools. Teachers gain some educational experience in the city and then move to the neighbouring towns where work is less demanding and life more relaxed. The urban education centre aims to support teachers and thereby keep them in the city. This is important since Manchester wants to raise its profile in education. MMU wants to track the developments in East Manchester to learn from these experiences;
3. the Centre for Regeneration Studies offers modern apprenticeships and masters degrees in regeneration studies. It is involved in several regeneration areas in Manchester;
4. the Centre for Aviation Transport and the Environment (CATE) is a multidisciplinary research centre. CATE's mission is to facilitate the integrated social, economic and environmental sustainability of the aviation industry through critical research and analysis, and through knowledge transfer between the academic, industry, regulatory and NGO sectors.[4]

MMU also boasts strong links with the media sector through a strong supply chain of pre-entry and development programmes in media (Westwood and Nathan, 2002). Through these links an attempt is made to turn knowledge directly into business. These and other activities of MMU show that HEIs can have close links with the local society and businesses. Such relationships are central to the MKC vision.

Campus Ventures Campus Ventures (CV) is a non-profit organisation established in 1995. Located on the UM campus, CV offers office space, legal and financial support, and provides access to university facilities such as labs or powerful computers (van den Berg and van Winden, 2002). CV is funded through European funds (European Regional Development Fund (ERDF)), national schemes and royalties: start-ups have to pay royalties to CV in the form of a certain percentage of their turnover after three years of profitable operation. Other financial contributors are the University of Manchester, the North West Development Agency and Airport Ventures Ltd. Campus Ventures provides support for anyone with a good business idea and the right attitude – it is not restricted to graduates only. However, CV predominantly serves UM, UMIST and MMU students and graduates. From June 1995 to October 2002, CV has incubated some 100 firms, of which about 87 per cent survived. Approximately 772 jobs have been created. The turnover of the active companies was approximately £9.9m (€14m) in the year for which there are the most recent statistics.[5]

One of the stumbling blocks for incubators is the shortage of seed capital after the dot.com bubble burst. With the global economic downturn, stimulating new companies has become even more difficult. As a result, CV has had to limit its support for new business ideas. At the moment, only activities that are very 'close to the market' (and appear to be promising) are being subsidised within CV.

CV now has several branch offices in other parts of the northwest region (West Lakes, South Cumbria, East Lancashire). A prime example of one of CV's successful companies is Neural Response (now: KMS); a company specialised in neural networks. The share value and employment of this firm have grown rapidly.

For CV networking is essential and it is therefore very desirable for CV to be more embedded in the knowledge capital project. It can contribute in a substantial way to keeping high-level skills in Manchester.

One concern is that large firms are not very involved in CV (with the exception of Unilever), although there are contacts with some other large concerns. However, it is hard to convince these companies to become involved, since they mostly want to keep intellectual property in-house.

The MKC strategy aims to create more incubators and workspaces, stating that venture capitals will be increased and there will be fresh strategies for growth in the creative, cultural, digital and biotech industries.

MMU and CV are good examples of how Manchester is turning knowledge into business. As a university with a strong vocational and business focus, it is crucial for MMU to encourage and develop knowledge activities which are close to the market. Several programmes have been developed by MMU for which there was already a broad economic as well as knowledge base in Manchester, such as modern textile activities and regeneration processes. Moreover, the Centre for Urban Education can clearly address the pressing requirement to provide high quality education at all levels in the city and thereby help to build up the knowledge base for the future.

CV has been successful in helping to start up a range of new knowledge intensive businesses and thereby create highly-skilled jobs. The current economic situation, however, has reduced the availability of sufficient venture capital and it is therefore expected that the number of new ventures supported will decrease on the short term. Whilst there have been strong economic growth trends in Manchester over recent years, particularly in knowledge sectors, it is as yet unproven whether the economic base is sufficiently broad and diverse to withstand severe global economic recession. The need to develop stronger links between knowledge institutions and large companies is one way of developing competitive advantage through sharing and applying knowledge, which would help to weather a severe macroeconomic downturn.

Making new combinations Partnership working and cooperation between stakeholders are central to the MKC, as is the setting up of strategic networks to exchange information and knowledge. MKC intends that the Manchester of the future will be renowned as a place where knowledge is disseminated through cutting-edge knowledge transfer initiatives, usage of the latest communication technologies and sophisticated knowledge-sharing processes.

The MKC initiative therefore hopes to contribute to setting up new and supporting existing networks, especially those between the city, HEIs and businesses of all sizes. One of these networks is Pro-Manchester (http://www.pro-manchester.org.uk/), a network of private sector representatives who meet regularly with the HEIs to discuss knowledge/business collaboration.

An important role in setting up and attracting new businesses is played by the Manchester Investment and Development Agency Service (MIDAS). As part of the Manchester Enterprises Group, MIDAS is a partnership organisation covering the local authority areas of Manchester, Salford, Tameside and Trafford. MIDAS, financed by predominantly public funds (including ERDF funds), offers a package of advice and assistance tailored to the needs of national and international companies that are considering setting up, relocating or expanding into the Manchester area. In 2002/03 MIDAS, employing 22 people, helped to safeguard or create nearly 3,200 jobs and played a part in securing investments from almost 100 businesses. The aim of MIDAS is to secure significant levels of new investment and employment for the area through the global marketing of the Manchester region.

MIDAS has also been active in providing a 'soft landing' for companies wanting to relocate: UK firms setting up business in, for instance, Montreal, US companies starting activities in the UK, etc. These soft landings are organised by the International Links project (www.intlinks.net). It started with North America (Pittsburgh, Atlanta, Montreal), but cities elsewhere have joined (e.g., Melbourne in Australia). In choosing participant cities for the project, their size is considered (see Section 6.3.6); they should be comparable to Manchester, and they have to have a hub airport (good accessibility). The activities are mostly in the fields of biotech, ICT and creative (business) activities; MIDAS does not involve manufacturing businesses. International Links' business plan is made up of three categories of firms: 1) customer centres; mostly with a financial services focus, but also ICT; 2) knowledge-based firms (small ICT, creative industries); and 3) high-growth industries: niche industries, like Asian food and food processing. MIDAS looks for business brokers abroad, using linked websites targeted at the three focus areas. The weakness of the MIDAS contacts is that the large (tech) companies (like Fujitsu, Sharp and Siemens) are no involved, but MIDAS is working on this. Currently this programme is particularly important for indigenous companies, but also to attract new companies.

Manchester has a good knowledge base and offers quite some urban diversity, building blocks that we assume to be necessary for an urban region to be successful in making new combinations.

6.4.4 Developing New Growth Clusters

According to Westwood and Nathan (2002), promising economic clusters in Manchester are: health (including biosciences), media, cultural and sports industries, retail, hi-tech manufacturing (including biosciences) and financial services.

Public and private health services are a large and growing employer and there are a number of important and growing relations with the HEIs, such as the University Medical School. The key growth area for this cluster is the 'health campus' adjoining the main universities between Oxford Road and Upper Brook Street to the south of the city centre. This campus will provide employment for over 12,000 employees.

Table 6.5 An indication of jobs within different knowledge clusters in the Greater Manchester area (percentage of total jobs and absolute number)

	%	Number
Health	10.3	116,048
Cultural and creative	4.1	46,200
Retail	10.9	122,808
Hi-tech manufacturing	29.0	244,000
Financial	6.3	70,500
Total	54.3	599,556

Source: Westwood and Nathan (2002, pp. 37–44).

The largest UK media cluster outside London is located in Manchester. This cluster consists of local and regional media, such as Granada, *City Life* and the *Manchester Evening News*), new media companies and regional headquarters of national bodies (such as the BBC and *The Guardian*).

Manchester also has the largest concentration of cultural and sports industries in the UK outside London. In particular, the 2002 Commonwealth Games had a huge impact on the Manchester economy. These games involved £600m of public investment (€845m), 6,300 full-time equivalent jobs for a period of ten years and a year total of 300,000 visitors to the city.

In the last decade, the city has recognised that there is economic potential in an area which is not directly visible in the traditional economic figures – in its cultural and creative industries (van den Berg, Braun and van Winden, 2001). Manchester TEC has included the arts, culture and media among the 14 key sectors in the TEC area and credits 2 per cent of the TEC area's employment to that sector (Manchester TEC, 1998). According to their figures the sector was quite stable in the early 1990s and holds promise for the future. More recent extensive quantitative research in the Manchester Training Enterprise Council area has adopted a wider view on the cultural industries and estimated that 3.6 per cent of the working population (18,058 jobs) is (self)employed in cultural enterprise (MIPC & DCA, 1998). In comparison with other UK cities, Manchester's employment figures stand out positively. Only smaller towns with a few large employers (Cardiff) do better (see Table 6.6).

The ongoing reurbanisation of the city centre has strengthened Manchester's already-strong tradition in retail business: at the time of writing 10.9 per cent of Greater Manchester jobs were in the retail sector. This sector has been stimulated by major city redevelopment programmes after the IRA bombing of the city centre in 1996. These programmes were predominantly focused on retail, public realm and cultural/leisure facilities. New flagship stores opened recently in Manchester include Marks & Spencer, Selfridges and Harvey Nichols.

A high-tech manufacturing cluster has grown from the traditional manufacturing base. Manufacturing still plays a huge part in the city's new economic success. The sector still employs 244,000 people in Manchester and 576,000 in the northwest region in total. The region's manufacturing sector invests over £2.2b (€3.1bn) annually in buildings, plant and machinery and generates £14.5b (€20.4bn) in value-added products (Westwood and Nathan, 2002).

Manchester is the principal centre of ICT activity in the northwest of England. The region hosts some leading ICT firms' European headquarters; the city also hosts around 1,000 smaller ICT firms active in software development, Internet and telecoms; their number has been rising sharply in recent years, as is reflected in the development of the Manchester Science Park (van den Berg and van Winden, 2002). Furthermore, Manchester is hoping to become a test centre for new Nokia products. This might contribute to the knowledge development in Manchester and to the image of the city as a high-tech area. In earlier times it was important to attract factories; nowadays, knowledge and image seem to be more important. The Manchester Digital Development Agency (MDDA) has been established to attract and sustain further investment in ICT and e-commerce across all sectors.

And finally, Manchester is the regional hub for a wide range of national and international financial and professional services organisations, including major law and accountancy firms, all major UK banks and a range of overseas and merchant banks. According to Westwood and Nathan (2002), Manchester has generated more jobs in financial and professional services in the last three years than the rest of the Core Cities put together. Between 1991 and 1999, the number of jobs in financial and business centres in the Greater Manchester area has grown from 56,600 towards 70,500 (see Table 6.5).

The MKC strategy is explicitly focused on supporting knowledge-intensive economic clusters: the aim is to contribute to the creation of 100,000 new jobs by developing strategies in creative, ICT/digital, bio, professional and financial, education, health and nanotechnology sectors. Currently, with the global economic recession, these aims seem to be very ambitious. However, with an economic upswing, particularly American companies are expected to be attracted to the Manchester region.

Table 6.6 Manchester cultural enterprises in UK perspective

	Share of employment
Manchester	3.6%
London	2.3%
UK average	2.3%
Cardiff	4.3%

Source: MIPC &DCA, May 1998, 'Cultural Production Strategy for Manchester', draft.

The Manchester region appears to have a number of promising economic clusters. Considering its limited urban scale (it is and will remain a secondary city within Europe) it will be crucial to focus on and further develop certain niche clusters. In the framework of the knowledge economy, some activities need high-grade knowledge: for example, the health and the biotechnology clusters. For both clusters, there seems to be a good knowledge as well as economic base in Manchester. Furthermore, for the media, cultural and sports industries, Manchester has competitive advantages, due in part to its improving quality of life (and related image) and its urban diversity. It appears that the city is doing well in the UK perspective, but there is significant potential to enlarge these economic activities and achieve greater levels of growth.

6.5 Conclusions and Perspectives

6.5.1 Knowledge Economy Strategy

Several strategies have been developed over the previous decade for stimulating new and enhancing existing economic activities in the Manchester region. These strategies, aimed at a transition process towards a knowledge economy, have been articulated in a range of documents, including the City Pride Prospectus and, recently, the 'Manchester: Knowledge Capital' (MKC) Prospectus. These publications have helped to encourage cooperation between the actors involved, to widen the support for the strategies and to attract funds (European, national and local public and private funds) for necessary activities. One of the essential elements of the MKC strategy is to strengthen the links between higher education, business and local communities. At the moment, the universities play a central role in the stimulation of the knowledge economy, with growing recognition of the need to maximise the transfer of that knowledge into local businesses and thereby achieve competitive advantage. In order to facilitate this process, however, greater autonomy and decision-making powers are required at the local government level, so that locally determined funding and strategic priorities can be pursued effectively.

6.5.2 Foundations of the Knowledge Economy

In Manchester, key generators for the transition towards a knowledge economy appear to be: the airport (international accessibility), the regional-economic centres (its economic base) and the higher education institutes (its knowledge base). An indication of the value of the knowledge economy foundations in Manchester is given in Table 6.7. This is explained below.

Table 6.7 An indication of the value of knowledge economy foundations in Manchester

	Foundations	Score in Manchester
1	Knowledge base	++
2	Economic base	☐
3	Quality of life	+
4	Accessibility	+
5	Urban diversity	++
6	Urban scale	+
7	Social equity	--

-- = very weak; – = weak; ☐ = moderate; + = good; ++ = very good.

The Manchester region's knowledge base is considered to be very good. There are four universities and several other high-performing education institutions. These HEIs have differing specialities varying from science to a vocational focus. Due to a reduction in the availability of decreasing public funds, the HEIs are starting to adopt a more market-oriented approach, which is desirable for a city that wants to stimulate stronger relationships between knowledge institutions, businesses and local communities.

The economic base of the city is growing in strength but there are still relatively high unemployment levels. Manchester has a rich manufacturing past, which on the one hand has been a breeding ground for high-tech manufacturing activities; but on the other hand, the legacy of the manufacturing past has contributed towards the significant job losses over the last 30 years. The city has a vibrant cultural sector, helping to boost new economic activities. Significantly, the knowledge institutions have been modern job generators: the universities are now amongst the largest employers in Manchester.

The quality of life as well as the image of Manchester as a place to live and work in have improved substantially in the last decade. Previously Manchester had a relatively bad image as an unsafe and unattractive city. Regeneration initiatives, for example in the city centre, Hulme and East Manchester, have, however, contributed to the development of a more attractive city and a better image. Nowadays the city centre functions as a magnet for young urban professionals. Furthermore, the many creative and cultural activities and sports and the success of the Commonwealth Games in 2002 have been very important in changing the image and perception of the city.

Connectivity and accessibility are considered to be essential for attracting knowledge activities. These aspects of Manchester are judged as being good, but some elements could be improved. The growing international airport ensures international connectivity and greatly enhances the attractiveness of the region as a location for business and investment. The quality of the ground transport (rail and road) is important to guarantee good connections to the big magnet London and other major UK cities. However, car accessibility could be improved by reducing congestion problems and tackling parking issues in the city centre. The light-rail system, Metrolink, and bus networks are crucial to maximising accessibility across the city-region and helping the regeneration process in deprived neighbourhoods, by offering them better links with the core of the conurbation and its job opportunities.

Related to the high urban quality of life, Manchester also has a rich urban diversity. The large variety within the built-up area, the urban activities and the inhabitants all help to make the urban region attractive in the knowledge economy.

The urban scale of Manchester is considered to be good for competing successfully with other knowledge cities. As a non-capital city, it has to develop a series of competitive advantages in specific niche markets. It is, however, still unclear which niches these should be. The merger of UM and UMIST and the implementation of the MKC vision should help the city to gain weight and profile as a knowledge city.

Currently, there are relatively wide skills and wealth imbalances between resident communities. This could be a threat for the development of the city as a balanced knowledge city, a fact acknowledged by the actors involved. The economic and social regeneration of deprived areas plays a central role in the knowledge capital initiative. The idea is that all areas should benefit from new economic opportunities in order to successfully function in a competitive knowledge economy. Within the framework of the MKC strategy, therefore, explicit skills support and regeneration programmes will be implemented in deprived areas such as East Manchester.

6.5.3 Activities of the Knowledge Economy

The dramatic growth in the number of residents in the city centre appears to illustrate the start of a successful transition process in Manchester. Within one decade, the number of residents in the city centre grew by more than a factor of six! The large number of building activities and the high prices for housing evidence the expectation of further residential growth in the city centre.

Many of the new residents in the city centre work in knowledge-intensive activities. A large proportion of them studied at one of the Manchester HEIs. In the past, most graduates left Manchester because of a lack of high-grade jobs. Though there is still a shortage, the situation has improved. The founding and attraction of additional knowledge-intensive businesses and job opportunities seem to be the most urgent activity required to retain knowledge workers. The capacity to attract and retain knowledge workers can therefore be judged as good, but improvements are required (see Table 6.8).

The merger of UM and UMIST is crucial for Manchester to compete with other knowledge cities. By creating a larger university, economies of scale can be achieved and more public and private funding can be attracted, along with high quality students and staff. A new geopolitical situation could arise; instead of the – currently – three leading knowledge cities (London, Cambridge and Oxford), there will be a 'golden quadrangle' (also including Manchester). Investing in technical disciplines such as bioscience is considered essential to obtain a better reputation as a knowledge city. Manchester already has a reputation as a city where new creative ideas and skills are being developed ('creating knowledge'). All the universities and other knowledge institutions

Table 6.8 An indication of the value of knowledge economy activities in Manchester

	Activities	Score in Manchester
1	Attracting knowledge workers	+
2	Creating knowledge	+
3	Applying knowledge/making new combinations	+
4	Developing new growth clusters	+

-- = very weak; – = weak; ☐ = moderate; + = good; ++ = very good.

must contribute to sustaining and developing this position in creative and cultural knowledge development.

Incubators like Campus Ventures (CV) have helped to start up quite a number of new companies. Currently, however, the growth of new companies has declined because of the reduction of available venture capital. CV is considered to be a good strategic vehicle for turning knowledge into business ('applying knowledge'). For the near future, it is expected that incubators and other business support agencies will play an important role in developing Manchester as a knowledge city, as outlined in the MKC strategy.

Stimulating the establishment of strategic networks between public and private agencies, universities and businesses is considered to be important to further the transition process to a knowledge capital. In our opinion, the knowledge activity of 'making new combinations' could be strengthened. Sub-optimal building blocks for this activity are probably the economic base, which could be improved, and the low critical weight of the city as a knowledge city (urban scale). However, good building blocks in this respect are the knowledge base and the urban diversity.

Where the city is still relatively weak is in relationships and cooperation. In particular, relationships between knowledge institutions and larger companies could be improved, with mutual benefit. However, the large companies seem to be reluctant in this respect: they seem to prefer to keep their knowledge development in-house.

Promising 'growth clusters' for Manchester appear to be health, sports, culture and media, and high-tech manufacturing. For the health and high-tech manufacturing sectors high-grade knowledge is required. To this end, Manchester should benefit from the merger of UM and UMIST. In particular in sports, cultural and media activities, Manchester has achieved high growth rates. These could become important niche clusters for the city to specialise in. Some specialisation appears to be advisable considering the limited urban scale.

6.5.4 Perspectives

Although at the time of writing there is a global recession, it is expected that those cities with a good base of knowledge activities, strong strategic networks between key actors, an attractive environment and good quality of life will directly benefit from a forecasted upswing in the economy. In Manchester, it is expected that, when macroeconomic fortunes improve, there are sufficient assets and strategies in place to attract further investment and growth in the knowledge economy.

In this study, we have tried to analyse the foundations and activities of the knowledge economy. It was discovered that within Manchester two foundations were relatively weak: the economic base and social equity. The economic performance of the region is still below the UK average and there are relatively high unemployment rates. Strengthening both the economic base as well as social equity therefore have to be critical aims of the knowledge economy strategy.

Furthermore, the development of certain niche activities is considered crucial for the Manchester region. More research is needed to determine where specialisation should be concentrated, by examining the size, strengths and weaknesses of existing clusters within Manchester and also in competitor cities.

The success of the knowledge capital strategy will depend on ensuring that programmes and activity do not lose momentum. At the moment, all relevant partners seem to be convinced of the importance of the current strategies towards the knowledge economy. But a large number of projects are still to be implemented to strengthen and consolidate the position of Manchester in the knowledge economy. To achieve sufficient weight as a knowledge city and to create and attract more knowledge-intensive activities it will be important to invest in the knowledge quarter, the merger of the universities, and in the improvement of the connectivity and accessibility. Continuous attention to such investments will be crucial.

Most of the discussion partners were quite optimistic about the prospects for the knowledge economy of the Manchester region. Notwithstanding the global recession, the UK economy and regional real estate development are doing reasonably well. With its current regional developments, the MKC strategy and related plans, Manchester could enjoy relatively high economic growth rates over the coming decade.

Notes

1 The Manchester TEC area includes Manchester, Salford, Trafford and Tameside. The population in the TEC area is estimated to be just over 1.1 million people.
2 These figures are drawn from 2001 Census results, which are currently the subject of a challenge between Manchester City Council and the Office of National Statistics. Quotation of these figures does not imply the Council's acceptance of the Census results.
3 See http://www.hollings.mmu.ac.uk/faculty/cdt/index.php.
4 See http://www.docm.mmu.ac.uk/EXT/cate/.
5 See http://www.campus-ventures.co.uk/history.asp.

Discussion partners

Ms Sarah Adolph, Central Policy Unit, Manchester City Council.
Ms Kathryn Carr, Central Policy Unit, Manchester City Council.
Mr Dave Carter, Acting Head of Economic Initiatives Group.
Mr Peter Fell, Director of Regional Affairs, University of Manchester.
Mr Richard Jones, Manchester Investment and Development Agency Service (MIDAS).
Cllr Richard Leese, Leader of Manchester City Council.
Ms Sarah O'Donnell, Head of Regional Affairs, External Relations, Manchester Metropolitan University.
Mr Mike Radcliffe, Campus Ventures, University of Manchester.

References

Berg, L. van den and W. van Winden (2002), *Information and Communications Technology as Potential Catalyst for Sustainable Urban Development, Experiences in Eindhoven, Helsinki, Manchester, Marseilles and The Hague*, Aldershot: Ashgate.
Berg, L. van den, E. Braun and A.H.J. Otgaar (2002), *Sports and City Marketing in European Cities*, Aldershot: Ashgate.
Berg, L. van den, E. Braun and W. van Winden (2001), *Growth Clusters in European Metropolitan Cities*, Aldershot: Ashgate.
City Pride Partnership (1994), 'City Pride: A focus for the future'.
City Pride Partnership (1997), 'City Pride 2: Partnerships for a successful future'.
Crime and Disorder Team (2003), Crime and Disorder Audit, Manchester City Council, http://www.manchester.gov.uk/crime/index.htm.
Manchester City Council and the Knowledge Capital Partnership (2003), 'Manchester: Knowledge Capital' (MKC), 'A Place for Inspiration, A World of Opportunities', Manchester.

Manchester TEC (1998), 'Economic Assessment 1998 – Part 3: Business and the Economy & part 4: Sectors'.

MIPC & DCA (1998), 'Cultural Production Strategy for Manchester', draft.

Nevala, A.M. (2003), 'Urban Physical Environment and Safety of Cities', Masters thesis MEMR, Rotterdam.

Phelan, L. (1997), 'Manchester: A city on the move; sports as a tool for city marketing', paper presented to the ICSS Workshop on the Economic Impact of Sport, International Centre for Sports Studies, Neuchatel, Switzerland.

The Working Group of Government Departments, the Core Cities and the Regional Development Agencies (2003), 'Cities, Regions and Competitiveness', June.

Westwood, Andy and Max Nathan (2002), *Manchester: Ideopolis? Developing a Knowledge Capital*, London: The Work Foundation.

Chapter 7

Munich

7.1 Introduction

Munich is the capital of the German Free State of Bavaria and, with a population of 1.3 million, the third-largest city in Germany, after Berlin and Hamburg. The city is the centre of an urban region with over 2.4 million inhabitants. The urban region comprises eight Landkreise with populations varying from 97,000 to 278,000 people.

Table 7.1 Population and GDP, 2001

	Population	GDP in millions of €
Bavaria	12,330,000	357,400
Region of Munich	2,483,000	107,220
City of Munich	1,228,000	64,000
Periphery	1,255,000	43,220

Source: Landeshauptstadt München (2003).

Munich's unemployment level is relatively low compared to other major cities in Germany (see Figure 7.1). Also, purchasing power and GDP per capita rank among the highest in Germany.

7.2 Knowledge Strategy

The city of Munich does not have an explicit and comprehensive 'knowledge economy strategy', but a number of its economic development policies are related to the promotion of a knowledge economy. This can be seen in the intensive cooperation between the city and the universities, firms and research institutions in the city. Examples are initiatives to promote promising economic sectors, to promote innovation in the city, to help SMEs in innovation or to support starters.

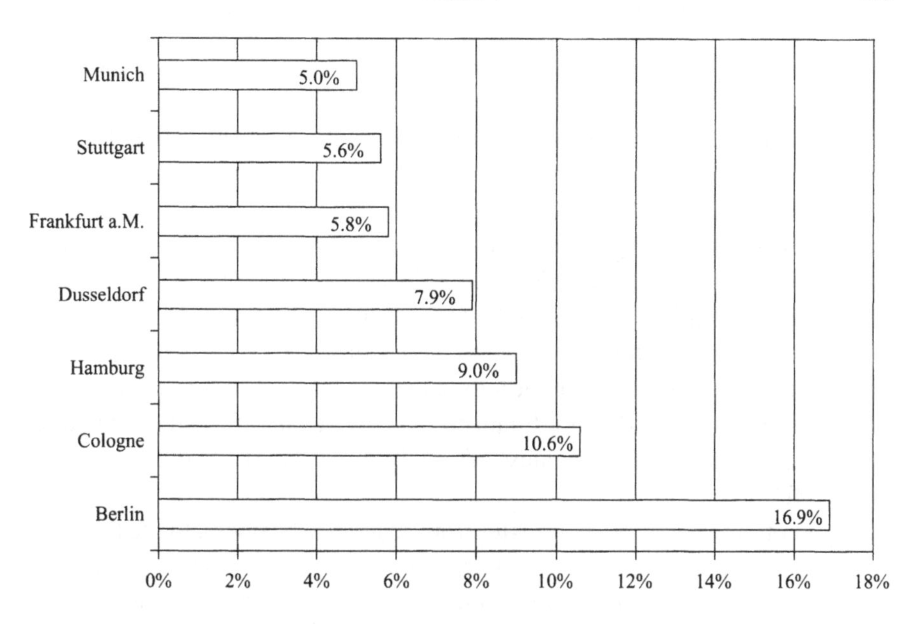

Figure 7.1 Unemployment rates in large German cities, 2002

Source: Landeshauptstadt München (2003).

It could be argued that the city does not need an explicit knowledge strategy, because, as we will see more in detail later in this case study, the Munich economy is already highly knowledge-intensive. Nevertheless, Munich recently commissioned a study to reveal all Munich's knowledge activities. The resulting report (2002), 'München, Stadt des Wissens' ('Munich, City of Knowledge') describes the enormous strength and variety of the city's knowledge base.

In the Federal Republic of Germany, the regions (Länder) are largely responsible for R&D policy, universities and education policy, and they have a large influence on some sectors, among which are the media and health/biotechnology. Thus, in the area of knowledge economy strategies, the Freestate of Bavaria is a key player with a huge impact on Munich. The Freestate has for decades very actively promoted the transition of the region into a knowledge economy. Since the 1960s, the Freestate of Bavaria has invested substantially and systematically in its knowledge infrastructure. Five new universities were established, and a number of other knowledge institutes were also founded. Concerning investments in research and development, Bavaria is among the top regions in Germany, with 2.9 per cent of GDP. Only

Baden Württemberg (3.9 per cent) and Berlin (3.7 per cent) have higher rates (Freistaat Bayern, 2002, p. 30).

The Freestate undertakes the following activities (Freistaat Bayern, 2002):

- the creation of 'centres of excellence', in cooperation with universities and business. One example is the relocation of the university faculties (machine building, IT and mathematics) of the Technical University out of the city centre into Garching, to the north of the city. Another example is the creation of a biotechnology cluster, Martinsried, in the southwest of Munich, where Bavaria financed the new construction of the faculties of Chemistry and Pharmacy of the Ludwig Maximilians University (LMU);
- promotion of technology transfer from science to business. In the Freestate of Bavaria, there are over 100 centres for technology transfer. In 1995, Bavaria founded a company named 'Bayern Innovativ' to increase the efficiency and transparency of technology transfer;
- financial support for research and development. Initiatives in this field include the BayTP Programme, which financially supports the transition of new technologies into profitable products and services, and BayTOU, supporting starters and young technology firms to develop new products with a high risk profile. Companies that develop new materials in cooperation with a university institute can receive state support: in 1990, Bavaria created a 'research foundation' to promote applied research in the fields of life sciences, ICT, new materials and environmental technologies. Finally, the Freestate has operated a venture capital fund since 1995, through which Bavaria takes shares in young promising companies (but only if there are other shareholders as well). This has greatly improved the ability of start-ups to obtain capital. The other shareholders, in some instances, receive an option to sell their share to the Freestate within a certain time period;
- the high-tech offensive Zukunft Bayern project, which began in 2000 and aims to support the development of promising technologies. Its budget – an impressive €1.18b – was generated by the privatisation and sale of some public utility companies. The project focuses on the following technologies: life sciences, ICT, new materials, environmental technologies and mechatronics. The Freestate aims to disseminate the 'spearheads' over the territory: every region in Bavaria has to develop its own strengths.

Although its efforts were directed towards the entire Bavarian territory, most investments have benefited the City of Munich, which is the undisputed knowledge capital of Bavaria. To intensify these positive impulses the city of Munich itself developed some projects to improve the knowledge base in Munich. Here the MTZ Munich Technology centre, or other start up centres could be named, as well as the MEB (Münchner Existentzgründungsberatung) a consultancy office for start-ups, organised by the city of Munich together with the chamber of industry and commerce. On the other hand increasing competition from other Bavarian cities must be expected, for a number of reasons – for example, 80 per cent of the very high budget of the Bavarian High-tech Offensive is going to other Bavarian cities, closing the gap with Munich.

7.3 Knowledge Foundations

In this section, the knowledge foundations – as discussed in the theoretical chapter – will be assessed for Munich. They include the knowledge base, the economic base, the quality of life, the accessibility, the urban diversity, the urban scale, and the social equity in the city.

7.3.1 Knowledge Base

The city's knowledge base is very strong. In the first place, the educational level of the population is high. Some 18.5 per cent of the employees in Munich hold a university or applied sciences degree (Landeshauptstadt München, 2003), which is a relatively high percentage compared to the German average. Secondly, the city has a large number of knowledge institutions. The city has two large universities, a polytechnic, and a number of academies (see Table 7.2). In total, the institutions have over 80,000 students, which makes it number two among German cities.

The universities have a good reputation, in both science and in education, although they are not the leaders in Germany in all fields. There are numerous other public and semi-public research institutes in Munich, such as the Max Planck Institutes, the Fraunhofer Institutes (both including the headquarters) and GSF (The National Research Centre for Environment). It is estimated that Munich has over 50,000 people working in research and development (see Table 7.3). Note that the majority is in private companies.

In the last few decades, several new university establishments have been built outside the city of Munich. Examples are the university establishments in

Table 7.2 Institutes of higher education in Munich

Institute	Number of students
Ludwig-Maximilians University	41,943
Technical University	18,864
Polytechnic University	12,494
University of the Bundeswehr	2,685
Foundation Polytechnic University	1,528
College of Music and Theatre	724
College of Political Sciences	587
Academy of Fine Arts	648
College of Philosophy	415
College of Television and Film	340
Total	80,228

Source: Landeshauptstadt München (2002).

Table 7.3 Researchers in Munich

	Number of researchers
Companies	31,500
Medical research	9,000
Others (Max Planck, Fraunhofer, Deutsches Zentrum für Luft und Raumfahrt (DLR), National Research Centre for Environment (GSF), research groups in universities)	10,000
Total	50,500

Source: Landeshauptstadt München (2002).

Freising, Weihenstephan, Garching and Martinsried. The institutions located at Martinsried, Weihenstephan and Garching are 'Centres of Excellence' employing scientific staff with a high potential in the fields of bio-technology and new materials.

The decentralisation trend is a natural result of the growth of the universities. So far, the location strategy has been opportunity-driven, with some leading actors setting the main trends. When the Max Planck Institute moved out of the city centre in the 1970s, the bio-chemistry faculty of the LMU was more or less compelled to do the same. However, at a strategic level, it should be considered that the competitiveness of the university system in Munich depends

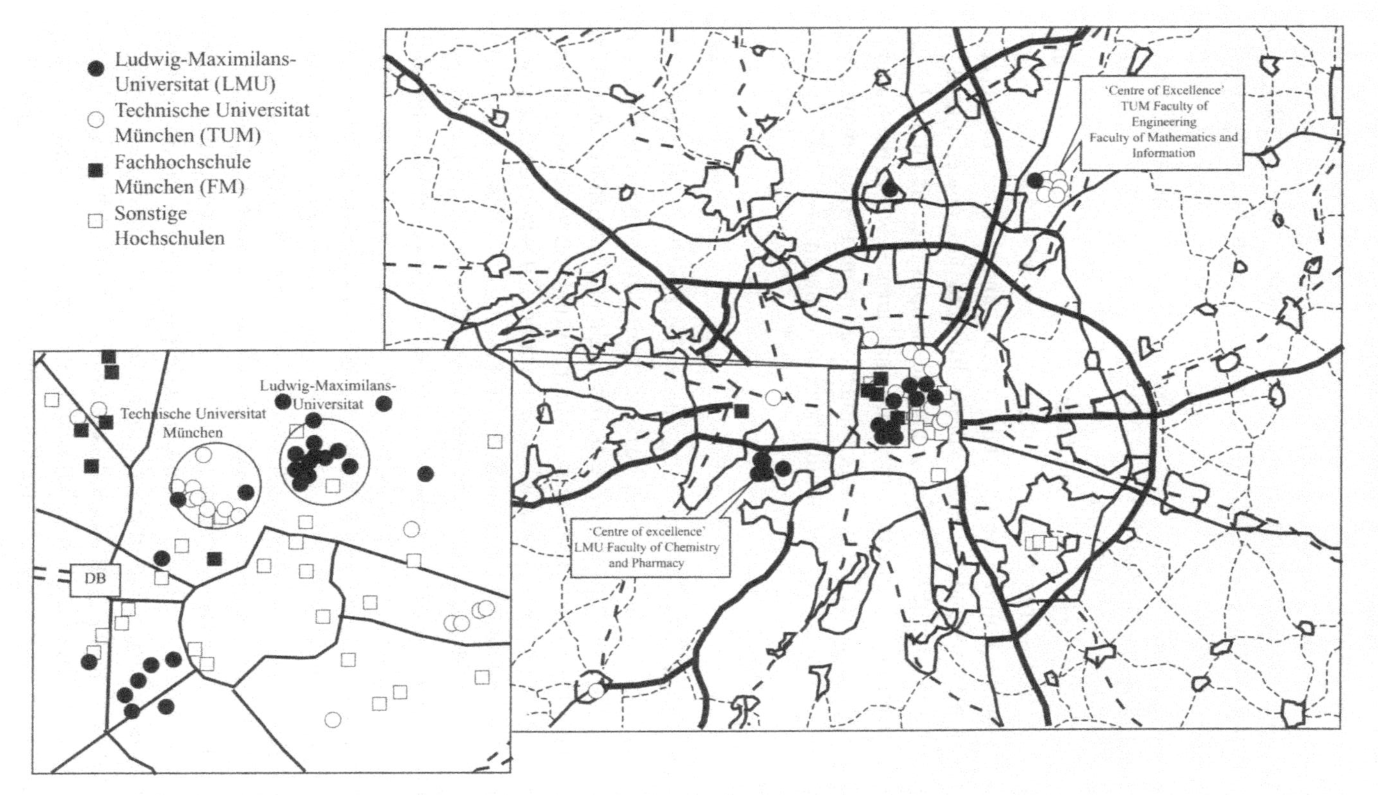

Figure 7.2 Higher education establishments in the Munich region

Note: Each point on the map stands for a university institute, a professorship or a university department.

very much on the quality of life and on its association with a pleasant, safe and wealthy city. If the education sector becomes disconnected from the inner city, in a rather isolated neighbourhood where no real urban facilities are to be found, this may be a serious problem for them. The idea of creating a 'campus university' following the Anglo-Saxon model – proposed by the leadership of the TU, for instance – is difficult to put into practice. The combination of living and working would be an interesting solution, but there are no such projects at this stage. That is why LMU is resisting the trend to move away; the human and social sciences studies are particularly integrated into urban locations where they find a favourable cultural context. This might change in the future, as new generations of lecturers and students express different preferences, but at the moment there is no such trend to be seen. The campus area of the technical university in Garching is at the time of writing still an area where the students go for lectures, but they then leave the campus to go home to their living areas in and around the city. The same situation can be found in Martinsried.

7.3.2 Economic Base

Munich's standing as a key economic engine in Germany is relatively recent. Before World War II and for some years thereafter, the city was far from where it is now in economic terms. It had some light industries and functioned as a regional capital for its predominantly agricultural hinterland. Four key factors have played a role in Munich's favourable economic development since the 1960s. The first is a determined effort on the part of the Freestate of Bavaria to develop the regional economy into a modern knowledge economy (more on this later). Secondly, after World War II, many large companies fled from Eastern Germany to Munich. A prime example is Siemens, which used to have its main operations in Berlin. Many publishers moved from Leipzig (which was Germany's prime publishing city) and other major Eastern German cities to Munich, to escape from the communist regime. But some knowledge institutes also came to Munich: the headquarters of the Max Planck Institute is an example. After German reunification in the 1990s, it was widely feared that Munich would lose many companies or establishments to Eastern Germany again, but this certainly did not happen on a large scale. Thirdly, Munich, unlike regions in North Rhine Westfalia, does not carry the burden of old industrial sectors that went into decline from the 1970s on. Munich's industrial development dates from after World War II, which means that traditional 'heavy' industries are virtually absent from the city. This makes it easier for the city to adapt to the requirements of the global knowledge economy. Finally, Munich's high quality of life has been a key driver of its economic success. This will be elaborated in the next section.

Nowadays, Munich's economic base can be characterised as highly diversified and knowledge intensive. The city has a very modern industry and hosts the headquarters and research bases of key companies such as Siemens and BMW. The banking and insurance sector has grown strongly during the 1990s and is one of the key sectors in the city. Key companies are Allianz, Munich Re and the HypoVereinsbank. Therefore, the city in second place in Germany as regards banking, behind Frankfurt, and in first place in insurance in both Germany and Europe. The city has a relatively high share of sectors with strong growth potential, such as ICT. The economic area of the city is indeed the major IT location in Germany, employing over 155,000 people in 8,000 firms engaged in ICT business and research. But other growth sectors are World War II strong in Munich. Examples are the (new) media, biotechnology and life sciences.

Many foreign companies have their major German or even European operations in Munich, attracted by the high educational level of the workforce, the attractiveness of the city and the strongly developed knowledge infrastructure.

The balanced mix of Munich's economy is often referred to as 'Münchner Mischung', the 'Munich Mix'. The mix refers to the presence of a variety of sectors, a variety of firm sizes (global players as well as a large number of SMEs), and the mix of foreign and domestic companies. The mix has at least two benefits. In the first place, it decreases the city's vulnerability to fluctuations in certain sectors. Secondly, the large variety of sectors offers scope for the creation of 'new combinations', or intersectoral marriages that are the source of innovation and new business opportunities (e.g. bio-informatics). In a later section this will be discussed in more detail.

7.3.3 Quality of Life

Munich is a city that offers a wide variety of cultural and leisure facilities. Furthermore, the surroundings of the city also have a lot to offer: The Alps are nearby, and there are lakes in and near the city. Many highly-skilled Germans feel attracted to this unique mix of culture and leisure possibilities at such a small distance. There is a price to be paid for Munich's attractiveness: it is one of the most expensive cities in Germany. Rent levels and house prices rank among the highest in the country.

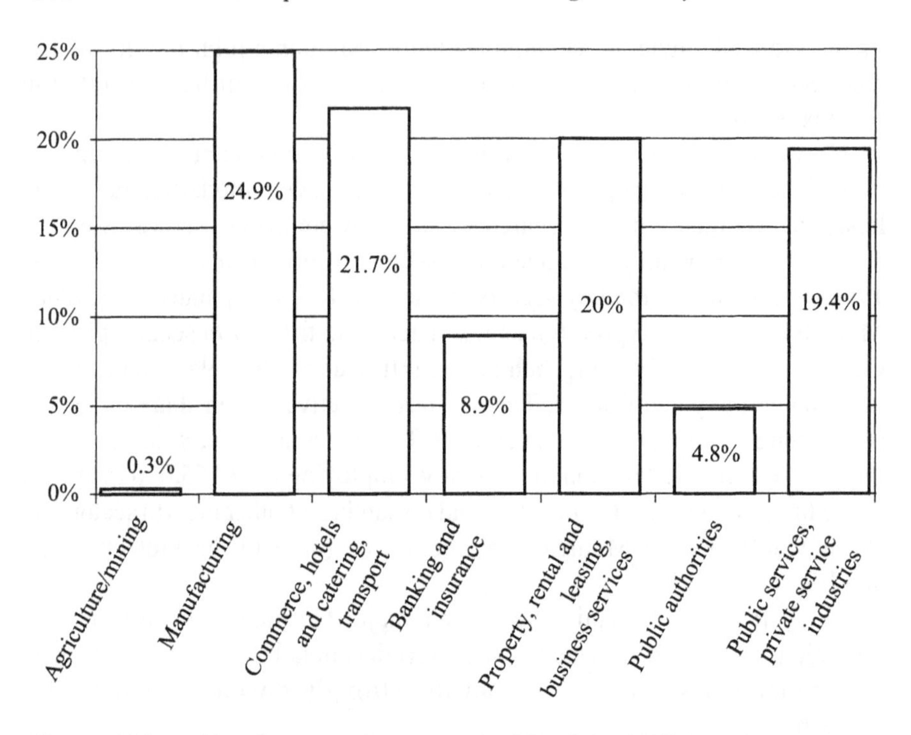

Figure 7.3 Employment by sector in Munich, 2001 (total employment = 695,650)

Source: Landeshauptstadt München (2003).

As a pleasant, clean and elegant city, Munich is also highly attractive to visitors. Munich's hospitality function can be split into two broad categories: leisure and culture, and congress and business. In 2003 the Munich Tourist Office reported 3,462,300 arrivals or 7,057,500 overnight stays. The total turnover directly or indirectly from tourism is estimated at €3.5b in 2003.

However, the strength of Munich as a tourist destination is mainly due to the traditional cultural-tourist market, enjoying an enviable palette of cultural institutions, as well as intangible elements related to its celebrated atmosphere and quality of life, and countless festivals and events, the most famous of which is the popular 'Oktoberfest'. On account of this, Munich is among the European 'stars' for short breaks and cultural visits. Munich's cultural speciality is music. There are three large orchestras are in the Bavarian capital, with international superstars such as Levine, Maazel, Mehta and Thielemann as conductors. In the mid-1990s the city developed a programme of high-calibre open-air pop and jazz events. Munich is well endowed with formal 'high-culture' institutions, but the development of underground and/or alternative cultures is much less obvious. As a result, artists and other people in the creative industries sometimes tend to rank cities like Berlin and Hamburg higher than Munich.

7.3.4 Accessibility

Munich's external accessibility is excellent. Munich's new airport (opened in 1992) is the second largest airport in Germany after Frankfurt, with 24.2m passengers in 2003. The city is also well connected to the high-speed railway system. Furthermore, the city has an efficient public transport system, with an extended metro network and a dense tramway system. However, car traffic congestion is an increasing problem, in particular during peak hours: since the 1970s, Munich has witnessed a growth of residential settlements on the outskirts of the city, in its immediate surroundings and further outside the city. At the same time, there has been a concentration of jobs in the city. This has resulted in a substantial growth of the traffic load in the Munich region (City of Munich, 1999).

One of the challenges for the future will be to expand and upgrade the public transport system to a higher quality level. Since the late 1970s, no major investments have been made in the S-Bahn, so that in 2000 the Freestate of Bavaria began a €260m investment programme for new wagons, stations

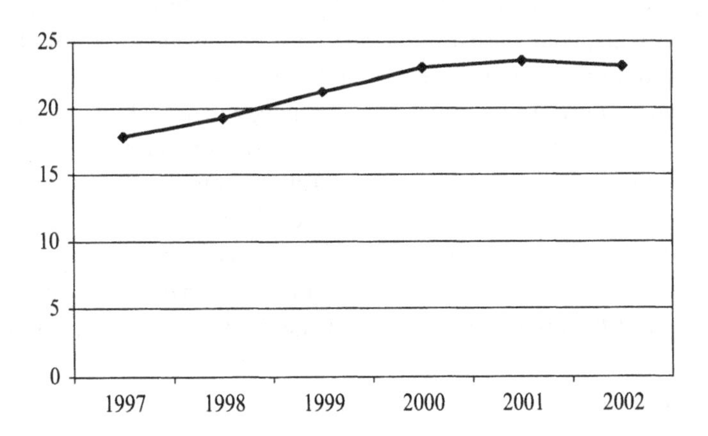

Figure 7.4 Number of passengers (in millions) travelling through Munich Airport, 1997–2002

Source: Landeshauptstadt München (2003).

and a second tunnel through the city. On the other hand, the city itself is permanently improving the underground, bus and tram system to offer an attractive alternative to the private car. Some international benchmarks, which rank the Munich public transport system among the best, show the success of these efforts.

In recent discussions, the question has been raised of whether to connect the airport with the city with a magnetic levitation system connection or an express railway. Which one of these alternatives will be realised is an open issue between the Freestate of Bavaria – in favour of the 'transrapid' connection – and the City. A 'transrapid' connection is likely to reduce travel time to the airport from 45 to 15 minutes. The Express-S-Bahn would take 20 to 25 minutes to reach the airport.

7.3.5 Urban Diversity

As already noted, Munich's urban *economy* is highly diverse: the Munich Mix stands for diversity in terms of economic sectors and firms' size. Nevertheless, the city does not have the image of being a very 'diverse' city from a cultural point of view. It has an image of being wealthy but not particularly 'funky'. The city is more known for its traditional Bavarian cuisine and beerhalls than for its vibrant nightlife, although it is certainly there. On the other hand, 23 per cent of Munich's inhabitants do not have German citizenship, which also could be seen as a sign of diversity. The kind of people who bring diversity to many cities (immigrants, young artists, hip and trendy youngsters) in form of cultural creation do not form a very visible part of the city's street image. This has to do with the high rent levels and prices of real estate, which make it difficult if not impossible to find cheap locations for artistic experimentation. Nevertheless, areas like Kunstpark Ost and Kunstpark Fröttmaning and the clubbing scene show that Munich has also established a specific and vital culture for young people. And although Munich has the image of an economic city where success is highly valued, positive less tangible factors are also part of Munich's image: high quality of life, a big variety and high standard of leisure time activities.

7.3.6 Urban Scale

The literature review showed that cities need to have a certain scale to support economic specialisations, and a number of amenities and infrastructures for which a demand threshold needs to be exceeded. The region of Munich, with approximately 2.6m inhabitants and a large economic base, is certainly large enough to qualify as a city with sufficient scale to have all the metropolitan amenities and infrastructures that support the development of a knowledge economy.

7.3.7 Social Equity

As a result of structural economic change, many people have difficulty in adapting to the new circumstances. The 'knowledge economy' entails growing demands on vocational qualifications and personal skills, and not everyone is in the possession of these assets. The official poverty ratio in Munich rose from 65 per 1,000 in 1986 to 111 per 1,000 in 2000, in 1995 it reached its peak at 125 per 1,000 (City of Munich, 2000, p. 27), and the 'poor population' in the city is now over 140,000. This trend is reinforced by the high cost of housing in Munich. Inward migration also puts a strain on Munich's social fabric. Although immigration may compensate for an ageing population, the influx of low-educated immigrants especially also brings tensions.

Nevertheless, Munich can be characterised as a socially balanced city. Unemployment levels are low, and social exclusion is not much of a problem compared to other major German cities. Some immigrant groups have difficulty in integrating into German society, and/or underperform at school. However, unlike virtually every other large European city, Munich has no clear spatial concentrations of social and economic problems, or 'problem neighbourhoods'. But, equally, as in other cities some areas can be found with temporary concentrations of ethnic minorities and some related problems.

7.4 Knowledge Activities

In the last section, the 'foundations' of Munich's knowledge economy were discussed. Now, we turn to knowledge activities and ask how well the city (and the actors in it) manage to: 1) attract and retain talented people; 2) create new knowledge; 3) apply new knowledge and make new combinations; and 4) develop new growth clusters.

7.4.1 Attracting and Retaining Knowledge Workers

How well does the city manage to attract and retain knowledge workers? Munich is considered an attractive city for high-skilled workers. Not only does it offer a large variety of jobs for which high qualifications are required, but also, the city's good quality of life and its cultural and leisure amenities make the city an attractive place for working and living.

The large number of IT specialists who came to the city in recent years might illustrate Munich's attractiveness for foreign highly skilled workers. In Germany by the late 1990s there was an increasing shortage of IT professionals. To address this problem, the German federal government decided to give temporary work permits to non-EU foreigners with high-level ICT skills, in the 'Green Card' programme. From August 2000 until the end of 2002 some 10,000 people arrived in Germany from all over the world. The largest contingents came from India and Central and Eastern European countries (Schreyer and Gebhart, 2003). In Germany, the city of Munich was by far the largest recipient of Green Card holders: 1,532 in total. They were mainly attracted by the booming ICT sector in the city. Munich received more specialists than all the cities in North Rhine Westfalia together.

Figure 7.5 shows the composition of the foreign population of Munich's economy. Although many of the foreigners are in low-skilled jobs, it can safely be assumed that a large percentage of the foreigners are highly skilled. This holds notably for people from the EU and the US.

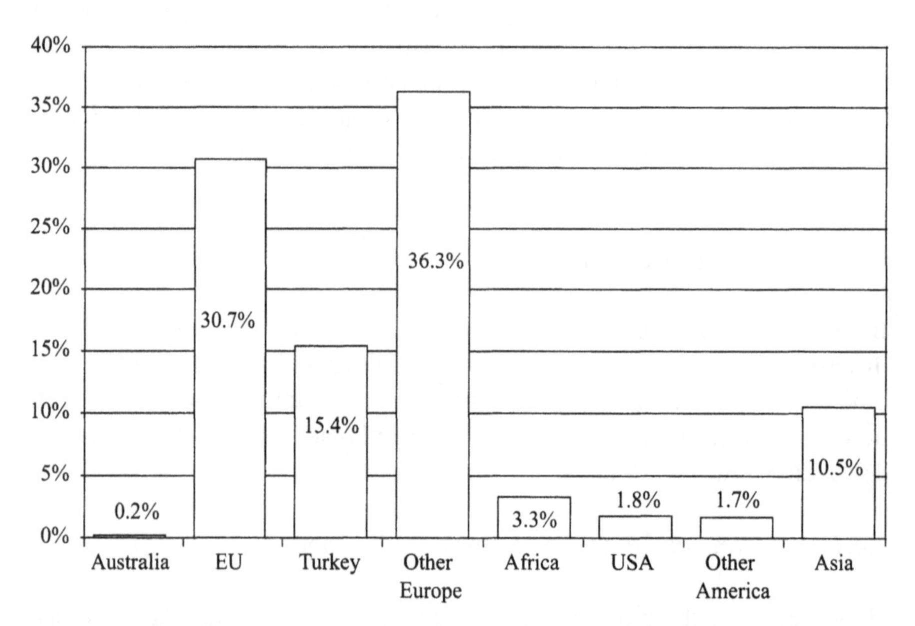

Figure 7.5 Foreign population in Munich, 2002 (total = 289,263)

Source: www.muenchen.de.

7.4.2 Attracting and Retaining Students

Attracting and retaining students is vital for any local knowledge economy: students can contribute to urban diversity, they are cultural and social innovators, they have substantial purchasing power, and, no less importantly, they are the 'human capital' of the urban economy of the near future. Munich has a student population of over 80,000 but it has problems in housing them appropriately. There are considerable shortages in student housing. The rents paid in Munich are among the highest in Europe – up to €1,000 for a flat. Many students choose to commute from their hometown and save on housing costs, but then they are then less likely to be integrated into the community. At the time of writing, a complex with 550 apartments is being constructed in the northern part of the city, an investment of approximately €37m. Some 2,500 new apartments are planned, but the first will not be in place until 2004. This means Studentenwerk (the main student organisation) has to improvise solutions for the housing shortage in the short term, and creative solutions are needed. At the moment there is a waiting list of 3,500 students who are now staying somewhere else, and the situation looks likely to continue. Studentenwerk offers furnished placements to first year students, a joint project with the University of Arts, which may cater for 2,500 students in the shortest time schedule, placed in a former military area. They also offer 'first aid' of a mattress on the floor; in practice, everybody gets something in their first weeks, but the quality of these solutions is questionable.

In the meantime, though, the national and international potential of Munich's higher education institutions is reduced as a consequence of the housing market constraint. Even at a regional level, Munich increasingly faces competition from more accessible student cities. Moreover, the city cannot fully internalise in its economy the expenditure generated by its more than 80,000 higher education students, who continue to use the city as commuters rather than citizens. In Munich, these problems are more dramatic than in other cities that compete with Munich as poles of economic development. For instance, in Berlin housing is much cheaper (and the housing stock bigger), and it is easier for students from different origins to be integrated into the community, with the result that the creative input of young people is strong and very visible in the 'Berliner' atmosphere. To what extent Munich may be losing out in terms of its reputation and economic strength to more 'open' and student-friendly cities remains a long-term issue that the city has to address more proactively than at present.

The Munich universities are not known for their international orientation regarding student exchanges. The number of exchange students from other countries is still relatively low, and the internationalisation of the universities has yet to take off. An improvement in this respect could further help Munich to attract talent to the city. Nevertheless, international orientation should not only be measured by the number of students. Munich universities take part in a number of multinational research projects and share their research results with other universities. To complete these international activities there are interchange projects for international researchers and exchange of lectures.

One impulse for the attraction of the best foreign talent comes from the Max Planck Institutes: recently, they have set up top-level international research schools in cooperation with universities. The Institute attracts students from China, Eastern Europe and India. A further example might be the cooperation of the Max Planck Institute, the Technical University Munich, the University of Augsburg and the George Washington University Law school, founding together the Munich Intellectual Property Law Centre.

7.4.3 Creating Knowledge

New knowledge is being created in Munich on a massive scale, both by the knowledge infrastructure (the formal knowledge institutes) and by firms.

An analysis of the number of patent registrations suggests that Munich is one of the leading innovative regions in Germany. With 3,091 registrations in 2000 the city is second in Germany after Stuttgart. Figure 7.6 shows that companies are the main sources of patents, followed by individuals. The research institutes (universities plus public research institutes such as the Max Planck Institute and the Fraunhofer Gesellschaft) only have a small slice of the cake. However, it should be noted that Munich's research institutes take the first place in Germany where the absolute number of patent registrations is concerned (Greif and Schmiedl, 2002, p. 27). The figure shows that the number of patents grew substantially from 1995 to 2000.

Patent data should be interpreted with caution. First, the patent data refer to the location where the patent has been registered; this is not necessarily the location where the invention happened. This may overestimate Munich's innovativeness, with its headquarters of BMW and Siemens. Furthermore, innovations in services (which make up over 70 per cent of the economy) are normally not patented, although they make a major contribution to productivity and quality improvements. Secondly, it is not the number of patents registered which is important for the economic development and innovative character of a city, but the number of patents which are transferred into production or new processes. This is the important difference between invention and innovation.

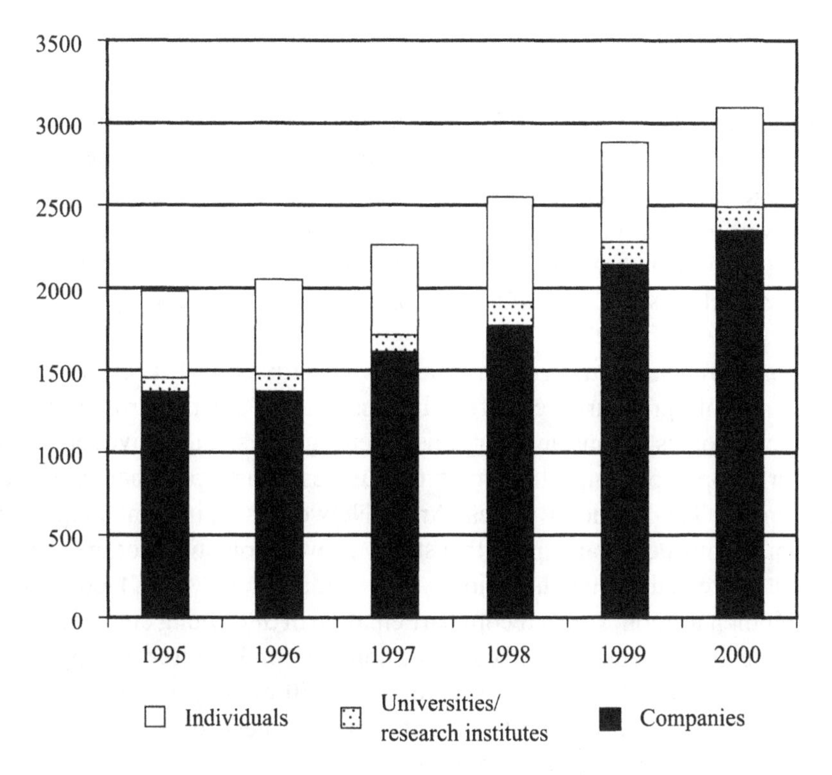

Figure 7.6 Patent registrations in Munich, 1995–2000

Source: Greif and Schmiedl (2002).

Nevertheless, the fact that the German Patent Office, the European Patent Office and the Munich Intellectual Property Law Centre are all situated in Munich makes the city one of the most important clusters in Europe concerning questions of patents and intellectual property rights.

The Freestate of Bavaria plays a key role in the promotion of knowledge creation, as university and science policies are determined on that level. In the last few decades, Bavaria was the region with the highest share of investments in higher education and R&D in Germany. As already mentioned, the Freestate of Bavaria has invested strongly in the creation of new knowledge. In 2000, they co-financed the new building of the Max Planck Institute, not only because of the importance of keeping such a high-quality research institute, but also because the presence of the Max Planck Institute attracts other knowledge institutes as well.

7.4.4 Applying Knowledge/Making New Combinations

The Freestate of Bavaria and the City of Munich together deploy an impressive variety of policies that are directed to boost entrepreneurship, to commercialise knowledge and to create networks of innovation. Figure 7.7 provides an overview of the measures supported by the Freestate of Bavaria as well as by the City of Munich in a number of fields. These policies, sometimes single actions, sometimes part of a wider strategy, are discussed below.

Raise entrepreneurial awareness Various initiatives have been taken to nurture an entrepreneurial culture. The Association for the Promotion of New Technologies, for instance, organises get-togethers, interactive seminars and workshops covering all aspects of business start-ups, financing and management. The Munich Business Angel Network is an informal, regional platform for business start-ups with a strong growth potential. For the fourth consecutive year, the Munich Business Plan Competition (MBPC) took place in the Munich Region, with a record participation of 386 young entrepreneurs and 126 business plans. In four years the competition has already supported the foundation of 103 companies with some 750 employees, financed by approximately €80m. About 60 per cent of the business ideas derived from the IT sector and 20 per cent from life sciences.

The FLÜGGE programme (initiated by the Freestate of Bavaria, and part of the Bavaria High-tech Offensive) aims to increase the number of spin-offs from the universities. It supports, for example, the technology transfer centre at the LMU, which forms a link between business demand and university supply of knowledge. It also offers university researchers the possibility to start their own company while remaining university employees on a half-time basis. This reduces the risk of starting up a company in the earliest stage. At the business school of the LMU there is a special centre for entrepreneurship and other faculties offering extra-curricular entrepreneurial courses. The Münchener Business Plan Wettbewerb (MBPW) is a business plan competition. It offers starters the possibility of having their business plan checked by experienced business and financial experts. Secondly, it offers start-ups the possibility to come together with business consultants and venture capitalists. In this way, although a new business idea might not win a prize, it can come to realisation because of the contacts made during the competition.

Promote innovative networks Many policy initiatives – most of them from the Freestate of Bavaria – aim to improve the cooperation between firms and

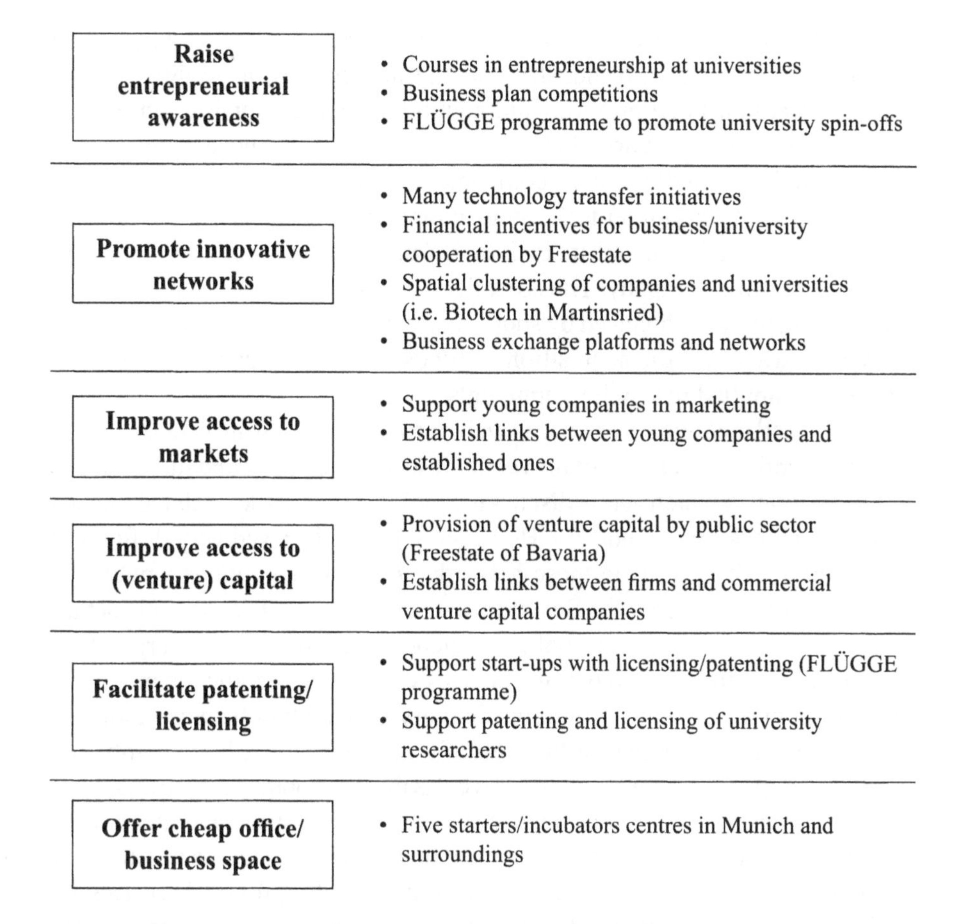

Figure 7.7 Policies for knowledge application and new combinations

universities, in order to make more of the regional knowledge base. The idea behind it is that universities and companies too often inhabit separate worlds: they have different drivers (profit making vs academic prestige), different time horizons, different work attitudes, and in practice, find it difficult to cooperate.[1] However, there are many synergies and interdependencies that could create room for fruitful cooperation: some knowledge developed at universities is commercially very valuable, or may become so in the longer run; university students and researchers are potential entrepreneurs; universities are the source of new staff for companies; university laboratories may be used by start-ups; universities may benefit from applied research assignment from the industry, and vice versa.

With this in mind, the Freestate of Bavaria and the City of Munich, often in close cooperation with industry and the universities, have developed a number of policies that promote interaction between universities and firms. One example is IRC Bavaria (Innovation Relay Centre), whose mandate is to support small and medium-sized companies and research institutes in Bavaria in establishing cross-border technology cooperation projects. Another policy is the provision of financial incentives for projects in which the university cooperates with a company. The so-called Bonus Programme (a programme encouraging research financed by sponsors from commerce and industry) pays financial rewards for the acquisition of funds from third parties for carrying out specific applied research assignments.

More 'soft' measures are taken as well: through its 'software offensive', the Freestate supports a network of university researchers, young and more mature software firms, in which actors discuss technological or market-related themes concerning the software industry. However, it has still proved very difficult to involve university people in these networks. *Bayern Innovativ* is a publicly-held company initiated by the Bavarian state government. It was jointly set up in 1995, by representatives of politics, science and industry as a corporation for innovation and technology transfer. This Centre has become the motor for stimulation and extension of interdisciplinary cooperation between industry and science. It is the spider in the web of a network with a large number of companies and scientific institutions as well as the development of professional, customer-oriented services. It is financed by the Bavarian state government with a fixed annual budget amounting to some €4m. Additional income is generated through the acquisition of strategic projects and fees for provided services.

There are indications that the quality of the knowledge and technology transfer system is a location factor for (foreign) companies: Munich's good score in this respect was one of the reasons for General Electric to choose Munich as a base for its European headquarters.

Improve access to markets Many start-up companies have developed a good product but do not have the business relations and/or expertise to sell their product successfully. To help companies with their marketing, a variety of support measures are available. Several starters' centres help companies to present themselves at international exhibitions. Another policy is to link start-ups to networks of established firms in the regions.

Facilitate patenting and licensing In the commercialisation of research, patenting and licensing are key elements. Patenting can be costly. Costs amount to €5,000 to get a patent from the German patent agency, and up to €15,000 for an international patent, and it also involves a lot of administrative work. Understandably, many start-ups and individuals can get lost in these domains and consequently many business opportunities are foregone. With this in mind, the Bavarian Patent organisation – in action in Munich since 2000 – helps university researchers when they want to takeout a patent. Box 1 explains how this works. Since 1955, Munich has hosted the 'Patentstelle für die deutsche Forschung' of the Fraunhofer Gesellschaft located in Munich, which does a similar job.

Box 7.1 Steps in the patenting process for university researchers

1 Researcher makes an invention.
2 Informs the discovery-notification agency.
3 Patent agency analyses whether the invention is really new.
4 If so, Fraunhofer Institute looks which company or investor could be interested in the patent.

If nobody is interested: no formal patent demand.
If there is an interest: patent will be created and licensed.

In the system, one-third of the revenues of the patent accrues to the individual scientist and two-thirds to the research group of which he/she is part. The Fraunhofer Institute checks whether the patent can be commercialised. However, if the researcher wants to start his or her own company, they has to buy back the licence, a process which is currently under discussion. The Ludwig-Maximilians University Munich does not operate its own starters' centre: it argues that there are already a lot of starters' centres in the Munich area. However, it has some pre-incubator facilities: starters can use university facilities (laboratories, etc) in their very early stage, and use the university address as postal address.

It is important to note that the number of patents emerging from the universities is still low, and it increased little in the 1990s. Only 1–2 per cent of the patents are from universities.

Venture capital Munich is rich in venture capital. Over 20 internationally operating venture capital companies provide about €500m in risk capital, mainly channelled to start-ups in the IT and biotech sectors. However, several of our interviewees indicated that venture capital is scarce, as investors have become a lot more cautious in recent years. In the booming years of the late

1990s many companies located in Munich because of the abundant presence of venture capital, but now some firms leave for the lack of it.

Start-up centres In Munich start-up support takes many forms. In the LMU faculties of Physics and Chemistry, there is a 'pre-incubator', where entrepreneurial scientists pay very little to make use of the laboratories and other facilities. The university can take a share in an eventual start-up. Next, there are a number of dedicated starters' centres in the city. It is both the policy of the City of Munich and of the Freestate of Bavaria to promote the start-up of new companies, and in Bavaria, there are more than 30 starters' centres. Five of them are in Munich. The largest of them – the MTZ, Munich Technology Centre – is owned and operated by the City of Munich (see Box 2), the others are initiated by the Freestate of Bavaria. In general, starters centres offer business space for start-ups at rents below the market levels. On top of that, many offer additional support such as administrative support, access to broadband networks, conference rooms, etc.

It is important to note that it is not only public agencies that invest in start-ups: large companies also invest in start-up activity, hoping to boost innovation on the fringe. BMW is an example.

Box 7.2 Start-up centre: The Munich Technology Centre

The MTZ (Munich Technology Centre) was founded in 1983 by the City of Munich and the Chamber of Industry and Commerce. It offers office space for start-ups at below-market rents, but also administrative services like telecom, office management and other services. The management of the centre also helps companies to build their client network, to get access to finance, or to market their products at trade exhibitions etc. The MTZ offers 11,000 m^2, and currently hosts 52 small firms in a variety of sectors. Since 1983, 130 companies have started their operations from the MTZ. Companies stay there for two years on average.

Box 7.3 Start-up centre: The Gate (Garchinger Technologie und Transfer Zentrum)

Garching is a medium-sized town to the northeast of Munich, close to the city border. Recently, a large technology park has been created there: the area is a concentration of technology institutes and faculties of the technical university, and also the Max Plack Institute for Plasma Physica, and an atomic reactor for research only, are there. Since 2002, there has also been a centre for start-ups called 'The Gate', offering some 8,000 m^2. The centre is primarily financed by the Freestate of Bavaria. It is focused on start-ups in information technology and mechatronics, two of the 'spearhead clusters' of the region. By 2003, there were some 23 companies located in the building, 30 per cent of which were start-ups from the university, and 70 per cent from outside. However, the outside firms also maintain links with the university. Rent levels are at €10/m^2, somewhat below the market price of €12–14. Some 10 per cent of the space is dedicated to production facilities; the rest is office space. The centre cooperates with the 'Software Offensive Bayern', a state-supported programme to boost the software sector in Bavaria. Every third Monday of the month, there is a meeting for start-ups where they can meet key people from established software firms or university professors and discuss themes. Furthermore, the Gate can use some of the technical university's facilities (laboratories, etc), and also they are linked to the IT network.

New combinations The 'Munich Mix' offers a favourable environment for creating new intersectoral combinations. One example can illustrate this. Arri-Richter, the film-equipment company (part of the media cluster), cooperates with a producer of medical technology equipment (another strong sector in Munich) to develop high-precision cameras that are integrated in medical equipment.

7.4.5 Developing New Growth Clusters

The Freestate of Bavaria, through its 'High-tech Offensive', invests heavily in developing new growth clusters in Bavaria, and in the city. The spearhead clusters are ICT, media and biotechnology.

The ICT cluster Set up in 1998 with a common effort of industry, science and politics, the Software Initiative is part of the Bavarian government's high-tech initiative. By investing the proceeds of the privatisation of public shareholdings (amounting to approximately €1,18bn) into R&D, training, and infrastructure, this project aims at making Bavaria an even more attractive location for the high-tech industry. The Software Initiative receives a substantial share of these investments. It concentrates on the following key areas:

- education and training – the IT industry faced a dramatic lack of skilled personnel. New university courses and training initiatives were set up to help to improve the availability of software experts;
- research and development – excellent results from R&D are the basis for globally successful products. The Software Initiative supports joint research and development projects between academia and industry;
- start-ups (see Box 3). The Software Initiative aims to stimulate high growth start-ups in the software industry, making Bavaria a top location for dynamic young businesses in the information and communications sectors.

An interesting development in the software cluster is the reconversion of an area of 45 ha., where the largest Siemens establishment within the city is located. Siemens will (in cooperation with the city) create a much more open Siemens-quarter, with housing, work and research. The whole concept is a little like a campus area. This can become a new 'hotspot' for the software cluster, as other firms will also be allowed on the site.

The biotechnology cluster Munich is one of Europe's biotech hotspots, and is rapidly growing: the number of people employed in the sector has risen from about 300 in 1996 to some 2,200 by the year 2000 (Bio^{M} AG, 1999; interview). By November 2000, 120 biotechnology and pharmaceutical companies were located in Munich, 100 being SMEs. Five out of the 15 German biotech firms with a stock exchange quotation are located in Munich. The sector is spatially concentrated in Martinsried, a village just outside Munich, where 850 people are already employed in the sector. Here the faculties of biology and physics of the LMU are also located, as well as a large hospital. In the near future, the medical faculty of the LMU will move to Martinsried as well. It is expected that the spatial clustering of complementary and competing knowledge industry will provide scope for (cross-disciplinary) the face-to-face contacts that are the basis for innovation.

The growth of the cluster accelerated in the late 1990s. However, before 1995, the ingredients for successful cluster development were already in place. The Gene Centre, for instance – a research institute for gene research – was already developing very well. From 1996 on, things speeded up. One factor was Bavaria's victory in a national competition between the German regions (organised by the federal government) to become a favoured 'biotech cluster'. The reward was substantial financial support (€25.5m in four years) from the Federal Ministry of Education and Research. These funds are invested in young biotech companies to help them develop and commercialise their R&D activities. A new organisation, Bio^M AG, has been set up, to distribute the grants and support the start-up companies. Starters can submit a business plan to be evaluated. Bio^M AG has been set up as a public-private partnership, owned by banks, venture capitalists, the Freestate of Bavaria, the pharmaceutical industry and several private investors. Each of these partners had an interest in creating a strong biotech cluster in Munich. The Free State of Bavaria has been very cooperative and has a positive impact on the cluster; it strongly supported the development of Martinsried, the small town just outside Munich where most biotech activity takes place. Bavaria invested €23m in the biotechnology park and strongly promotes and supports science and R&D activities, which it regards as principal growth activities for the knowledge economy. In the development of Bio^M AG a few key people with expertise and experience in both science and entrepreneurship have played a catalysing role. The city of Munich has also supported the development of the cluster.

With the life science park Freiham, the City of Munich is developing an area of around 50 ha., which should generate about 8,000 jobs. It is located some 12 km from the heart of the city, and it offers business locations in several sizes.

Other clusters Other strategic clusters in Munich are the media cluster, environmental technologies, and new materials. For the last two there are several research funds available and a number of technology transfer activities focus on these branches.

Munich is already an important media cluster, with several TV broadcasting and production companies: see http://www.bayern.de/Wirtschaftsstandort/Medien_und_IuK/Medien/Medienstandort/ for more data. To promote the cluster, a starters' centre for new media firms (Gründerzentrum für Neue Medien, GZM) was set up by the Freestate of Bavaria, the Landkreis München and the village of Unterföhring (very near the city) in cooperation with large media firms and investors. The centre is dedicated to start-ups in multimedia, Internet, TV and ICT. Other promotion policies include the operation of a film fund and the organisation of festivals.

7.5 Conclusions and Perspectives

This section contains conclusions and unfolds perspectives for Munich's position in the emerging knowledge economy.

Table 7.4 An indication of the value of knowledge economy foundations in Munich

	Foundations	Score in Munich
1	Knowledge base	++
2	Economic base	++
3	Quality of life	++
4	Accessibility	++
5	Urban diversity	□
6	Urban scale	+
7	Social equity	++

-- = very weak; – = weak; □ = moderate; + = good; ++ = very good.

7.5.1 Foundations of the Knowledge Economy

The foundations of Munich's knowledge economy are very solid. After World War II, the city benefited greatly from the in-migration of large companies and talented people from Eastern Germany. During the 1960s and 1970s, with the help of strong and persistent regional policies, Upper Bavaria has developed into one of the strongest knowledge regions of Europe. Munich, as the regional capital, drew most of the benefits and is the flagship of Bavaria's success.

It can be concluded that Munich has a very high score on almost all the foundation indicators. Its *knowledge base* is very strong by many degrees: the city has a high percentage of workers with high qualifications, and hosts an impressive share of knowledge-intensive companies and knowledge institutes. The *economic base* is highly diversified: the 'Munich Mix' comprises many growth sectors and few declining industries. The *quality of life* is high in many respects, with all the high-level amenities in the city and the mountains and lakes nearby. The other side of the coin is the high cost of living in the city. *Accessibility* is rather good. Munich is excellently linked with the rest of the world through its airport, a good railway and highway connection and an efficient urban public transport system. The *urban scale* of Munich is large enough to support a highly specialised labour market and many high-level amenities, that in turn attract talent and companies. *Social problems* are very moderate, especially compared to most European major cities. As the weakest point of Munich, we nominate the *urban diversity*. In particular, the city does not have many experimental artistic and underground scenes, which negatively impacts on the liveliness and 'metropolitan' atmosphere and cultural creativity of the city. The weak development of these sectors is a result of Munich's image as an 'economic' city, but is also related to its very high cost of living, which drives artistic talent to other cities, notably Berlin. However, this weak point is of limited importance for the knowledge base in the areas of economic and natural science: it is more related to the fine arts and the image of a city that wants to attract researchers. Future developments are potentially more important. There is a slight tendency in the production sector to leave the city, moving just outside the city borders, because of the expensive premises in Munich, or to leave the region completely for areas where the wages are lower. This development might break the link between research and production in the long run, so that the city also might lose its research reputation.

7.5.2 Activities of the Knowledge Economy

In accordance with our frame of reference, we have considered four types of activities of a knowledge city. It is clear that Munich scores quite highly here, too. The city is strong in *creating new knowledge*, in *applying knowledge* for commercial purposes, and in *developing new growth clusters*. For this, the city owes much to the Freestate of Bavaria's strong and sustained efforts to invest in science, higher education, knowledge transfer and start-ups, as this case study shows. Also, the Freestate's targeted policies to boost certain high-level growth clusters have benefited Munich, witness the media cluster and biotech clusters that emerged in the 1990s and made the 'Munich Mix' stronger.

Table 7.5 An indication of the value of knowledge economy activities in Munich

	Activities	Score in Munich
1	Attracting knowledge workers	+
2	Creating knowledge	++
3	Applying knowledge/making new combinations	++
4	Developing new growth clusters	++

-- = very weak; – = weak; □ = moderate; + = good; ++ = very good.

In the last few decades, Munich's economic success has *attracted knowledge workers* to the city, who further contributed to its performance. However, careful policies can make improvements in this area. For instance, we feel that the city could do more to attract (foreign) students. The city management can make a contribution by supplying adequate housing, but also more indirectly by urging the universities to become more outward-looking. As important as it is to get innovation and to create a creative climate by integrating foreign students and researchers it is as important to keep highly-qualified students and researchers in the city and stop the migration to other cities of knowledge. In this aspect the city image, the research conditions and the quality of life in a city are the essential assets, which can at least partly be influenced by a city strategy. An attractive city for students, offering good housing conditions might attract further talents.

7.5.3 Threats

Considering all this, Munich has very good prospects to thrive as a successful city in the knowledge economy in the years to come. However, we also identify some threats, some coming from outside and some from within the city.

- Competition from the developing world. Some activities might leave Munich. In the past decade we witnessed a displacement of routine jobs to lower-cost locations in Eastern Europe and Asia. This did not harm Munich too much, specialised as it is in R&D, high-level production and services. Increasingly, however, knowledge intensive activities can be also carried out in these countries (*Financial Times*, 7 and 20 August 2003). Recent research shows that Munich's vulnerability in this respect is higher than that of other cities. A thorough analysis is needed of how these developments might affect Munich's knowledge base on the longer run, and which adequate policy measures can be taken at the urban level. A strategy to counter this possible development might be to strengthen the described ability to create new knowledge by bringing together different partners. This is an advantage of the very diversified city of Munich compared to other low cost research cities. Intensifying research projects or supporting the High-tech Offensive by communal measures to develop new fields for production or innovation could be a possible action for the city. An other possible strategy might be to support niche branches and knowledge sectors.
- Competition from European cities. Munich faces competition from several European cities. In our interviews, Vienna was mentioned as a key competitor, not only as a knowledge city but also as gateway to the growing market of Eastern Europe. Furthermore, there are indications that

Munich loses creative people to Berlin or other European metropoles that have a better image, climate and infrastructure for artists and underground cultures. If it is true that the arts and the industries are becoming more interdependent – and there are good reasons to believe so – then this development could imply that new opportunities for growth are foregone. Munich could consider improving its image for the creative industries and creating conditions for it to flourish. For instance, it could subsidise the rent of older buildings and dedicate them to the creative industries, give other financial and non-financial incentives to artists who settle in Munich, or promote/support the artistic education infrastructure, although there are some successful examples where temporary use of areas/buildings for this is realised: for example, former military areas or old Stadtwerke centre.

- Competition from the US. German cities face the threat of losing talented people to the US. Each year, 100,000 well-educated Germans leave to work abroad, and reverse migration is much less pronounced.
- A further important measure, and at least partly the responsibility of the cities, should be the support and creation of a framework within which it is possible for families to combine their family life with their job. Comparing the European employment rates of women, a large spread can be seen, from about 40 per cent in Spain to 60 per cent in Germany and more than 70 per cent in the Scandinavian countries. Economic development of society as well as of the single enterprise is dependent on the qualifications of the working force. Losing talented or qualified people because of family changes should be avoided.
- Currently, the high rent levels in Munich imply a threat concerning the student population. Unless substantial investments are made in affordable student housing, students could increasingly choose other cities with lower costs of living and housing, which undermines Munich's knowledge potential in the longer run.
- Despite many success stories, we also have indications that some growth clusters are coming under pressure. This holds in particular for the biotechnology and media clusters. In biotechnology, expectations have been very high, but the actual results of biotech companies – with some exceptions – are discouraging. It will take a long time before real growth will be visible. In the media sector, the collapse of the Kirch imperium has given a substantial blow to the media cluster. A current survey of the media sector in Munich shows that the branch has been consolidated, but at the moment a new positive development can be seen.

Note

1 It should be noted that polytechnics more easily and frequently cooperate with the business sector. The reason is that often the professors and researchers came back to university after making a career in the private sector. Therefore they have very good connections to the firms.

Discussion partners

Dr C. Beck, Press and Public Relations, Max Planck Society.
Dr F. Glatz, Director, Gate Garchinger Technologie- und Gründerzentrum GmbH.
H.P. Heidebach, Economic Development and Employment Strategies, Department for Labour and Economic Development, City of Munich.
C. Mann, CEO, Münchner Technologiezentrum.
Monika Monat, Planning, Coordination and Control, Department of Education, City of Munich.
Dr R. Obermeier, Economic Promotion/Site Consultation, Chamber of Industry and Commerce for Munich and Upper Bavaria.
S. Reiss-Schmidt, CEO, Department of Urban Development Planning, City of Munich.
Dr J. Schade, President, Deutsches Patent- und Markenamt.
U. Schramm, Department of Urban Development Planning, City of Munich.
R. Vilsbeck, Referat für Presse- und Öffentlichkeitsarbeit, Deutsches Patent- und Markenamt
D. Wiedmann, Project Manager, ConM Gesellschaft für Marktforschung und Regionalanalysen mbH.
C. Zinser, contact office for research and technology transfer, Ludwig Maximilians-University Munich.

References

City of Munich (1999), 'The Munich Perspective. A Summary of the 1998 Urban Development Strategy', Department of Urban Planning.

Freistaat Bayern (2002), *Invest in Bavaria, Investorenatlas*, Bayerisches Staatsministerium für Wirtschaft, Verkehr und Technologie.

Greiff, S. and D. Schmiedl (2002), *Patentatlas Deutschland 2002, Dynamik und -Strukturen der Erfindungstätigkeit*, München: Deutsches Patent- und Markenamt.

Landeshaupstadt München (2002), 'München, Stadt des Wissens', Department for Labour and Economic Development, City of Munich.

Landeshauptstadt München (2003), 'Munich. The Business Location Department for Labour and Economic Development', City of Munich.

Schreier, F. and M. Gebhart (2003), *Green Card, IT-Krise und Arbeitslosigkeit*, Institut für Arbeitsmarkt- und Berufsforschung, Bundesagentur (vorher Bundesanstalt) für Arbeit.

Chapter 8

Münster

8.1 Introduction

The city of Münster has approximately 280,000 inhabitants and is located in the north of the German region of North-Rhine Westphalia (NRW) near the border with The Netherlands (City of Münster, 2003d). Münster is one of the five seats of government in NRW. The Münster region – Münsterland – houses around 1.5 million inhabitants (Münster Marketing, 2003). The city lies close to the industrial Ruhr area, where bigger cities such as Dortmund (50 km from Münster) and Düsseldorf are located (see Figure 8.1). Münster is a town that is

Figure 8.1 Münster's position in the region

Source: Wirtschaftsförderung Münster (2003a).

characterised by its big university and high number of students. Furthermore, Münster is known as a city of civil servants and law institutions.

As a result of the high number of students, the Münster population is relatively young: about 38 per cent of all inhabitants are younger than 30 years (City of Münster, 2004a). The city's per capita GDP is high: in 2001 Münster ranked 16th (€38,149) among the top European cities (Office of the UK Deputy Prime Minister, 2004). In 2001 the Münster purchasing power ratio was 111.8 compared to the German level (Wirtschaftsförderung, 2003b). The 2003 level of unemployment was about 8 per cent; this is a rather good position when compared to some major German cities (see Figure 8.2).

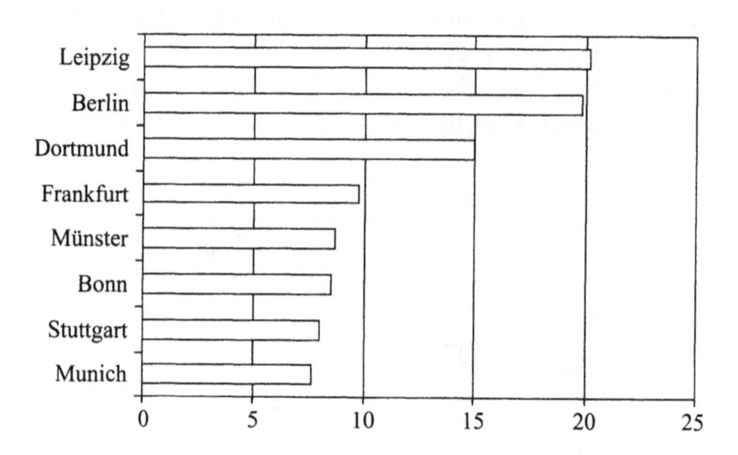

Figure 8.2 Unemployment rates in large German cities, 2003 (%)

Source: Wirtschaftswoche and IW Consult (2004).

8.2 Knowledge Strategy

The city of Münster does not have an explicit and comprehensive knowledge economy strategy, but several policies are related to the promotion of a knowledge economy.

'Multitalent Münster – Wege der Stadt der Wissenschaft 2005' ('City of Science' (City of Münster, 2003a)) is a policy that aims to increase the interaction between, on the one hand, the university, polytechnics and other research institutes and, on the other hand, the companies in the region, the various sorts of art institutes and Münster inhabitants. According to various interviewees such strengthening of interaction and linkages is needed because the Westfälische Wilhelms-Universität (WWU) is currently seen as an 'island' by some of the citizens and several firms. The City of Science 2005 will have a turnover of €3.8m; the municipality would contribute €700,000 of the total amount, and about €2m from other organisations has so far been committed. Some 65 projects were identified within the policy: 37 projects were to be executed whether Münster won the City of Science competition (set up by the 'Bundesforschungsministerium', the national ministry for research) or not, while 28 projects would only be executed if Münster won the national contest.

The projects include: Science Biking Tours (experts will guide citizens and visitors around scientific institutions in Münster by bike), Wissenschaftbox (a sea container will be placed in various parts of town to exhibit scientific themes through artists and visual craft makers), Berufe mit Megahertz (female school pupils will be informed via a website about science and technological studies and jobs in the Münster region), Münster gründet! (economic and scientific partners support small and medium-sized enterprises (SMEs) in the region by giving information, advice and by extending networks via fairs and conferences etc) and Stammbaum der Wissenschaft (scientists together with firms give various presentations for citizens about the ways scientific disciplines have developed in the past). About 75 per cent of all projects are aimed at the entire Münster population, 10 per cent is aimed specifically at experts in certain fields and the remaining projects are targeted at specific groups like the elderly, women and female pupils (City of Münster, 2003a).

In the City of Science competition for 2005 Münster scored well, with a place among the top ten, getting acknowledgements for the contributed concept (for press release see http://www.presse-service.de/static/56/567105.html). Bremen was selected 'City of Science 2005' by the Stifterverband. Münster will again participate in the competition in 2006, also because of the successful and fruitful collaboration with local players that the 2005 competition yielded.

The municipality of Münster also stimulates the knowledge economy by financially supporting the Centre for Nanotechnology CeNTech, Technologiepark Münster, Technology Initiative Münster (TIM) and Nano2Life. CeNTech promotes the formation of start-ups and firms expanding in nanotechnology; the Technologiepark offers space and services to firms in information and communication technology (ICT) and life sciences; TIM supports start-ups in all fields of high-tech activities; and Nano2Life bioanalytik-muenster, together with the Atomic Energy Commisssion (CEA) in Grenoble, coordinate a EC Network of Excellence, with the goal of strengthening the European scientific and economic leadership in nanobiotechnology (for these initiatives see section on applying knowledge).

Münster is not very well-known abroad. The city recently tried to become the European Cultural Capital of 2010 (City of Münster, 2004a). This would have contributed to enhancing the international profile. The city hoped that it would help to attract multinationals and high-grade workers. In the end, Münster was not selected to represent NRW as a candidate city. Most of the project proposals will be realised in any case, again because of the close and fruitful interaction between the actors involved. The projects also add to the integrated city marketing process of Münster (see below).

Recently Münster also presented a city marketing plan. Among others the plan aims to enhance the city's visibility for firms from other places. Goals of the marketing plan include: 1) to become a leading European city in education, science and research and development; and 2) to strengthen Münster as a place for public and private services (Münsterischer Anzeiger, 2004).

The policy process in Münster is organised quite swiftly: the municipality involves companies and other actors (e.g., the insurance firm Westfälische Provinzial Versicherung AG and the organisation of industrial companies Industrie in Münster) in the strategy and implementation phases by organising interactive sessions and keeping in close contact with all the actors involved.

In the Federal Republic of Germany, the regions (Länder) are mainly responsible for R&D and education policy. The regions have a large influence on some sectors like media and health/biotechnology. The region of North-Rhine Westphalia is implementing the Technology and Innovation Programme (TIP). This is the follow-up to the Technologieprogramm Wirtschaft (TPW); in TPW most life sciences money went to the Rhineland (70 per cent) and Ruhr area (17 per cent); Westphalia (which includes Münster) only received 13 per cent. The new TIP programme stimulates the development of future technologies and promotes technology transfer in the economy. TIP aims to support SMEs, among others: this provides good opportunities for Münster, since most firms in this region are SMEs. The 'Modellversuch Inkubator', for example, aims to stimulate top researchers to start up firms in the field of biotechnology in particular. Since Münster has few old industrial activities, the city does not benefit from the extensive financial regional help to old industrial cities ('Objective 2' regions) in the Ruhr region (Forschungszentrum Jülich GmbH, 2003a and b). The region of North-Rhine Westphalia (NRW) runs a bureau of the Life-Science-Agency, formerly known as Bio-Gen-

Tec initiative. According to NRW nanotechnology, analytics and molecular medicine are the key Münster competencies. In 1998, a regional bureau was set up for the Münsterland in the city of Münster. Since then 14 start-ups have been founded; four projects in 2001 totalled €10m in turnover. An important partner of the Life-Science Agency is CeNTech (see Section 8.4.3 on applying knowledge). The Life-Science Agency also contributes, for example, to the annual NanoBioTec conference in Münster, organised by CeNTech.

8.3 Knowledge Foundations

In this section, the knowledge foundations – as discussed in the theoretical chapter – will be assessed for Münster. They are the knowledge base, the economic base, the quality of life, the accessibility, the urban diversity, the urban scale, and the social equity in the city.

8.3.1 Knowledge Base

Münster has a strong knowledge base: it houses the highest number of students per inhabitant in Germany (almost 200 students per 1,000 inhabitants) (Cash, 2004). The city has the fourth biggest university in Germany regarding student numbers (see Table 8.1).[1] The biggest departments in the Westfälische Wilhelms-Universität Münster are philology (6,798 students), economics (5,498), law (5,425), education and social science (5,314), history/philosophy (4,152) and medicine (3,298).[2] Other areas are: theology, psychology, mathematics and informatics, physics, chemistry and pharmacy, geo-science and biology (City of Münster, 2003b). The Westfälische Wilhelms-Universität (WWU) Münster has a good reputation both in science and in education. Some 580 professors and over 3,000 other academic staff work at WWU. In the fields of biology and medicine, the university excels in neuro- and molecular biology.

Furthermore, Münster is home to the Muenster University of Applied Sciences where students study economics (1,815), welfare (1,230), structural engineering (1,034) and architecture (858), and to other polytechnics (see Table 8.2). The Muenster University of Applied Sciences (MUAS – Fachhochschule Münster) is one of the largest universities of applied sciences in Germany regarding student numbers. Some 240 professors work at the MUAS. The main focus of research projects at MUAS involves: chemical environmental technology, fuzzy technology in technical engineering, quality management,

Table 8.1 Students at German universities

University	Number of students 2002/03
Cologne, University	61,101
Hagen, 'Fernuniversität'	45,661
Munich, University	44,128
Münster, University	42,828
Berlin, 'Freie Universität'	41,691
Hamburg, University	39,896
Frankfurt am Main, University	39,855
Berlin, Humboldt University	36,493
Berlin, Technical University	28,673

Source: City of Münster (2003b).

applied materials science, applied laser engineering, micro systems engineering, multimedia, superconductor technology transport and medical laboratory technology. Affiliated institutes of the MUAS are: the Institute of Waste and Sewage Control, Institute of Textile Engineering and the Institute of Business Administration for SMEs (Fachhochschule Münster, 1999).

Table 8.2 Polytechnic students in Münster

Polytechnic (Fachhochschule – Fhs)	Number of students in Münster 2002/03
Fachhochschule Münster	6,721
Catholic Fhs North-Rhine Westphalia	695
National Fhs for Public Management: Finance	600
Fhs for Public Management	549
Art Academy	311
Fhs for Music	212
Philosophy-theology Hochschule	80

Source: City of Münster (2003b).

Münster accommodates a multitude of public and semi-public research institutes. For instance, the Centre for Molecular Biology of Inflammation (ZMBE) at the university and the Max Planck Institute for Molecular Biomedicine. The latter is a foundation financed by the national government (50 per cent) and the state of North-Rhine Westphalia (50 per cent) and was established in Münster in 2001. The institute currently employs about 90 people; this number will rise to 150–180 employees in 2006. The academic hospital also is a prominent part of the Münster knowledge base: more than 160 professors and 900 academic staff are employed there. Other knowledge base strengths of Münster are the Competence Centre for Nanoanalytics and the German-Chinese Centre for Nanoscience (CeNTech GmbH).

8.3.2 Economic Base

Münster is known as a city of civil servants and public management. Many government institutions are in Münster, because it was once the capital of Westphalia. The city is known as the 'Schreibtisch' (writing desk) of Westphalia. However, many government organisations have moved to Düsseldorf in recent years. Münster is also known as a 'Bildungsstadt' (education/knowledge city). Furthermore, Münster is highly oriented towards services (see Table 8.3): the percentage of services further increased in the period 1991–2001. The total GDP rose from €6.7m in 1991 to €8.7m in 2001. This is an increase of 31 per cent for Münster compared to a GDP growth for the state of NRW of 25.5 per cent. GDP per capita in Düsseldorf is significantly higher than in Münster, but the latter city shows a significantly larger GDP than the state of NRW average (Amt für Stadt- und Regionalentwicklung, Statistik, 2004). Table 8.4 illustrates the composition of the Münster economy. The percentage of services in Münster (84.1 per cent) is higher than the NRW average (71.2 per cent) (City of Münster, 2003d). The university (WWU) is the biggest employer in Münster with some 7,000 employees (City of Münster, 2003c).

Table 8.3 Economic development of Münster in the period 1991–2001

Economic activity	1991 absolute numbers	1991 percentages	2001 absolute numbers	2001 percentages
Agriculture, forestry and fishing	26.8m	0.4%	34.8m	0.4%
Industry	1366.8m	20.4%	1348.5m	15.5%
Services	5306.4m	79.2%	7316.7m	84.1%

Source: Amt für Stadt- und Regionalentwicklung, Statistik (2004).

Table 8.4 Structure of employees in Münster liable to contribute to social security in 2002

Private services	39.7%
Trading	13.8%
Processing industry	12.8%
Public bodies, social insurances	9.7%
Credit and insurance companies	8.3%
Social facilities	6.3%
Traffic, news services	3.5%
Construction industry	3.5%
Public utilities	1.4%
Agriculture and forestry	1.0%

Source: Wirtschaftsförderung Münster (2003a and b).

The city has few industrial companies – BASF Coatings (2,500 employees) is the biggest industrial employer: the company produces car paints for all major car manufacturers. Oevermann (2,000 employees) played an important role in the reconstruction of the city centre which was 90 per cent demolished by bombs inWorld War II. Winkhaus is a multinational producer of security systems (window and door technology and automation); most of its R&D is located in Münsterland (IiM, 2003). Large service firms in Münster are the credit and insurance companies Sparkasse, Provinzial and LVM (Landwirtschaftlicher Versicherungsverein). Münster has many SMEs; several excel in certain niche markets (e.g. TOF-SIMS and ION-TOF in biotechnology).

Münster ranks fifth out of 40 large German cities with regard to the ratio of newly-established companies per firms that leave/close down (1:24). The city ranks low (34th position) as regards the number of start-ups per 100 already-established companies. Regarding the number of insolvent firms per 100 already established companies (4), Münster ranks 17th. The city has quite a high tax yield per capita (€535): it ranks 11 out of 40 cities. The high tax yield per capita enables the city to invest more. The company tax rate in Münster is relatively low – the city ranks 25th. The average increase in employment in Münster is very high (2.42 per cent): the city is second behind Regensburg. Overall, Münster is ranked fourth in the list of most attractive German cities concerning real estate (Cash, 2004).

Münster has a GDP per capita of €38,149 compared to €61,360 in Munich, €26,548 in Dortmund and €25,209 as the German average (Office of the UK

Deputy Prime Minister, 2004). The level of unemployment in Münster (about 8 per cent) was a little higher than in Munich, but significantly lower than in Dortmund (15 per cent) (Wirtschaftswoche and IW Consult, 2004).

8.3.3 Quality of Life

Münster is a relatively small city in a European perspective. It is considered to be quite a green city (various parks, many trees and shrubs, etc): Münster ranks 13th out of 40 German cities in this respect (Cash, 2004). Münster is also considered a rather safe city: in the period 1990–2002 the number of reported crimes (excluding traffic offences) grew only moderately from 25,719 to 28,647 and the percentage of solved crimes rose from 39.9 per cent to 45.4 per cent (City of Münster, 2003d).

The city has what seems to be a very old city centre: this is a deceptive appearance, since the centre was for the most part destroyed by bombs in World War II. About 90 per cent of the buildings were destroyed, but Münster was reconstructed after the war ended (Münster Marketing, 2003). The seemingly old city centre is supplemented by various modern architecture buildings (e.g., the city library). Münster is known as the bicycle city of Germany: for many years the city has been rated the best German city for cyclists. Münster also received this title in 2004: the city has many separate cycle roads, good signposts for cyclists and a low number of cycling accidents (City of Münster, 2004b).

In the field of culture Münster has several festivals (e.g., the wine festival, the annual Christmas markets) and 26 museums (e.g., the Pablo Picasso museum, the Science and Planetarium Museum and the Mühlenhof-Freilicht Museum). High international attention is drawn to Münster by the 'Skulptur Münster' exhibition, which has been held every ten years since 1977, attracting more visitors than, for instance, the famous Documenta in Kassel. In 2002 the museums welcomed 628,855 visitors (City of Münster, 2003d). There are various cinemas in Münster (18 screens in total) and the city also has several theatres. The city theatre experienced sharply decrease in the number of visitors in the period 1994–2001, from 220,044 to 172,362 per year. Visitor numbers in other Münster theatres also declined in this period, from 8,050 to 4,633 people. Furthermore, the city orchestra saw its number of visitors decline. The nightlife in Münster cannot compete with vibrant scenes like in Berlin, but the number of bars and restaurants (1,277) is quite high, in part because of the big student population. Münster has 84 hotels, guest houses, etc., with 7,140 beds; this number has increased gradually since the 1990s.

The city counted more than 1,054,326 overnight stays in the year 2002. The majority of the guests are German (984,642); most foreign guests come from The Netherlands, the United Kingdom, the United States and Denmark. Each year the city hosts about 200 conferences, including more than 30 with over 300 participants. The Münsterland Halle annually attracts around 600,000 visitors to all kinds of exhibitions (City of Münster, 2003d).

In part of the harbour complex a new city district called the 'Kreativkai' (Creative Quay) is being developed. Since the 1970s port activities in the city harbour of Münster (1 km west of the inner city) have been declining. Therefore, several port buildings and pieces of land have fallen out of use. Since the 1990s the area is being restructured for both harbour activities and activities in other fields. The latter include the Creative Quay in city harbour 1: the area involves a small-scale mixture of cultural activities, gastronomy and bars and innovative service companies (Modellstadt Münster, 2004).

In recent years the supply of housing in Münster has increased in accordance with a gradually increased population. The amount of square metres per Münster citizen has increased from 34 in 1990 to 38 in 2000. Housing construction is expected to decrease, but this will not have much effect on the housing market in the short term, since there is a housing surplus. In general, the Münster housing market will transform from a supply-led to a demand-oriented market. This implies that the attractiveness of the neighbourhood and the quality of the housing are becoming more important (Amt für Wohnungswesen, 2002).

Taking into account the aspects discussed above, the quality of life in Münster is rather good for a city of this size. Münster does not offer the extent of cultural facilities that metropolises offer, but its cultural supply is more than sufficient. In addition, the city does not have some of the negative quality of life aspects that bigger cities tend to have (traffic congestion, high crime rates, expensive housing, etc).

8.3.4 Accessibility

Münster's internal accessibility is good. The city has an extensive bus network. The excellent cycling opportunities in the city have already been mentioned in the section on quality of life. Cyclists make up 35 per cent of all Münster traffic (City of Münster, 2004b). Pedestrians make up over 13 per cent of all journeys. Almost 11 per cent of the modal split consists of public transport and little over 40 per cent of travelling is done by motorised vehicles. Figure 8.3 shows the modal split in 2001 compared to 1994. The shares of public transport, motorised traffic and cycling have increased significantly at the expense of walking. Overall mobility of the Münster citizens has increased by 7.4 per cent between 1994 and 2001 (City of Münster, 2002). Still, Münster does not have a serious problem of traffic congestion. Compared to some other German cities (Bonn, Heidelberg, Gütersloh and Essen), Münster has a low percentage of motorised traffic and high percentage of cyclists (City of Münster, 2002).

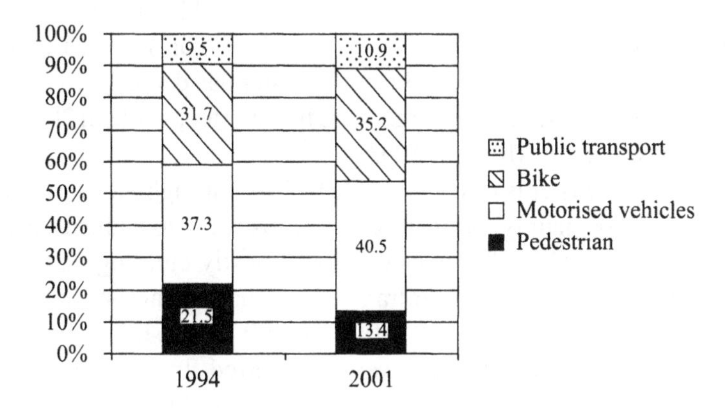

Figure 8.3 Modal split in Münster

Source: City of Münster (2002).

Münster's external accessibility is reasonably good. The Münster Osnabrück International Airport (FMO) in Greven is some 25 minutes by car from Münster city. FMO has direct flights to among others Berlin (three times a day), Frankfurt (four times a day), London (once a day), Mallorca, Munich (six times a day), Rome (once a day), Stuttgart (three times a day) and Vienna (once a day). The bigger Düsseldorf airport is about 100 km from Münster. Annually, FMO has about 50,000 lift-offs and touchdowns: this number has remained stable since 1990. The number of passengers has increased sharply: in 1990 around 307,000 passengers used FMO, while this number was 1,500,000 in the year 2002. Since 1999 the number of passengers is fairly stable. In the period 1990–2000 air cargo increased from 935 to 13,987 tonnes, but this subsequently dropped to 10,443 tonnes in 2002 (City of Münster, 2003d). Nevertheless, for FMO airport growth seems to be desirable to improve connections with other big cities and thereby attract more companies to Münster. It is planned to extend the runways, starting in 2005, to manage direct intercontinental flights from and to Münster Osnabrück International Airport. To attract firms that need very good international accessibility, the cities of Münster and Greven and the Steinfurt district are developing a business area called AirportPark next to the FMO airport. In the long term 200 ha. is available for the business park construction; in the short term 40 ha. will be used (Wirtschaftsförderung Münster, 2004).

Münster has good accessibility by rail: ICE, IC and EC trains stop at the Münster central station. Several large German and other European cities can be reached relatively quickly: e.g., Amsterdam (3h20m), Berlin (3h30m), Brussels (4h30m), Hamburg (2h20m), Munich (6h20m) and Paris (6h00m). The number of cargo trains to and from Münster has decreased sharply since 1990: in that year 35 tonnes of cargo were transported, while in 2002 this was only 2 tonnes (City of Münster, 2003d).

The city of Münster is also accessible by boat through the Dortmund-Ems canal. Some 250 (mainly German) cargo ships came to Münster harbour in 2002. In that year 247,000 tonnes of cargo (mostly building material) were transferred in the harbour. Cargo transfer in Münster harbour has decreased substantially over the years: in 1990 some 660 cargo ships still visited the harbour and 518 tonnes of cargo were transferred. In 2002, almost 13,000 loaded cargo ships passed Münster along the Dortmund-Ems canal: this number has also decreased significantly since 1990: at that time nearly 21,000 loaded cargo ships passed by (City of Münster, 2003d). At the time of writing the canal is being upgraded for larger vessels (Wirtschaftsförderung Münster, 2003b).

In the field of ICT infrastructure Münster has an advantage: the presence of the worldwide leading international Internet exchange NDIX. NDIX B.V. is a public company that was founded in 2001. The University of Twente and the regional development company OOM N.V. are the Dutch partners involved, while Stadtwerke Münster GmbH has been involved as a German partner since 2003, with a 50 per cent stake in initial capital. NDIX has connection points in Enschede (The Netherlands) and Münster (Germany). NDIX leads to lower Internet traffic costs for the customers compared to the former situation when large Internet traffic customers in Münster were connected to the Internet exchange in Frankfurt/M. Companies that want to benefit from the proximity of NDIX can buy premises and offices in the business park Loddenheide, where the German NDIX connection point is situated. The German-Dutch cooperation project NDIX is financially supported by the EU within the framework of the Interreg-IIIA programme from Structure Fund for Regional Development money (City of Münster, 2003c).

8.3.5 Urban Diversity

The state of North-Rhine Westphalia (NRW) has been a region of high inflow of foreigners for many decades: immigrants from the former DDR, labour migrants, etc. In 1976 NRW counted 1.2 million foreign inhabitants (6.9 per cent). Today, almost 2 million out of 18 million NRW inhabitants are of foreign origin: this is about 11 per cent. This figure is higher than the German average of 8.9 per cent (7.3 million inhabitants of foreign origin in a total population of 82.5 German million people) (Statistisches Bundesamt, 2004). Inhabitants of Turkish origin make up for most of the population with a foreign origin: immigrants from Yugoslavia and Italy also make up a significant part of the foreign-origin NRW population (Ministeriums für Frauen, Jugend, Familie und Gesundheit NRW, 2000).

Münster has almost 24,640 inhabitants of foreign origin: this is 9.2 per cent of the population. Over the years this number has steadily grown: in 1980 it was only 4.4 per cent. Most foreign-origin inhabitants come from Yugoslavia (3,016), Turkey (2,298), Portugal (1,895) and Poland (1,350). The presence of a significant number of foreign-origin inhabitants can contribute positively in several ways: it can add to the cultural climate in a region, make intercultural 'new combinations' possible and help solve labour market frictions. Currently, however, among inhabitants of foreign origin in NRW unemployment is significantly higher than for native German people: 16 per cent and 7.6 per cent respectively (Landesregierung Nordrhein-Westfalen, 2004; City of Münster, 2003d).

In addition, the Westfälische Wilhelms-Universität (WWU) adds to the urban diversity of Münster. More than 3,800 students at WWU are from abroad (almost 9 per cent of the WWU student population): most of them come from Turkey, Eastern Europe and Southeast Asia. There are also some 200 foreign guest lecturers working at WWU (Westfälische Wilhelms-Universität, website 2004; City of Münster, 2003d).

Münster's urban economy diversity is moderately positive. The region has few large industrial firms; most companies are in the service industry and the largest employer in Münster is the WWU university. The city's image is rather conservative and cautious: Münster has not experienced any major economic crises in previous decades and many inhabitants like the situation the way it is. According to an interviewee, Münster would be truly boring if it did not have the university whose students provide the city with some sense of dynamism and night life.

8.3.6 Urban Scale

Compared to other German and European cities, Münster is quite small (the number of inhabitants in the city is 280,000 and in the region, 1.5 million). The number of Münster inhabitants has been fairly stable since the 1980s (City of Münster, 2003d). The size of the city region is 302 km^2 and the Münster workforce numbers about 150,000 people (Wirtschaftsförderung Münster, 2003a; City of Münster, 2004c). Münster houses 17,594 companies; this number has grown considerably since 1990 when only 13,659 firms were present (City of Münster, 2003d). The city attracts many students from the region, but with regard to workers Münster witnesses a large number of outward commuters, mainly to Dortmund.

The university and polytechnics in Münster already participate in multiple networks with universities, polytechnics and other institutes in other German and European cities. For instance, universities in Münster, Osnabrück, Dortmund, Twente, Leuwen and Nijmegen are cooperating in research and technology transfer via the Network of Euregional Universities (NEU). The Fachhochschule Münster cooperates in, among others, the Euregio project Neuro Fuzzy Technology (making unclear situations mathematically understandable for computers): Fachhochschule Münster, Twente University, 16 companies and two municipal economic development agencies working together are implementing fuzzy technology in concrete products. This Euregio project is 50 per cent financed by the EU, 30 per cent by The Netherlands, the states of NRW and 20 per cent by companies in the project (Fachhochschule Münster, 2004).

Another international cooperative effort is found in the field of bio analytics: bioanalytik-muenster works together with the Atomic Energy Commission in Grenoble (France) to coordinate the EU Network of Excellence in nanotechnology: Nano2Life (bioanalytik-muenster, 2004).

8.3.7 Social Equity

The level of unemployment was 8.1 per cent, or 10,112 people, in Münster in 2002. While the percentage of inhabitants of foreign origin in Münster is only 9.2 per cent, 17.4 per cent of the unemployed consisted of people of foreign origin. This implies that unemployment in this group is higher than the Münster average. A similar trend can be discerned on the NRW level. Twenty-five per cent of Münster unemployment consists of the long-term unemployed (2,530 people). Münster's level of unemployment is structurally a little lower than in

NRW and the NRW unemployment rate is structurally lower than the German average (Wirtschaftsförderung Münster, 2003a; Landesregierung Nordrhein-Westfalen, 2004). In 2002, almost 15,000 people in Münster (including 2,500 asylum seekers) received social security benefits: this is 5.4 per cent of the total population. Münster has 662 homeless people. While the percentage of homeless people declined in NRW between 1996 and 2002 from 2.9 to 1.2 per cent, the figure increased slightly in Münster from 2.3 to 2.5 per cent (City of Münster, 2003d).

8.4 Knowledge Activities

In the last section, the 'foundations' of Münster's knowledge economy have been discussed. Now, we turn to knowledge activities and ask how well the city (and the actors in it) manage to: 1) attract and retain talented people; 2) create new knowledge; 3) apply new knowledge and make new combinations; and 4) develop new growth clusters.

8.4.1 Attracting and Retaining Knowledge Workers

How well does the city manage to attract and retain knowledge workers? Münster is a real 'student city': it has the highest student ratio in Germany: 198.5 students per 1,000 inhabitants (Cash, 2004). Most students come from the region (about 60 per cent from Münsterland); but many come from German locations between Münster and the North Sea. Although the overall number of WWU students decreased in the period 1997–2002 (from 45,647 to 43,800), the number of foreign students increased from 2,942 to 3,855. Most of them come from Turkey, Eastern Europe and Southeast Asia. The Fachhochschule Münster also has a considerable number of foreign students: 782, or 9 per cent (City of Münster, 2003d). In addition, some 200 foreign guest lecturers work at WWU (Westfälische Wilhelms-Universität, website 2004). Another trend is the increasing number of students coming from other places in Germany.

In Münster there is an active organisation for foreign students called 'die Brücke' (the Bridge). Among other things, this organisation arranges German language courses, social activities such as excursions and parties and helps students find accommodation and fill out forms. Die Brücke also transmits radio broadcasts, runs its own bar/restaurant where international meals can be bought and takes care of the Alumni International network. The last-mentioned was started in 1995 with financial support by DAAD (the German Academic

Exchange Service). Alumni International maintains contacts with former foreign students that want to stay in contact with Münster. Die Brücke is in close contact with, among others, the Fachhochschule Münster, the university, the Erasmus/Socrates bureau, AIESEC, the municipality and the Chamber of Commerce.

WWU is active in attracting foreign students – this is necessary in the light of the upcoming introduction in 2006 of the international bachelor-master system. WWU advertises in magazines and representatives attend international fairs to promote WWU abroad. Some courses at WWU are taught in English. The theological education at WWU is quite popular with students from South America.

Because of the presence of WWU and the polytechnics in Münster, the city is quite successful in attracting students. However, retaining them is much more difficult because the number of jobs in Münster is fairly limited: each year approximately 5,000 WWU students graduate, while the total Münster labour force only amounts to some 150,000 workers.

In contrast to the well-established Alumni International network, the alumni efforts by the university regarding German Münster graduates are not yet at an advanced stage. Student housing in Münster is not very abundant and therefore rent levels are rather high. The tight student housing market has improved a little in recent years because Studentenwerk is actively building new student apartments.

Attracting and retaining knowledge workers in the Münster region is stimulated by the presence of the university, several polytechnics and various scientific research institutes, e.g., the Max Planck Institute for Vascular Biology, the Centre for Molecular Biology of Inflammation (ZMBE) and the Centre for Nanotechnology within its networks such as the Competence Centre Nanoanalytics in Münster. For instance, the good reputation of WWU in the field of medical science and biology combined with the presence of the Max Planck Institute attracts excellent PhD students to Münster.

Whether all these knowledge workers are thoroughly rooted in the region is doubtful: at WWU so-called DiMiDo professors are a well-known phenomenon. These professors work three days a week in Münster (*Dienstag*, *Mittwoch* and *Donnerstag* –Tuesday, Wednesday and Thursday) and live and work for the rest of the week outside the Münster region (for example, in Munich). One cause of this phenomenon is the difficulty other members of their families have in finding appropriate jobs in Münster; this has to do with Münster's small urban scale. The municipality is not actively trying to solve this specific job bottleneck.

The positive side of Münster is its attractiveness for families: the city is quite safe and peaceful and rural surroundings are very nearby. Moreover, the city does not experience serious traffic congestion.

8.4.2 Creating Knowledge

The creation of new knowledge is quite strong in Münster. The university (WWU), polytechnics and scientific institutes in particular contribute to this. Spearhead areas are biomedicine, nanotechnology, ICT, laser technology and life sciences. In 2001, the university's research activities were supported with €56.4m from external parties. The amount of third-party funds at WWU has increased in recent years, especially in the faculties of Science, Information Technology and Medicine (Competence Centre Nanoanalytics in Münster, 2002). The state of NRW strongly influences the university policy. At the time of writing, NRW considers the University of Dortmund's activities too broad; it will try to have the university focus on a smaller number of fields.

With 594 patent registrations in 2000, the city ranks 19th in Germany; the number of patents per inhabitant is 0.002. Cities in NRW that have more patent registrations are Düsseldorf (1,901), Cologne (1,090), Aachen (697), Bielefeld (691), Bochum/Hagen (689) and Duisburg/Essen (654). While the number of patents in NRW increased by 24 per cent from 6,418 to 7,965 in the period 1995–2000, Münster showed an increase of 49 per cent from 400 to 594 (Greif and Schmiedl, 2002). Figure 8.4 shows the development of the number of patents in Münster. It is important to note that the suggested time scale is not optimal: if patent registrations from 2003 had been available, important changes because of the introduction of the 'Arbeitnehmer-Erfindungsgesetz' (the employee invention law) would erase distortions from the past: the number of patent registrations from universities is much higher in reality. In line with the overall German pattern, most patents in Münster are requested by private firms. The number of patents requested by universities and research institutes is low because, in most cases, the researchers apply for the patents on an individual basis: this is possible because German universities are lenient towards such activities. German universities have acknowledged that relatively few ideas are being patented. The national government has recognised this and therefore changed the law concerning inventions by employees. Polytechnics and universities are also trying to improve the incentives to register patents. Since 2003 the Patent Offensive Westphalia and Ruhr area (POWeR) has been active: the universities of Münster, Dortmund, Paderborn and Bielefeld try to stimulate a strong patent consciousness with university employees by organising information and training events. Also, AFO (Arbeitsstelle Forschungstransfer at WWU) is trying to stimulate a patenting culture through idea-mining projects in which good patentable ideas are being sought (Arbeitsstelle Forschungstransfer WWU, website 2004). CeNTech is in charge of analysing, filing and prosecution of ideas that emerge from the research field of nanotechnology.

The number of patent registrations per 100 R&D workers in Münster firms (22.9) is high compared to the German average (9.9). Münster is in second place (behind the neighbouring city of Osnabrück) regarding patents in the field of agriculture in Germany.

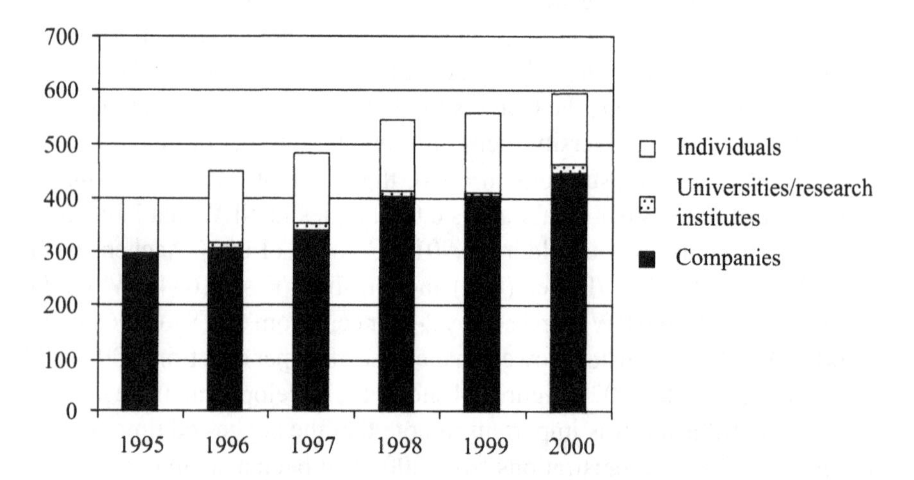

Figure 8.4 Patent registrations in Münster, 1995–2000

Source: Greif and Schmiedl (2002): data from pp. 144–5.

The patent data should be interpreted with caution. First, the data refer to the location where the patent was registered; this is not necessarily the location where the invention occurred. Furthermore, innovations in services (which make up 84 per cent of the Münster economy) are normally not patented, although they make a major contribution to productivity and quality improvements.

The state of NRW plays an important role in the promotion of knowledge creation, since university and science policies are mostly determined on that level. Münster has received relatively little money for science – Dortmund received much more. This might prove to be an advantage for WWU since it will be easier to make decisions independently from the NRW government. The Muenster University of Applied Sciences (third party research budget in 2003: €7.2m) obtains more funding from NRW while the polytechnic is successful in attracting financial means from other sources. Furthermore, NRW funds the Max Planck Institute for about 50 per cent; the other 50 per cent comes from national government.

8.4.3 Applying Knowledge/Making New Combinations

The city of Münster houses a rather extensive number of institutions in the field of applying knowledge. These organisations are mostly set up and stimulated by the Münster municipality, the state of NRW and the Westfälische Wilhelms-Universität (WWU). The application of knowledge by private companies is quite limited because Münster has little industry and few large companies. Therefore ideas have to be transferred into SMEs and/or companies from outside which have to be attracted to the Münster region. Several firms that have been set-up are spin-offs from the university or research institutes: for example, Cilian is a biotech spin-off, Denovo Biolabels (founded in 2001) and XanTech Bioanalytics (founded in 1997) are spin-offs from the Institute for Chemical and Biochemical Sensor Research (ICB) and ION-TOF (founded in 1989, 50 employees) is a university spin-off that produces ion mass spectrometers for a global market (Bio-Gen-Tec-NRW, 2002; Gesellschaft für Bioanalytik Münster e.V. 2002). The application of knowledge also takes place in those hospitals that cooperate with the Max Planck Institute and the WWU. Münster's initiatives in the field of applying knowledge include raising entrepreneurial awareness, promoting innovative networks, improving access to markets, facilitating patenting/licensing and offering business space. Table 8.5 shows the core activities of the various initiatives. Figure 8.5 shows a map of several knowledge institutions in Münster: the organisations are geographically quite near to each other. Next, the institutions are discussed.

CeNTech The main aim of incubator CeNTech GmbH is to promote the setup of start-up firms originating from university research and to support the expansion of companies in the nanotechnology sector. CeNTech is located in the western part of Münster next to the Technologiehof, Technologiepark and in the vicinity of the university institutes. CeNTech provides entrepreneurs with technical and business know-how and business space. Since 2003 about 2,400 m^2 of business and research space (a cleanroom and laboratories with vibration-free foundations) are available to researchers in nanoanalytics

Table 8.5 Overview of Münster initiatives in applying knowledge

Initiative	Main activities
CeNTech	Nanotech incubator for start-up firms and promoting knowledge exchange and cooperation
Technologiepark Münster	High-quality business space for young firms (up to 6 years) in life sciences and ICT
Technologiehof Münster	ICT and life sciences incubator for start-ups and other young firms
Technology Initiative Münster	Public-private cooperation to profile Münster as a technology location, increase application of knowledge and stimulate employment in biotechnology
Bioanalytic muenster	Regional non-profit organisation to enhance communication to promote teamwork in nanobioanalytics research
AFO at WWU	Centre for research transfer at the university aimed at transferring scientific knowledge to private companies
GDF (society of benefactors at the polytechnic)	Stimulates students who want to found start-up firms through financial support
Polytechnic transfer agency	Strengthen cooperation between polytechnic and businesses
Wissensregion	German-Dutch cooperation between municipalities, universities, polytechnics, etc. to enhance cooperation between knowledge institutes and private firms
Industrie in Münster	Organisation of Münster's industrial companies that among others tries to enhance their image and visibility

and nanobiotechnology. The researchers mostly come from WWU faculties (chemistry, physics, biology and medicine). Research topics include optimisation of scanning probe microscopy, studying new materials and nanoscale structures, analysing biophysical effects in cellular processes and utilising biological and biochemical processes in nanotechnology. At the time of writing around 60 people are working at CeNTech; the professors come mainly from the WWU and make up their own working teams. Three start-up firms are operating in CeNTech; there is room for about five other start-ups. CeNTech looks for firms that need to be close to the university researchers. Twelve research groups are active in CeNTech; ten are led by university

1 Technologiepark
2 Leonardo Campus
3 Technologiehof
4 ICB
5 CeNTech
6 University of Applied Sciences
7 Centre for Dermatology and Centre for Molecular Biology of Inflammation (ZMBE)
8 Institute of Physics
9 Institute of Microbiology
10 Institute of Pharmaceutical Technology
11 Institute of Pharmaceutical Chemistry
12 Institute of Botany
13 Administration of the University (WWU)
14 Institute of Physical Chemistry
15 Interdisciplinary Centre of Clinical Research (IZKF)

Figure 8.5 Knowledge institutions in Münster

Source: Technologiepark Münster GmbH (2001; 2003).

professors and two by former employees of the Max Planck Institute and Caesar.

CeNTech supports start-ups by advising on financing and licensing trajectories, giving education and training, stimulating cooperation with industry and research institutes and organising conferences and workshops. CeNTech also promotes knowledge exchange and cooperation, for example, through the EU 6th Framework Programme Nano2Life, the Competence Centre Nanoanalytics and the German-Chinese Centre for Nanotechnology. CeNTech has some Chinese employees; these workers learn how to build and run a centre like CeNTech so as to export this know-how at a later stage. It is hard to attract experts: the salaries in Münster are lower than in the US and UK and the city has few facilities aimed at an international community.

CeNTech is financed by the state of NRW (€4.8m), the city of Münster (€2.3m), the university (€664,000) and the Sparkasse Münster bank (€511,000). The owners are the Technologiepark Münster GmbH (79 per cent), Sparkasse Münster (20 per cent) and the University of Münster (1 per cent). Although the university owns only a small part of CeNTech, its involvement is substantial because it takes care of the CeNTech facilities. Apart from the money that is earned by renting business space, CeNTech makes some money by organising workshops and annually receives around €150,000 from the municipality. The CeNTech business plan aims to be profitable in 5–7 years. The city of Münster is willing to cover the losses of CeNTech for this time. Competitors of CeNTech are centres in Saarbrücken, Munich and Berlin in Germany and Grenoble, Lund, Cork, Barcelona, Vienna and Eindhoven in Europe (Technologiepark Münster GmbH, 2001; 2003).

Technologiepark Münster GmbH Technologiepark Münster GmbH started in 1985 and is owned by the city of Münster (88 per cent), the Sparkasse Münster (6 per cent) and the energy company Stadtwerke Münster GmbH (6 per cent). The Technologiepark Münster GmbH is active in three sectors: 1) managing the Technologiehof Münster as an incubator for technology-oriented companies; 2) marketing of the properties in the Technologiepark Münster, together with the Wirtschaftsförderung Münster GmbH; and 3) promotion of innovations and marketing of the technology location Münster (Technologiepark Münster GmbH, 2001; 2003).

Technologiehof Münster The Technologiehof is spatially a little removed from the university (some 500 m) and is a technology centre, focusing on defined technologies. The Technologiehof can be found just across the road from CeNTech (in the northwestern part of Münster) and covers an area of 10,000 m^2. It offers offices, laboratories and workrooms on a rental basis to start-ups and young companies in the fields of ICT and life sciences. The Technologiehof also offers services like business plan support, finding financing partners and organising trade show presentations. Moreover, joint use of a reception office, conference rooms and other infrastructure facilities help to minimise the costs of companies located there. The centre also includes a transfer office of the university and several industry initiatives. Over the years, more than 200 firms have started at the Technologiehof. The centre houses around 40 firms (Technologiepark Münster GmbH, 2003): one-third in life sciences (mainly nanobiotechnology and biomedicine) and two-thirds in ICT (mainly data mining, financial sector ICT and geographic information systems). Companies are allowed to stay in the Technologiehof for 3–5 years. In general, the life science firms need a little more time (3–6 years) than the ICT firms. About 90 per cent of the Technologiehof firms come from the university. The Technologiehof looks for innovation in the university and subsequently helps start-ups by organising regular entrepreneurship workshops for students. Interaction between the Technologiehof and WWU is extensive: the university has a knowledge transfer department located in the Technologiehof.

The Technologiehof is one of 17 technology centres in NRW. Around 90 per cent of firms in the technology centres have survived up to the time of writing, compared to 50 per cent of the firms outside such centres. There is some competition between the technology centres: Dortmund gives financial support to firms, but Münster cannot do this. As a former industrial city Dortmund (an Objective-2 region) recieves funding to change its economic structure: Münster receives no such funds. Münster tries to attract firms by having good networks, contacts and huge range of experts. The city is a first-class location for nanobiotech firms, because it has an excellent physics institute and an also excellent medical department (Technologiepark Münster GmbH, 2001; 2003).

Technologiepark Münster The park encompasses 66,000 m^2 meant for firms in the fields of ICT and life sciences. The park is next to the Leonardo campus of the university, which houses the Institute for Information Systems, the Institute for ICT and Media Law, the Institute for the Establishment and Development of Companies and the Art College. The Department of Computer Science, WWU institutes of natural science, the WWU Centre for Information Processing, the medical faculty, the University of Applied Sciences Münster, the Technologiehof and the CeNTech can be found within a radius of 2 km.

About one-third of the 66,000 m^2 have been sold; the property has a medium-level rent. Only ICT and life science firms are allowed in the Technologiepark (Technologiepark Münster GmbH, 2001; 2003).

Technology Initiative Münster (TIM) and bioanalytik muenster TIM is a joint initiative originally started by the city of Münster, WWU, the Fachhochschule Münster, the Chamber of Commerce, the Handwerkskammer Münster (an association for craftsmanship), the Sparkasse bank, the Technologiepark Münster and the Institute for Chemo- and Bio sensors (ICB). Later, TIM was enlarged by the Bio-Gen-Tec-NRW and the Competence Centre Nanoanalytics. TIM aims to profile Münster more strongly as a technology location, to increase both the application of scientific knowledge and employment in biotechnology. TIM's results include the foundation of a regional Bio-Gen-Tec-NRW office in Münster, the international NanoBioTech Congress and CeNTech. Furthermore, TIM finances a WWU professorship in Company Establishment and Entrepreneurship – Management for SMEs. TIM also founded the association bioanalytik-muenster, a regional network consisting of members from science, economy and public institutions that aims to strengthen the local structures in the field of nanobioanalytics by promoting teamwork in research through enhancing communication (Technologiepark Münster, 2001; 2003). Bioanalytik-muenster is a non-profit organisation founded in 2000. The bioanalytik network contains about 30 firms employing some 1,350 people. Since early 2004, bioanalytik-muenster, together with the Atomic Energy Commission (CEA) in Grenoble, has coordinated the EU Network of Excellence Nano2Life (Bioanalytik-muenster, website 2004). Nano2Life includes 23 partners all over Europe (Barcelona, Lund, Saarbrücken, Twente, etc). Nano2Life aims to develop a European nanotech centre that can compete with the United States.

Centre for Research Transfer at WWU (AFO) At the university in Münster, the Arbeitsstelle Forschungstransfer (AFO – Centre for Research Transfer) functions as the main centre for contact and coordination between research and business. AFO has been active since 1984; it directs outside parties towards the right departments and employees within the university and advises university employees about potential business partners. It also provides free legal advice and used furniture to WWU start-ups. In addition, AFO organises and supports the presentation of scientific results at industrial fairs and arranges meetings (Competence Centre Nanoanalytics in Münster, 2002). In 1999, the state of NRW appointed AFO as a 'Gründerhochschule' (a founding academy), because it had helped so many start-up firms: by then over 100 firms had been created from the university. AFO cooperates with the Fachhochschule Münster to help start-up firms. Also, AFO informs university employees about European research and technology programmes. Through the Europa-Kompetenz-Netzwerk in NRW, the AFO director maintains contacts at both national and international level (Arbeitsstelle Forschungstransfer Westfälische Wilhelms-Universität, website). The university is an important partner for the Sparkasse bank, because the bank requires good mathematicians for their insurance activities: cooperation in this field could be intensified.

Society of benefactors of the Münster polytechnic (GDF) In 1977 the polytechnic (Fachhochschule) of Münster started 'Gesellschaft der Förderer der Fachhochschule Münster' (GDF), an initiative to stimulate and support start-up companies from the polytechnic and its graduates (Fachhochschule Münster, 2002). Today, GDF gives scholarships to students who want to start a firm: non-refundable amounts of almost €750 per month for a maximum period of one year. Students in the end phase of their study at the polytechnic can apply for the financial support. If they receive financial support, they are obliged to report about the status of their venture at the halfway point and write a short report after their project is finished. Afterwards they are obliged to undertake a learning assignment and actively participate in GDF and polytechnic founding activities. GDF is financially supported by the Münster Chamber of Commerce (IHK), the polytechnic and some sponsors from the region.

Polytechnic transfer agency The 'Transferstelle' (Transfer Agency) of the Fachhochschule Münster aims to strengthen cooperation between the polytechnic and private firms. The Transferstelle tries to enhance the visibility of the polytechnic in and outside the Münster region and tries to link technical and scientific expertise at the polytechnic to technical and economic problems in the 'real world'. The focus of the Transferstelle is SMEs, since, unlike larger companies, they mostly do not have their own extensive R&D departments. However, associations, municipal and other public organisations and individuals can also apply for expertise at the Transferstelle. The Transferstelle offers information about the R&D expertise areas of the polytechnic, stimulates ties between companies and scientists, gives advice about national and international research programmes, takes care of cooperation projects between the polytechnic and firms, presents polytechnic R&D discoveries at fairs and exhibitions, organises workshops at the polytechnic about practical problems and gives information to start-ups (Fachhochschule Münster, 2004).

Wissensregion The regions of Münster, Osnabrück and Enschede/Hengelo work together to strengthen the networks between research and education institutes and companies. In the fields of logistics, surface technology, healthcare services and ICT, the Knowledge Region ('Wissensregion') supports and enhances contacts. Partners involved are three polytechnics, three universities, six municipalities, CeNTech, Technologiehof and the Economic Development Agency of Osnabrück (www.wissensregion.de).

Industrie in Münster Several large industrial companies in the Münster region founded Industrie in Münster (IiM) to make their economic activities more visible and to create a universal focal point for Münster industrial companies. Members include Armacell, BASF, Westfälische Nachrichten (a media firm) and Winkhaus (www.ihk-nordwestfalen.de/industrie_in_muenster/). IiM firms cooperate with university institutes and polytechnics in Münster in several fields, such as training engineers (apprenticeships), solving specific company problems through university research and obtaining good personnel (graduates and professors). IiM cooperates with the university not only concerning technical disciplines but also in other fields of research and education. IiM does not have strong linkages with the Technologiepark yet, but it would like to strengthen such ties with regard to know-how and know-who.

Venture capital Münster has no large venture capital funds. Three years ago Sparkasse set up a small venture capital fund. In the field of life sciences the lack of local venture capital funds is not a real problem, since such funds work globally. Up to the time of writing, the Münster market for capital funds has been small, since most companies were in the start-up stage. Now, Technologiepark Münster is trying to establish links with venture capital funds because an increasing number of Münster firms might have a demand for such funding.

New combinations Currently, the formation of new combinations is scarce in Münster: most sectors (life sciences, ICT, etc.) are organised mainly according to the traditional fields of science and economic activity. However, the 'Multitalent Münster – Wege der Stadt der Wissenschaft 2005' ('City of Science' (City of Münster, 2003a)) project has proved to be an impetus for starting up new combinations: people, companies and institutions from different fields have been brought together to develop new ideas and products. The basis of forming new combinations is the strengthening of relationships between different actors: virtually all the City of Science projects aim to do precisely this. An example is the strengthening of ties with polytechnic alumni by founding a comprehensive Münster polytechnics alumni network and enhancing linkages between companies and scientists in the Münster region in the 'KICK 8' initiative (City of Münster, 2003b).

8.4.4 Developing Growth Clusters

Münster has some economic fields that stand out. The main field is biotechnology (comprising of nanotechnology, nano-analytics and molecular medicine). Information and communication technology (including media activities) also is quite well represented in Münster.

Biotechnology In the first half of the 1990s the German government recognised that the country had fallen behind in the biotechnology industry compared to the United States and, in Europe, had been surpassed by the UK: only few patents were being registered and the number of biotech spin-off firms from academic research projects was very low. In 1996, the national government (the Federal Ministry of Education, Science, Research and Technology (BMBF)) launched the Bioregio competition: this initiative aimed to stimulate the biotech industry and make it the European leader. Three regions with strong biotech potential were to be awarded DM50m (about €25m) to build up an internationally competitive biotech cluster. On behalf of the NRW region, the Bio-Gen-Tec-NRW organisation submitted a proposal entitled 'BioRegio Rhineland'. This initiative, now called BioRiver, referred to the area between Aachen, Cologne and Düsseldorf. The other regions that won the Bioregio competition were Heidelberg/Mannheim and Munich (Bio-Gen-Tec-NRW, 2002; Zeller, 2001). In the years after Bio-Gen-Tec-NRW won the Bioregio competition, it also began to broaden its geographic scope outside the Aachen-Cologne-Düsseldorf area, including the Münster region.

The Bio-Gen-Tec-NRW organisation (now the Life-Science Agency) was set up in 1994 as a state initiative. Since the Bioregio competition, the organisation has remained important for biotechnology activities in the region. Bio-Gen-Tec-NRW functions as the contact agency for biotech entrepreneurs and start-ups. It offers, among other things, business services, consultancy advice and information about funding and takes care of public relations. The head office is located in Cologne, two branch offices are located in the East Ruhr region and Münster (opened in 1998). In the period 1997–2001 Bio-Gen-Tec-NRW provided business support for 89 start-ups; around €400m was invested, including over €150m from public funds (Bio-Gen-Tec-NRW, 2002).

In Münster, the main institutes for research and development are the university (WWU), together with its academic hospital, and the polytechnic (University of Applied Sciences – Fachhochschule Münster). In the field of life sciences the emphasis lies on nanoscale analytics, sensor technology and molecular medicine. In natural science and medicine, several collaborations take place. Other institutes that are important include the Centre for Molecular Biology of Inflammation (ZMBE) at the university, the Max Planck Institute for Vascular Biology, the Institute for Chemo- and Biosensors, the Competence Centre Nanoanalytics, the German-Chinese Centre for Nanoscience and the Centre for NanoTechnology (CeNTech). In Münster there are experts in the areas of scanning probe microscopy and surface mass spectrometry (TOF-SIMS – Time-Of-Flight Secondary Ion Mass Spectrometry). Coordinating institutions are TIM (Technology Initiative Münster) and bioanalytik muenster. ION-TOF GmbH (50 employees) is an important biotech company in Münster. Since the city has only a few large firms, the biotechnology cluster mainly consists of SMEs that originated from the university and polytechnic.

ICT Information and communication technology (including media) has quite a strong base in Münster, but the city does not house many large firms in this sector. Westdeutsche Rundfunk (WDR, a German broadcasting company) is the most important large ICT firm in the region. The ICT firm Nixdorf is also a significant player in Münster, but its number of employees is declining. The city also houses some outsourcing companies of banks, although one company is moving away. Deutsche Telekom is also reducing its staff. Apart from the big players, there is a significant number of small and medium ICT companies. The core of the ICT cluster consists of around 150 companies, some of them organised in the IDML e.V. (Internet Dienstleister MünsterLand, see below).

Other cities in NRW and other parts of Germany are much stronger in ICT and multimedia: e.g., Düsseldorf, Cologne, Hamburg and Munich. Even so, Münster has a strong ICT base. The university (WWU) has a large Faculty of Mathematics and Informatics (2,265 students) where topics like ICT implementation in organisations, web design, e-learning and data mining are taught and researched. Furthermore, Münster has the Technologiehof. The city may have a chance to build upon this ICT base if it stimulates entrepreneurship in this field. The university is also active in ICT-related fields like copyright and patent law. An institute active in this area is the Institute for Information, Telecommunication and Media Law (City of Münster, 2003b).

In order to put ICT more on the map in Münster and make the city more visible as an ICT city, some ICT entrepreneurs founded Internet Dialog

Münsterland (IDML) in 2003. IDML is a network for Internet providers, advertising firms and companies that have Internet-based business models. It organises 3–4 meetings per year for members where topics such as technological developments and new forms of marketing are presented and discussed. Other IDML activities include informing the press agencies about relevant developments regarding Münster-based ICT firms, promoting the ICT sector in other networks and organisations and trying to obtain support from the Münster municipality. Up to the time of writing, the municipality and the Chamber of Commerce (IHK) have not acknowledged the importance of IDML, since they do not think of Münster as an ICT city. Some 60 out of almost 400 ICT firms in the Münster region participate in IDML (Internet Dialog Münsterland, website 2004).

8.5 Conclusions and Perspectives

This section contains conclusions about the Münster case and unfolds perspectives for the city's position in the knowledge economy.

8.5.1 Foundations of the Knowledge Economy

The foundations of Münster's knowledge economy are quite strong. The largest employer in the city is the university and Münster has the highest ratio of students per inhabitant in Germany. Furthermore, Münster is known as a city of civil servants and law institutions. The number of large private firms is rather limited; in industrial sectors like manufacturing in particular, the region only houses a few companies.

An indication of the value of the knowledge economy foundations in Münster is given in Table 8.6. It can be concluded that the city has a good score on almost all foundation indicators. The *knowledge base* is strong: Münster has one of the largest universities in Germany and also one of the largest German polytechnics. In addition, there are several research institutes in the city, mostly working in the fields of medicine and biotechnology. The *economic base* is highly oriented towards services; Münster has very few declining sectors. The GDP per capita is high compared to the state average and other European cities. The city has few large companies and its economic base therefore mostly consists of SMEs. The low number of high-grade industrial firms in Münster is a disadvantage. The *quality of life* is high in many respects: Münster is a relatively green and safe city and has many bars and restaurants because of the student population. However, the city is not very well-known abroad and lacks an extensive dynamic and vibrant cultural climate. *Accessibility* is good in the Münster region, with little traffic congestion, excellent cycling facilities and an appropriate public transport system. External accessibility by train is good, but the number and range of flights at the international airport is rather limited presently. *Urban diversity* is moderate in Münster: about 9 per cent of the population and the university students are of foreign origin. The urban economy diversity is moderately positive; the city image is rather conservative and cautious. The *urban scale* of Münster is quite small compared to other German and European cities. A drawback of this small scale the small labour market: many people leave Münster because they cannot find an appropriate job. Because of the small urban scale, the importance of cooperating with companies from elsewhere is important for the city. *Social equity* is good: the percentage of unemployment is low. However, relatively many foreign citizens are unemployed.

Table 8.6 An indication of the value of knowledge economy foundations in Münster

	Foundations	Score in Münster
1	Knowledge base	+
2	Economic base	+
3	Quality of life	+
4	Accessibility	+
5	Urban diversity	□
6	Urban scale	□
7	Social equity	+

-- = very weak; – = weak; □ = moderate; + = good; ++ = very good.

8.5.2 Activities of the Knowledge Economy

In accordance with our framework of reference, Table 8.7 considers four types of activities in a knowledge city, and attaches scores for Münster. The overall score is moderately good. The city is successful in attracting both German and foreign students. Retaining graduated students is difficult since the Münster economy offers only few jobs. Also, the universities and several research institutes attract a multitude of good knowledge workers, but sometimes it is hard to retain them because it is often not easy for their partners to find suitable jobs. *Creating knowledge* in Münster is also quite strong: the university, polytechnics and research institutes contribute to this especially. Spearhead areas include biomedicine, nanotechnology and ICT. The *application of knowledge* by private firms is quite limited in Münster, because the city has few large companies. Yet the number of institutions in the field of stimulating the appliance of knowledge is extensive. Many of these organisations are supported by the state of NRW and the municipality. These intermediary institutes support SMEs that are developing various state-of-the-art products, particularly in the field of biotechnology (including nanotechnology and biomedicine). It will take several years before these companies become large, if they ever do. With regard to *developing growth clusters*, Münster is very much focused on biotechnology. The activities in this field are not dominated by certain cluster engines, since Münster lacks large firms in biotechnology. The biotech activities are concentrated in the university, research institutes and SMEs, most of which are university spin-offs. Besides biotech, the city also has a strong base in ICT, which consists of the large Faculty of Mathematics and Informatics at the university (WWU). Similarly to biotechnology, the ICT sector is dominated by SMEs. ICT activities in Münster might be strengthened by increasing entrepreneurship.

Table 8.7 An indication of the value of knowledge economy activities in Münster

	Activities	Score in Münster
1	Attracting knowledge workers	+
2	Creating knowledge	+
3	Applying knowledge/making new combinations	□
4	Developing (new) growth clusters	□

-- = very weak; – = weak; □ = moderate; + = good; ++ = very good.

8.5.3 Perspectives

As regards knowledge economy foundations and activities, Münster seems to be in a rather good position. However, there are some points of attention that the city needs to assess and/or keep in mind in order to successfully develop further towards a knowledge economy:

- The university occasionally seems to be at some distance from the city psychologically speaking. A number of ordinary citizens do not know what happens at WWU, while some WWU employees are not very interested in Münster. The preparation of the project City of Science 2005 has already partly bridged this gap between the university, polytechnics, research institutes and citizens and laid the basis for further and close-partnered collaboration and knowledge exchange. The success influences the possibilities of new combinations: all organisations and people involved are becoming more conscious of the problems, possibilities, weaknesses and opportunities that play an important role in the city and in the university. This leads to new ideas, products and solutions at the university and in the urban society, but also in companies. The increasing connectedness of the university can be illustrated by its participation in developing the Technologiehof, its role of shareholder in CeNTech, as co-founder of TIM and also the university's various cooperation projects with elementary schools, BASF Coatings and others.

  The city of Münster is not very well-known abroad. By increasing the more vibrant aspects of the quality of life (the artists' scene, more extensive night life, etc.) publicity abroad could be increased: this could help in attracting more foreign students and knowledge workers. Münster's bid to become the European Cultural Capital of 2010 could help to stimulate such activities. Whether or not the city wins the contest should not prevent Münster from strongly stimulating vibrant cultural places and events.
- Münster also has to increase its visibility for firms both from abroad and from other parts of Germany. The urban scale of Münster is rather small and therefore it is essential to extend this scale by enhancing ties with firms and institutes from elsewhere, and perhaps try to attract some large firms to the region. The recently presented city marketing plan can help contribute to this aspect.
- In previous years various well-qualified professors left Münster because their partners could not find appropriate jobs in this region. The city can try to retain such professors by actively helping their partners get suitable jobs. Support to attract and retain scientific staff by some measures, e.g. developing 'welcome packages' including such services for partners, might be an interesting idea.
- The knowledge base in Münster is good, but the economic base is lacking since the city has few high-grade large industrial companies. There are several ways for Münster to move forward: it could attract companies from elsewhere by offering an excellent knowledge base and/or offering

knowledge created in Münster in other places. Endogenous growth of knowledge intensive firms in Münster is a good development, but this takes time. Furthermore, such companies need venture capital to facilitate further growth; currently, such financial means are scarce in the city.

- Organising capacity in Münster is quite well developed: various municipal policies are proactive (e.g., the city marketing approach and the way the city is involved in the development of CeNTech) and the networks between the municipality, companies and the university are solid.
- The city is aware of its present situation: there is no economic crisis and the economy is fairly stable. However, various government functions are moving to Düsseldorf and employment in the military sector in Münster is decreasing. This means that the city must try to reach feasible goals regarding new economic activities. The current focus on specific biotechnology activities can contribute to this.
- Not only do municipality initiatives play a key role in stimulating the knowledge economy in Germany, the federal states also play an important role. The promotion of biotechnology activities by the state of NRW has led to many programmes in various cities, including Münster.
- Finding and appointing top leading people in key positions is paramount: becoming successful as a knowledge city demands being the best in specific areas. In Münster, as elsewhere, it is observed that top professionals attract other high-grade workers: researchers like to work in research groups that are led by the best professionals in their field of work.
- Being a major city of education, Münster has several opportunities to strengthen this aspect. Now that the EU countries are harmonising their educational systems (bachelor-master structure), opportunities will evolve to attract students from other countries. One way that Münster can increase the number of incoming foreign students is by offering more English-language courses.

Notes

1 The actual numbers for the summer semester did slightly decrease because of the introduction of tuition fees in German universities which made some (mainly long-term) students leave the university.

2 Up to the time of writing only preliminary numbers have been published by the university (see http://www.uni-muenster.de/Rektorat/Statistik/D2s104aa.htm).

Discussion Partners

Mr Klaus Anderbrügge, President, Westfälische Wilhelms-Universität Münster.
Mr Michael von Bartenwerffer, CEO Winkhous Holding.
Mr Wilhelm Bauhus, Director, Arbeitsstelle Forschungstransfer (Job Research Transfer) Westfälische Wilhelms-Universität Münster.
Mr Werner Funcke, Pro-Rector for Research, Development and Knowledge Transfer, Münster University of Applied Sciences (Fachhochschule Münster).
Mr Stephan Hüwel, Managing Director, Technologiepark Münster GmbH.
Ms Anna-Maija Kasanen, Arbeitsstelle Forschungstransfer (Job Research Transfer), Westfälische Wilhelms-Universität Münster.
Mr Matthias Schmidt, Hochschulangelegenheiten, Wissenschaftsstandort, Technologie-Initiative Münster – City of Münster.
Mr Carsten Schröder, Managing Director, Transferagentur der Fachhochschule Münster GmbH (transfer agency of the Münster polytechnic).
Mr Frank Schröder-Oeynhausen, CEO CeNTech GmbH (Centre for NanoTechnology).
Mr Christian Sorge, Client Advisor Technology, Wirtschaftsförderung Münster GmbH (Economy Promotion Company).
Mr Claas Sudbrake, Scientific Coordinator, Competence Centre for Nanoanalytics CeNTech.
Mr Berthold Tillmann, Mayor of the City of Münster.
Mr Klaus-Michael Weltring, Managing Director, bioanalytic-muenster.

References

Amt für Stadt- und Regionalentwicklung, Statistik (2004), 'Wirtschaftsentwicklung in Münster', City of Münster, http://www.muenster.de/stadt/medien.
Amt für Wohnungswesen (2002), *Wohnungsmarkt Münster 2001*, City of Münster.
Arbeitsstelle Forschungstransfer Westfälische Wilhelms-Universität (2004), http://afo.uni-muenster.de.
Bioanalytik-muenster (2004), http://www.bioanalytik-muenster.de.
Bio-Gen-Tec-NRW (2002), *BTi April/May 2002*, http://www.bio-gen-tec-nrw.de.
Cash (2004), *Das grosse deutsche Standort-Ranking* (Large German Cities' Ranking), Nos 1–2.
City of Münster (2002), *Verkehrsbild Münster 2001* (Traffic Composition).
City of Münster (2003a), *Multitalent Münster – Wege der Stadt der Wissenschaft 2005.*
City of Münster (2003b), *Multitalent Münster – Wege der Stadt der Wissenschaft 2005 – Die Details.*
City of Münster (2003c), *48eins – Magazin rund um den Wirtschafts- und Immobilienstandort Münster*, October.
City of Münster (2003d), *Jahres-Statistik 2001/2002 Stadt Münster.*
City of Münster (2004a), *Kulturhauptstadt für Europa 2010 (Cultural Capital of Europe 2010) – Anlage 2*, http://www.muenster2010.de.
City of Münster (2004b), *Test: Radfahren in 22 Grossstädten*, http://www.muenster.de/stadt/stadtplanung/ms-_19929.html.
City of Münster (2004c), *Modellstadt Münster.*

Competence Centre Nanoanalytics in Münster (2002), *Where Nanotechnology Meets Life Science.*

Fachhochschule Münster (1999), *Shaping the Future*, http://www.fh-muenster.de.

Fachhochschule Münster (2002), *Gründerzeit an der Fachhochule Münster (GDF).*

Fachhochschule Münster (2004), *Jahresbericht Fachhochschule Münster 2003.*

Forschungszentrum Jülich GmbH (2003a), Geschäftsbericht 2002.

Forschungszentrum Jülich GmbH (2003b), Programme der Bundesländer – Nordrhein-Westfalen. http://www.ihk-nordwestfalen.de/industrie_in_muenster/.

Gesellschaft für Bioanalytik Münster e.V. (2002), *Münster biotech-region.*

Greif, S. and D. Schmiedl (2002), *Patentatlas Deutschland – Ausgabe 2002*, Deutches Patent- und Markenamt, München.

IiM (Initiative Industrie in Münster) (2003), *Industrie in Münster, gestern, heute, morgen.*

Internet Dialog Münsterland (2004), http://www.idml.de.

IW Consult (2004), 'Deutsche Großstädte im Vergleich', Cologne, http://www.iwconsult.de.

Landeshauptstadt München (2003), *The Business Location*, Referat für Arbeit und Wirtschaft, City of Munich.

Landesregierung Nordrhein-Westfalen (2004), *Zuwanderung und Integration in Nordrhein-Westfalen*, http://www.nrw.de.

Ministeriums für Frauen, Jugend, Familie und Gesundheit des Landes Nordrhein-Westfalen (2000), *Gesundheit von Zuwanderern in Nordrhein-Westfalen*, http://www.mfjfg.nrw.de.

Modellstadt Münster (2004), 'Gute beispiele zum Handlungsfeld "Wirtschaft"', City of Münster/Wirtschaftsförderung website.

Münster Marketing (2003), *Stadsgids Münster 2004 – Beleven en ontdekken*, http://www.tourismus.muenster.de.

Münsterischer Anzeiger (2004), *Sieben Ziele für die Zukunft – Stadtmarketingkonzept vorgestellt*, 17 March.

Office of the UK Deputy Prime Minister (2004), 'Competitive European Cities: Where do the Core Cities stand?', London, http://www.odpm.gov.uk.

Statistisches Bundesamt (2004), http://www.destatis.de.

Technologiepark Münster GmbH, Powerpoint presentation.

Technologiepark Münster GmbH (2001), *Your Location for Life Science Industries/Information and Communication Technologies.*

Technologiepark Münster GmbH (2003), 'The Technologiehof in Münster', http://www.technologiehof-muenster.de.

Westfälische Wilhelms-Universität (2004), http://www.uni-muenster.de.

Wirtschaftsförderung Münster (2003a), 'Standort Münster – Spielraum für Unternehmen', http://www.wfm-muenster.de.

Wirtschaftsförderung Münster (2003b), *The Location: Münster – Scope for Enterprise.*

Wirtschaftsförderung Münster (2004), *Oberzentrum Münster, Kreis Steinfurt und Stadt Greven schaffen für das Münsterland grosses interkommunales Gewerbegebiet*, press release, 5 March.

Wirtschaftswoche (2004), *Lichtjahre entfernt – Städtetest*, 15 April.

Zeller, C. (2001), 'Clustering Biotech: A recipe for success? Spatial Patterns of Growth of Biotechnology in Munich, Rhineland and Hamburg', *Small Business Economics*, No. 17, pp. 123–41.

Chapter 9

Rotterdam

9.1 Introduction

Rotterdam is the second largest city in The Netherlands with a population of 600,000 inhabitants. Only Amsterdam has more inhabitants (735,000); The Hague (458,000) and Utrecht (207,000) trail Rotterdam. Rotterdam's population has been growing since 1985 and this growth is expected to continue. It is forecast that by 2017 the city will have 635,000 inhabitants (COS, 2003a). The Rijnmond region, consisting of Rotterdam and surrounding municipalities, has 1.1 million inhabitants.

Within The Netherlands Rotterdam is the only big city where the average age of the population is getting younger: the percentage of young people is increasing while the share of elderly is decreasing. This development is caused by the increasing number of non-native inhabitants (mostly from Turkey, Morocco and Surinam). It also has an impact on the economy: already 45 per cent of the labour force is of foreign origin. Rotterdam houses relatively many people with low incomes compared to both other big cities in The Netherlands and the national average. The wider Rotterdam region has a relatively low number of low-income households; among other things, this has to do with the flow of many relatively wealthy Rotterdam inhabitants towards satellite towns (COS, 2003b). Figure 9.1 shows the comparatively high level of unemployment in Rotterdam.

Since 1994 unemployment decreased in Rotterdam and in The Netherlands in general but in 2001 the figures started to rise again. The unemployment percentage in Rotterdam has been higher than the national figure for many years: the difference ranges between 3–5 per cent (OBR, 2003a).

9.2 Knowledge Strategy

Both on the urban (Rotterdam) and provincial (South-Holland) level, strategies to promote the knowledge economy are in place. Furthermore, the national government aims to stimulate the knowledge economy via an innovation platform.

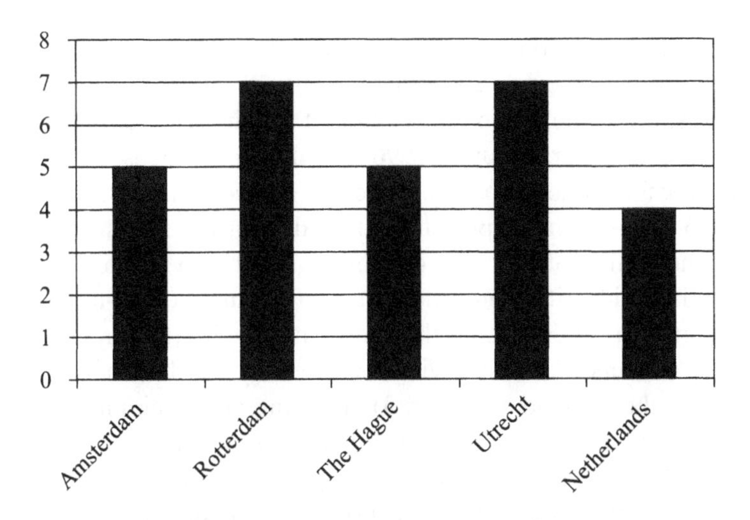

Figure 9.1 Unemployed labour force in 2002 (%)

Source: OBR (2003a).

9.2.1 City Strategy

In 2002 the newly-elected city council of Rotterdam formed a new policy programme, the aims of which included 'stimulating the knowledge economy'. The Development Corporation Rotterdam (OBR) has elaborated this theme: the 'Programma Kenniseconomie' ('Programme Knowledge Economy' (PKE)) was published in 2003. PKE stresses three main elements: quality of the education; cooperation and interaction between private firms, knowledge institutions and government; and quality of life (attractive surroundings to live in). The three main actors on the government side are the municipal agencies OBR (Development Company Rotterdam) and DSO (Department for Education) and the Port Company Rotterdam (which is no longer part of the municipality since 2004). These organisations should stimulate the knowledge economy in cooperation with businesses and knowledge institutions: the aim is to increase production and delivery of services and knowledge-intensive economic activities to reach higher employment numbers (Stuurgroep Kenniseconomie, 2003). In order to start up, promote and coordinate knowledge economy activities in Rotterdam, the Economic Development Board was founded: this group consists of leading people from those organisations and companies that are important in the urban region regarding knowledge-intensive activities.

The PKE elaborates on several earlier studies by the municipality: e.g., the Economic Vision Rotterdam 2002–2006 (OBR, 2002). This vision was based on a broad study of the economic position of the city. The starting point was the economic vision drawn up in 1996: that document focused measures on strengthening the economic basis of the city. The Economic Vision Rotterdam 2002–2006 stresses the importance of quality in the fields of economic structure and quality of life. This document is still relevant for Rotterdam's position today. It identifies three main ways to stimulate a long-term balanced economic growth: 1) enhancing the quality of the city (to attract and retain knowledge workers); 2) diversifying the economy via the cluster policy by focussing on the demand side; and 3) attracting and retaining knowledge-intensive research and education institutes and companies. In the period 1996–2002 the cluster-oriented economic policy was introduced. In recent years the municipality has focused on a smaller number of cluster projects to increase the yields of the cluster policy. Several criteria determine how much support clusters[1] receive:

- the contribution to accelerated diversification of the economic structure;
- the degree to which clusters bind knowledge to the city/region;
- the extend to which clusters increase the 'quality of the city';
- the contribution to improving the innovative climate.

The clusters that are supported by municipal policy include the AV (audio-visual) cluster, the ICT (information and communication technology) cluster and the medical cluster.

Three 'tracks' are categorised in the PKE to stimulate and support the growth of the knowledge economy. Within the tracks several sub-programmes have been identified. These programmes clarify the goals, the intended results and the primary responsible municipal agencies. The prime responsible agencies are in control of the sub-budgets.

1 **Developing the labour potential**

Strengthening vocational education and reinforcing the linkages to jobs
Goal: decreasing school dropouts and enlarging the number of people that go on to a higher education (prime responsibility: DSO)
Strengthening cooperation between Erasmus University Rotterdam (EUR) and Technical University Delft (TUD)
Goal: strengthening cooperation EUR-TUD and Rotterdam-EUR (prime responsibility: DSO).

2 **Reinforcing innovation power**
Knowledge and technology centres
Goals: founding a top knowledge institute linked to existing knowledge institutions and starting a centre of excellence in the field of transport (prime responsibility: DSO).
Lectureships/professorships
Goal: establishing at least three lectureships/professorships in ICT/logistics and industry (2003), metalelectro, process chemistry, urban/architecture (2004) and healthcare/medical/life sciences (2005) (prime responsibility: DSO).
Knowledge Circles ('Kenniskringen')
Goal: founding Knowledge Circles with stakeholders from private firms, education and government to strategically cooperate in the areas of research and development and lobbying in process chemistry, audio-visual, metalelectro, architecture and healthcare/medical/life sciences (prime responsibility: OBR).

3 **Improving infrastructure and facilities**
Site development (geographical focal points)
Goal: develop four locations for companies in promising clusters (old city harbours, Hoboken, area near the regional airport, Lloydkwartier) (prime responsibility: OBR).
Telecom infrastructure
Goal: develop citywide business case, implement two pilot projects (prime responsibility: OBR).
Retaining students in the city
Goal: improve housing for students and graduates, directing graduates towards companies in the city and region (prime responsibility: OBR).

The three tracks described above are inter-related. Better education will improve the quality of the labour force which will eventually increase the quality of the economic clusters. The clusters should improve the quality of education and attract knowledge workers. Good infrastructure and facilities are meant to increase the quality of the city by, among other things, attracting highly skilled workers; these people subsequently contribute to better education and clusters and a better living environment. The municipal funds are mainly aimed at strengthening the linkages between education, businesses and municipality (Stuurgroep Kenniseconomie, 2003).

9.2.2 Regional/Provincial Strategy

On the regional/provincial level the Knowledge Alliance South-Holland ('Kennisalliantie Zuid-Holland' – KA) is active. KA is a network organisation that is financed by (among others) the European Union, Senter and Syntens (both are agencies of the ministry of Economic Affairs), the province of South-Holland, the big municipalities (The Hague, Rotterdam, Delft, Leiden), the chambers of commerce in The Hague region and Rotterdam district, four polytechnics and the universities in Rotterdam, Delft and Leiden. In Figure 9.2 the working area of the KA is shown (Knowledge Alliance South-Holland, 2003/4).

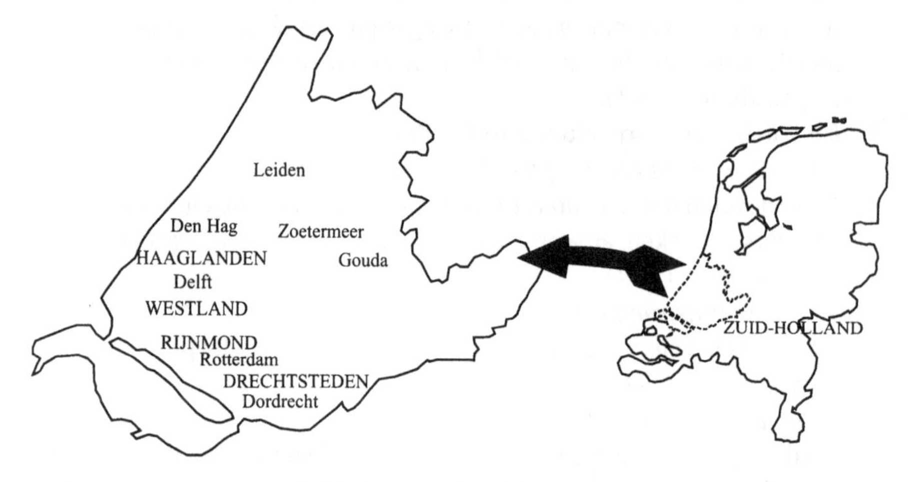

Figure 9.2 Working area of Knowledge Alliance South-Holland

Source: Knowledge Alliance South-Holland, 2003.

KA aims to increase cooperation between education, research, business and government. The organisation centres on nine 'knowledge pillars': 1) water technology; 2) ICT (including telecom); 3) aerospace; 4) life sciences; 5) shipping, transport and logistics; 6) sensor and nanotechnology; 7) international law; 8) cultivation under glass; and 9) process and petrochemistry. KA aims are: a) South-Holland should enter the top three of the most innovative regions in The Netherlands within five years; and b) more high-grade knowledge institutions and start-up companies will be present in the province. These aims should be obtained by exchanging knowledge and experiences (through real life meetings and using virtual networks), enhancing lobbying and promotion activities, using the KA organisation as a think tank for strategy development, and gearing policies, visions and action programmes to each other. The KA started its activities in 2003; four projects were defined (Knowledge Alliance South-Holland, 2003/4):

1. *knowledge to the market*: stimulating cooperation between firms and knowledge institutes to realise new product market combinations (PMCs). Activities include market research, feasibility studies, thematic meetings, scouting PMCs, an annual 'knowledge festival' (to be organised in Rotterdam in 2004) and publishing lessons learned;
2. *high-grade innovative graduation apprenticeship programme SMEs*: enhancing innovation by deploying polytechnic and university students in SME apprenticeships. This project also includes a contest for best graduation projects;
3. *strategic think tank for the regional knowledge economy*: inventorying best practices in The Netherlands and abroad by knowledge workers who are temporarily attached to SMEs, listing successful patent applications, supporting cooperation between organisations in the province;
4. *information point techno starters South-Holland (IPT):* improving the establishment conditions for start-ups. This includes monitoring, advising and scouting techno starters, organising network meetings.

According to the Knowledge Alliance, many initiatives have been taken and reports have been written about knowledge-intensive activities, but up to the time of writing, too little has been done to implement this knowledge and these experiences. In the previous period the KA has tried to get all relevant parties to take one line. Now KA is focusing in competencies and trying to find the right leaders (former entrepreneurs) to manage the activities in the nine 'knowledge pillars'. Within each pillar, about three sub-themes will be identified; for each of these sub-themes innovation workshops will be organised four times per year.

At the time of writing, the municipal knowledge economy strategy of Rotterdam and the Knowledge Alliance strategy for South-Holland have not been explicitly tuned to each other: this might be because both the programmes have only been running for a short period of time. Furthermore, the provincial policy regarding the knowledge economy seems to be rather top-down oriented.

9.2.3 National Policy

Besides the city and regional knowledge strategies, stimulating the knowledge economy is also an aim at the national level. The Dutch Innovation Platform was set up in 2003 to stimulate knowledge-intensive economic activities in The Netherlands. This platform was formed in accordance with the Finnish Science and Technology Policy Council, which started in the 1980s. Just as in Finland, the Dutch Innovation Platform is chaired by the prime minister. Up to the time of writing, the output of the platform has been minimal.

9.3 Knowledge Foundations

In this section, the knowledge foundations – as discussed in the theoretical chapter – will be assessed for Rotterdam. They include the knowledge base, the economic base, the quality of life, the accessibility, the urban diversity, the urban scale, and the social equity in the city.

9.3.1 Knowledge Base

The Rotterdam labour force has a relatively low education level compared to the Dutch average. In the Rotterdam region ('Groot-Rijnmond') 26 per cent of the labour force has a polytechnic or university degree and 34 per cent has a low education level (at most vocational education). The same picture applies to the Rotterdam municipality: the low education level is linked with the high proportion of non-native citizens. For high-grade jobs Rotterdam-based firms depend on commuters from outside the city: there are 97,000 high-grade jobs in Rotterdam, whereas only 36,000 of these jobs are filled by people who live within the city. This suggests that over 60 per cent of the Rotterdam high-grade jobs are filled by commuters. Most of these commuters live in the Rotterdam region, in adjacent municipalities like Capelle aan den IJssel and Spijkenisse. After that, many commuters come from The Hague region, Dordrecht and adjacent towns. It is striking to note that, while many commuters travel to Rotterdam for high-grade jobs (61,000), at the same time many Rotterdam inhabitants (31,000) commute to other places for high-grade jobs (OBR, 2003a).

In Rotterdam almost 190,000 people attend classes. About 27 per cent of these people are in a polytechnic/university; the absolute number of polytechnic/university students is growing. The Rotterdam percentage of people with third level education is in line with many other European cities (Office of the Deputy Prime Minister, 2004). Table 9.1 shows the number of students in different education types.

Table 9.1 Number of students in Rotterdam educational institutes

Type of education	Number of students (2000)
Primary education	53,900
Special education	6,300
Secondary education	77,000
Polytechnic	30,700
University	19,500

Source: OBR (2003a).

Rotterdam has several polytechnics; Hogeschool Rotterdam (HR) and InHolland are the biggest players. HR has a mainly regional function and focuses on seven fields of knowledge (applied technology, healthcare, city renewal and creative industries like architecture are important focal points), while InHolland is part of a national network of polytechnics that offers a wide range of studies. Furthermore, the Rotterdam region has two universities in its surroundings: the Erasmus University Rotterdam (EUR) and the Technical University of the neighbouring city of Delft (TUD). TUD has almost 13,500 students of whom most (over 3,100 students) study architecture (TU Delft, 2003). The EUR student population has grown fast in recent years: the university counted 17,400 students in 1998, rising by 26 per cent to 22,000 students in 2002. EUR has well over 10,000 students in economics and management; almost 4,000 law students and about 1,700 medicine students. The faculties of law and business in particular grew swiftly (2000–2001, www.eur.nl; COS, 2003d). EUR has a good reputation and track record in several fields. The medical faculty (Hoboken campus, west of the city centre) is involved in world leading research in the field of virology and also performs well in cellular biology and immunology. In the field of economics and business (Woudestein campus, east of the city centre) the Rotterdam School of Management belongs among the top institutes in Europe.

The labour force in Rotterdam is relatively young: the percentage of people under 35 is 47 per cent compared to 39 per cent for The Netherlands. The labour force is forecast to continue to grow: an annual increase of 1.6

per cent is expected. This is high compared to the Dutch average. Although it is true that in a quantitative way there are no longer any major frictions in the Rotterdam labour market, in qualitative terms there is a shortage of highly educated people. Workers with a technical background are especially in demand (OBR, 2003a). This can be seen, for instance, in the harbour and industrial complex. While employment in this sector is declining, an upgrading of functions is taking place: demand for people with higher levels of secondary education and polytechnic/university graduates is increasing. In the harbour, continuing bottlenecks in the labour market are expected: there are many non-native students but they mostly only follow lower vocational training. At the same time the workforce is ageing, which contributes to increasing labour costs (Port of Rotterdam, 2003).

9.3.2 Economic Base

The presence of one of the world's largest harbours in Rotterdam implies that industry, transport and logistics significantly contribute to the economy, both on a local/regional and national level. Although the harbour is by far the largest in Europe, in recent years it has lost market share to other European ports (Antwerp, Hamburg, Amsterdam): while trans-shipment in Rotterdam increased in absolute numbers, the growth in competing ports was greater. However, in previous years it is the service sector that has grown most rapidly; this development is expected to continue (OBR, 2003a). Table 9.2 shows the sector contribution to the gross regional product of the Rotterdam region.

Table 9.2 Contribution of economic sectors to Rotterdam region GRP

Economic sector	Contribution to GRP, 1990 (%)	Contribution to GRP, 2000 (%)	Change (% points)
Agriculture	3	2	–1
Industry	20	19	–1
Construction	6	5	–1
Trade	16	14	–2
Transport	15	11	–4
Financial services	6	8	+2
Business services	16	21	+5
Non-commercial services	18	20	+2

Source: OBR (2003a).

Employment shows a similar picture to GRP development. The largest employer in Rotterdam is business services: 58,000 jobs make up about 20 per cent of total employment. Transport, storage and communication comprise 12 per cent of all jobs. Compared to The Hague, Amsterdam and Utrecht (the other cities that are part of the Randstad or G4), Rotterdam has relatively many jobs in sectors that show a decreasing demand for labour – industry, transport and logistics. This makesRotterdam's economic base vulnerable, especially with regard to employment growth. In business services Rotterdam is catching up with the other G4 cities. While most sectors experienced a decline in employment as a result of the economic recession in 2001–2003, the number of jobs in government, education, healthcare and welfare increased in Rotterdam (OBR, 2003a).

The productivity of Rotterdam workers is high: €67,000 per labour year compared to €62,000 in The Hague and €58,000 in Amsterdam and Utrecht. This is a result of among others the capital-intensive petrochemistry. The productivity growth, however, is worrisome for Rotterdam: while productivity increased only moderately in the other G4 cities and in the country as a whole in the period 1996–2001, productivity declined in Rotterdam by almost 0.5 per cent per year. An important reason for this decline is the strong downturn in added value per job in the oil refinery business (OBR, 2003a).

A relatively large proportion of employment in Rotterdam (35 per cent) is found in medium-sized firms (10–99 employees) compared to the other Randstad cities. The big companies (more than 100 employees) contain most jobs: less than 2 per cent of all Rotterdam-based firms have over 50 per cent of all jobs. Rotterdam houses relatively many big firms compared to the national average. While in Amsterdam and The Hague it was mostly the large companies that contributed to employment growth in 2002, in Rotterdam the SMEs created more jobs. In 2002 the city had a significant number of start-up firms: almost 2,500. Table 9.3 shows the percentage of starters. However, in 2003 the number of start-ups decreased sharply; the decline was greater than the Dutch average. Also the relative number of bankruptcies was high in 2003 compared to the other G4 cities and the national average (OBR, 2003a).

The extent of 'driving' economic sectors (all economic sectors that contribute to the competitiveness of the Dutch economy on international markets and that are not mainly dependent on local/regional demand regarding their location choice) informs us about the degree to which the Dutch economy can compete in international markets. Especially in the G4 cities, driving sectors are strongly represented. Rotterdam underperforms compared to Amsterdam: the driving sectors are mostly industry and distribution, where

Table 9.3 Percentages of start-up companies compared in 2002

City	Percentage of start-ups
Amsterdam	8.5
Rotterdam	9.0
The Hague	8.8
Utrecht	13.0
Province of South-Holland	7.6
Netherlands	7.0

Source: OBR (2003).

relatively few people work. The growth in driving sectors also shows that Rotterdam is lagging behind: growth mainly took place in the eastern part of the city, while mainport Rotterdam (the harbour) created relatively few jobs in driving sectors (OBR, 2003a).

An international comparison shows that Rotterdam ranked 40th in GDP per capita among 61 European cities in 2001: €26,200 compared to €74,500 in Frankfurt am Main (leader), €38,200 in Amsterdam and €30,100 in The Hague. The percentage of employees working in high-tech manufacturing is relatively low compared to other European cities, while the percentage of employees working in high tech services is relatively high. Also, the percentage of employees working in knowledge-intensive services is relatively high (Office of the Deputy Prime Minister, 2004).

9.3.3 Quality of Life

Increasingly firms focus their location choice on their demand for (high-grade) workers. This implies that 'quality of life' factors (living environment, recreational facilities, safety, etc.) are becoming more important. Companies value Rotterdam highly compared to other G4 cities because of the presence of sufficient good company locations. Yet citizens value the housing conditions/ living environment the lowest among the G4 cities (although rating differences are small). Almost 50 per cent of Rotterdam inhabitants find that trash and vandalism occur frequently in their neighbourhood; around 8 per cent of the citizens claim to have been a victim of vandalism themselves and 20 per cent say they have been hit by car theft. The number of violent crimes is on the same level as in the other G4 cities. Social safety in public transport was rated 6.8 out of 10 in 2002; this is a little below the other G4 ratings (COS,

2003b). Thirty-seven per cent of city inhabitants feel their neighbourhood has deteriorated in recent years. According to the citizens of Rotterdam, safety and public policy are the most pressing urban problems. About 70 per cent of the Rotterdam citizens feel (very) healthy and happy: this is about the same as in Amsterdam, Utrecht and The Hague (COS, 2003c). The recreational facilities are on a level that competes with the other G4 cities and the (feeling of) safety is on a comparable level to Utrecht but somewhat below The Hague and Amsterdam (Ministerie van Economische Zaken, 2002a). There are relatively many shops in Rotterdam, but the diversity is limited (OBR, 2003a). The inhabitants of Rotterdam are not particularly proud of their city, but some subjects are frequently mentioned: the harbour, the skyline, the dynamic and modern city image, festivals and events (COS, 2003c).

Rotterdam has some 285,000 homes: 25 per cent are owner-occupied houses and 75 per cent are rental houses compared to a 50/50 division at the national level. Rotterdam inhabitants pay the lowest housing prices (€1,570 per square metre) compared to the other G4 cities; the Rotterdam figure is also below the Dutch average. The city has a qualitative housing mismatch: there is a surplus of cheap homes and a shortage of medium and expensive homes. The mismatch amounts to 50,000 homes (OBR, 2003a).

The city of Rotterdam offers a multitude of leisure activities in several fields: sports (recreational, but also a top premier league soccer club, an annual international marathon and an ATP tennis tournament), culture (multiple cinemas and museums, several theatres, the International Film Festival, the Gergiev classical music festival, summer carnival and annual dance parade, etc) and a vibrant night life (including several big discothèques such as Now & Wow, Hollywood and Off Corso). In the period 1998–2002 the overall number of visitors to Rotterdam increased by 8.8 per cent; museum visits showed severe decline (–22.5 per cent), while events (+20.3 per cent), theatres and pop concerts saw major increases in visitor numbers (+14.4 per cent). In the top five of most visited Dutch cities by the Dutch, Rotterdam ranked second behind Amsterdam. The top three crowd-pullers in Rotterdam are Ahoy' (a multi-facility hall for pop concerts, fairs and sporting events), the Pathé cinemas and the Blijdorp Zoo (OBR, 2003a). Rotterdam citizens visit cultural inheritance relatively little compared to other G4 inhabitants (COS, 2003c).

The people who visit Rotterdam are young (15–29 years) compared to other G4 city visits. Most visitors come from the province of South-Holland. In 2001, Rotterdam counted 867,000 overnight stays: most of these stays involved foreign businesspeople. Amsterdam and The Hague both welcomed more overnight visitors and their share of foreign tourists in the overnight stays was significantly larger. In the last five years Rotterdam experienced a high growth in the number of hotel beds: a growth rate (20 per cent) comparable to Utrecht, but higher than Amsterdam and The Hague (Ecorys, 2003). Employment in the leisure sector in Rotterdam increased 17.3 per cent in the period 1998–2002 compared to 12.6 per cent in The Netherlands as a whole. Currently, 5.4 per cent of Rotterdam's employment can be ascribed to the leisure sector; this is much lower than in Amsterdam (10 per cent) and the national average (6.5 per cent) (OBR, 2003a).

Increasing the quality of life indirectly also depends on the image of the city. In recent years Rotterdam has undertaken many activities to positively contribute to this image: for example, Rotterdam co-hosted the European soccer championships in 2000 and in the next year the city was Cultural Capital of Europe, together with Porto. To integrate the city marketing activities, the city founded the Rotterdam Marketing Agency (van den Berg et al., 2000).

9.3.4 Accessibility

Firms' accessibility by car in Rotterdam is somewhat better than in the other G4 cities; car parking facilities are also best in Rotterdam. Internal car accessibility is best in Rotterdam, but the city is hit relatively hard by traffic congestion as far as external accessibility is concerned. Accessibility by public transport is just as good as in Amsterdam and Utrecht; The Hague performs a little better (Ministerie van Economische Zaken, 2002b). The province of South-Holland experienced most traffic jams in 2001, but the province of North-Holland (where Amsterdam is situated) was hit harder with more severe traffic jams.

Rotterdam's international accessibility is quite good. The city has its own airport, although this has a limited capacity: in 2001 765,000 people used Rotterdam Airport, placing the city airport in 143rd position in the European rankings (Office of the Deputy Prime Minister, 2004). However, the number of destinations significantly increased recently and airport capacity also grew with the opening of a new terminal. Furthermore, Rotterdam is close to Schiphol Airport, which was ranked fourth in Europe in 2001 (39,309,000 passengers): currently travelling time by train or car from Rotterdam to Schiphol is about 45 minutes. In the near future travelling time will be shortened further when the HST-South (high-speed train) comes into use: the HST is expected to reduce travelling time to 20 minutes (Pol, 2002). Regarding trains, Rotterdam is also connected to the TGV network (which connects The Netherlands with Belgium and France) and eventually the Betuweline will improve the train connection eastward to Germany and beyond; the Betuweline is especially important to improve the hinterland connections of the Rotterdam harbour.

Besides physical accessibility, virtual access through ICT networks is becoming increasingly important. Since Rotterdam is a metropolitan urban region it is pretty well endowed with ICT infrastructure: several providers offer ADSL, cable and some other data networks. The city of Rotterdam, however, wants to speed up broadband Internet deployment: it has planned the construction of optic fibre networks in the restructuring area Lloydkwartier and the newly-built suburb Nesselande. These pilot projects should connect some 6,750 houses and multiple firms (van Winden and Woets, 2003).

9.3.5 Urban Diversity

Cultural diversity in Rotterdam is extensive: many citizens have a foreign background (46 per cent) and this number is expected to increase to 58 per cent in 2017. The biggest minority groups are Surinamese (52,100), Turks (42,700), Moroccans (33,100), Antilleans/Arubans (39,400) and Cape Verdians (14,800). The last four of these groups more often choose lower-grade follow-up studies than do native Rotterdam inhabitants (COS, 2003a). This contributes to friction in a labour market that increasingly demands high-skilled workers. A positive effect of the high proportion of non-native Rotterdam citizens is their rejuvenation of the population composition, which in the longer term can help lower the average age of the work force. In recent years some people who belong to specific minority groups in Rotterdam have caused the city significant damage in the fields of city image, safety and its perception, resulting in a decreased quality of life. Many of the problems seemed to be the result of a lacking integration of some minority groups: because several other Dutch cities (The Hague, Amsterdam, etc.) experienced similar problems, the subject has received national government attention. In spite of this negative publicity, cultural diversity has also proved to be a positive asset for Rotterdam: for example, the leisure scene benefits from having many different sorts of restaurants and several cultural events attract many people and contribute positively to the city image (for instance, the annual Summer Carnival). The cultural diversity also offers opportunities for innovative crossovers in several fields: not only cultural but also economic activities. To actively utilise the potential of the large minority groups, some specific policies are executed such as Non-native Entrepreneurship and encouraging non-native youths to be educated for, among other things, high-grade harbour jobs (OBR, 2003b; Port of Rotterdam, 2003).

9.3.6 Urban Scale

Rotterdam (city: 600,000; region: 1.1 million) is the second-biggest city in The Netherlands after Amsterdam (city: 730,000; region: 1.4 million). On the European level, Rotterdam is a secondary city in terms of number of inhabitants. However, the city is part of the densely populated area called the Randstad ('Edge City', which includes the four largest Dutch cities: Amsterdam, Utrecht, The Hague and Rotterdam); this area consists of some 7 million inhabitants (compared to a total Dutch population of 16 million people). Regarding economic activities and cooperation, a pattern can be distinguished that divides the Randstad into a northern wing (Amsterdam – Utrecht) and a southern wing (The Hague – Rotterdam). The northern wing is more of a service area compared to the southern wing: during the ICT boom in the second half of the 1990s this resulted in a higher economic growth for Amsterdam and Utrecht compared to The Hague and Rotterdam. Benefits of the northern wing include more high-grade office buildings and a better city image. A drawback of this focus on the service sector is that the northern wing is more susceptible to economic recessions in this sector (Het Financiële Dagblad, 2001). Regarding business start-ups, the Randstad divide is illustrated by Kloosterman and Lambregts (2001.) Although the northern wing has performed better than the southern wing, both have achieved good economic results in a northwestern European perspective (Bureau Louter, 2003). The large cities in The Netherlands are experiencing an increasing urban scale with regard to population and employment growth: growth is highest in the suburbs, which implies that the spatial scale of the cities is widening (VROM, 2001).

It is expected that the economic interaction between Rotterdam and Amsterdam will increase when the HSL railway link is ready in 2007, since this connection will reduce travel time from 64 to 35 minutes (van den Berg and Pol, 1998). For the Randstad area it will be important to keep upgrading its infrastructure linkages (HSL, Betuwelijn, roads and the mainports Rotterdam harbour and Schiphol Airport) in order to maintain a good position internationally: in view of the European Union's expansion towards the eastern European countries, the Randstad area will become a somewhat more peripheral location from the EU perspective. Extension and upgrading of infrastructure can contribute to expanding the urban scale.

9.3.7 Social Equity

Social equity in Rotterdam is quite poor. In the introduction to this chapter, it was noted that unemployment in Rotterdam and in The Netherlands in general decreased since 1994 but that the figures started to rise again in 2001. Rotterdam's unemployment percentage has been higher than the national average for many years; the difference ranges between 3–5 per cent. Also, within the Randstad, Rotterdam has the highest number of unemployed. In 2002–2003 unemployment in the city rose by 25 per cent. Over 60 per cent of the unemployed in Rotterdam have a foreign background. Most of the unemployed people in the city have a low level of education. A worrisome fact concerns the high percentage of long-term unemployed: more than 60 per cent have been out of a job for over a year. Most predictions show increasing unemployment numbers for 2003 and 2004: an annual average increase of 9.1 per cent is expected. From 2005 onwards, unemployment is expected to decline again. These figures are in line with the national development (OBR, 2003a). In the G4 cities (which include Rotterdam), more than 20 per cent of households have a low income compared to about 15 per cent on a national level; in Rotterdam 10 per cent of households received social security benefits in 2000 compared to a 3 per cent Dutch average (G4, 2002). Among the G4 cities, Rotterdam has the highest number of young people on social security: almost 8 per cent in the age group 15–34 (COS, 2003b).

9.4 Knowledge Activities

In the last section, the foundations of Rotterdam's knowledge economy have been discussed. Now, we turn to knowledge activities and ask how well the city (and the actors in it) manage to: 1) attract and retain talented people; 2) create new knowledge; 3) apply new knowledge and make new combinations; and 4) develop new growth clusters.

9.4.1 Attracting and Retaining Knowledge Workers

Rotterdam's performance in attracting knowledge workers is average. The number of polytechnic and university students in Rotterdam has risen by 22 per cent in the period 1997–2002. The city now has 22,000 university students and 32,000 polytechnic students. Over 44,000 pupils are in vocational education (COS, 2003d). Students are potential knowledge workers. Key reasons for choosing a place of residence after graduation are the job location and the attractiveness of the place of residence. The number of foreign students who have come to Rotterdam has increased. In particular, many foreign students are in graduate courses; there are some 550 international students at the EUR. Taking all Rotterdam educational institutes together the number is 2,000–3,000. Van den Berg and Russo (2002) conducted research into how to attract and retain students to a city: currently, many Dutch students continue living with their parents (outside Rotterdam). By improving information, utilising city marketing and offering 'living career' possibilities in the city and region, these future knowledge workers can be attracted and retained. The city should integrate students into the community by bringing them into touch with Rotterdam's cultural life: the polytechnics and Erasmus university are increasingly doing this both by organising introductory periods for newcomers and via a range of cultural activities (discount theatre, museum visits, et.) targeted at students through specific communication channels. Efforts to retain students from abroad might gain a lot, because currently the foreign student population is scarcely integrated into the overall student and city life. The efforts of the EUR to attract foreign students do not yet seem optima; promotional tools are needed to enhance this situation. Moreover, an integral city marketing policy has to be created that should also include this aspect of the city, in cooperation with polytechnics and the university. The city and the OBR urban development agency have recognised these challenges and started to act accordingly, but the current focus of embedding knowledge into the city is on the 'education chain' (van den Berg and Russo, 2002). Even so, EUR has acknowledged the increasing importance of international students and there are plans to build a 1,350-student campus near the university.

Increasing the labour potential is an important part of the Rotterdam knowledge economy programme. The educational department is directing most of these projects. A covenant was signed to increase the efficiency and effectiveness of vocational education: the aim is to increase the yield in number of graduated pupils by 50 per cent in 2006 and to help 4,000 dropouts get a job or an education. Another project which is already being implemented concerns the creation of 1,000 'learning jobs' in three years from the time of writing. Programmes on the verge of getting started are the Summer School and the Rotterdam Talent programme. The Summer School will be directed by polytechnic InHolland; it aims to offer courses to 3,000 higher-educated people; the Summer School should contribute to enhancing the image of the participating polytechnics and the image of Rotterdam as a knowledge and education city. The Talent programme aims to improve the match between talented polytechnic students and SMEs via apprenticeships and traineeships. The polytechnic InHolland opened an SME desk which should, among other things, stimulate the number of apprenticeships for polytechnic students in SMEs.

The harbour and industrial complex (HIC; industry, trade and transport) in Rotterdam mainly requires employees with a vocational education. In the discussion about the knowledge base, the increasing upgrading of job functions was identified; this also holds true for the harbour. There is a shortage in technicians both at lower and higher education levels. The municipality, together with the port authority and businesses (through Deltalinqs, the HIC representative organisation), are active in enlarging the intake of young people into harbour courses and jobs. General measures include improving the harbour image through the World Harbour Festival and more specifically organising days where potential students can take a look in HIC companies. Also, Deltalinqs (a network organisation of harbour related companies), together with HIC firms, finances and organisationally supports the Education Information Centre Mainport Rotterdam, which arranges tours through the harbour, HIC company visits, interactive exhibitions and provides learning materials to schools (www.eic-mainport.nl). A specific measure to increase the inflow of HIC-related courses is the Procescollege Mainport Rotterdam, a cooperative effort between HIC companies and two schools in Rotterdam to strengthen the chain of education and job-seeking in the Rotterdam process industry (www.procescollege.nl). At the high end of harbour education the Erasmus University Rotterdam (economists and managers) and the Technical University Delft (technicians) play an important role. The HIC complex, together with the municipality and the two universities, wants to strengthen this segment by founding an Academic Centre for Trade, Transport and Logistics; at the time of writing the set-up of the centre is still in the planning stage. Additionally, the Traineeship Rotterdam Harbour is aimed at the high educational end: this cooperation includes the Hogeschool Rotterdam (polytechnic), employment agency Yacht, the port authority and Deltalinqs. The traineeship aims to increase the number of polytechnic students that continue their education in the HIC field (Port of Rotterdam, 2003).

Company acquisitions Besides attracting knowledge workers via educational efforts, the municipality also draws knowledge workers by attracting companies. The municipal acquisition department helped to attract 39 firms to Rotterdam in 2003; this resulted in 1,200 additional jobs. The firms come from Asia (13), The Netherlands (11), the United States (10), Europe (3) and South

America (2). This number is stable compared to previous years. The number of extra jobs increased, especially because several companies established their Dutch branch office in Rotterdam and others moved/established European headquarters in the city. Foreign firms seem to be adding strategic activities in their Rotterdam offices. The number of Dutch firms that came to Rotterdam is lower than in previous years; in particular, fewer firms in the audio-visual (AV) sector have settled in the city. However, the AV companies that did come included some large ventures (OBR, 2003b).

Attractive housing For many years, Rotterdam has lacked sufficient attractive housing for medium and high-income groups. Recently the municipality has been actively reducing this qualitative housing mismatch: more high-grade apartments and houses are now available (e.g., Kop van Zuid area and the high-rise apartment building 'Hoge Heren') and construction activities in this medium and upmarket segment are still going on (large scale neighbourhood renovation, for instance, in Crooswijk).

9.4.2 Creating Knowledge

Universities are an important place for creating knowledge. The Netherlands is doing quite well in this field: seven Dutch universities are ranked in the European top 20 of universities with the greatest impact regarding scientific publications. Erasmus University is ranked ninth, the University of Leiden 16th, Technical University Delft 17th, but Eindhoven (in the province of Noord-Brabant) has the highest Dutch ranking of third (Stichting Nederland Kennisland, 2003).

In the province of South-Holland (together with Zeeland) €1.5bn was spent on research and development (R&D) in 2000 (compared to €1.7m in 1999); this makes the province the runner-up in absolute numbers after Noord-Brabant (which includes Eindhoven, the homebase of Philips). However, when the R&D expenditures are compared to the gross regional product, South-Holland only ranks eighth: about 1.6 per cent of GRP was spent on R&D in 2000 (compared to leader Noord-Brabant, with 3.1 per cent). Most of the R&D expenditure in South-Holland takes place in research and education institutes: 58 per cent. Regarding private businesses, R&D-intensive sectors (electrotechnics, manufacturing of machines, ICT, etc.) are not very abundant in South-Holland. Most R&D expenditure happens in the food industry (Unilever, Sara Lee) and chemical companies (Shell, DSM). Chemistry makes up for the largest part of South-Holland R&D: 13 per cent. The large firms contribute most to R&D spending: 1–2 per cent of the total number of firms contributes over 50 per cent of all business R&D expenditure. South-Holland ranks second in the number of patents applied for: 509 patents – 13.1 per cent of the Dutch total number. The leading province in The Netherlands is Noord-Brabant: 1937 patents – 50 per cent. The share of high-tech patents per total number of patents in South-Holland (14.9 per cent) is quite low compared to the Dutch average (29.3 per cent) (OBR, 2003a).

The Rotterdam region ('regio Rijnmond') performs relatively well concerning the number of firms with new product announcements. Main drivers behind these product innovators are the Technical University Delft and the high business density in the Rotterdam region. In R&D the region does quite well: Greater Rijnmond makes up 7 per cent of total R&D labour costs in The Netherlands (this ranks the region fourth of all Dutch regions). When the large cities in the country are compared, Rotterdam ranks third in R&D labour costs after Eindhoven and Amsterdam. Regarding R&D intensity and concentration, Rotterdam scores averagely. In high and medium-tech the city scores relatively well compared to the other G4 cities; Rotterdam has a high percentage of high- and medium-tech services, but a low share of high- and medium-tech manufacturing, compared to the Dutch average (OBR, 2003a).

The Rotterdam municipality aims to strengthen the knowledge creation capacity of the city and region: in the Knowledge Economy Programme (PKE) (Stuurgroep Kenniseconomie, 2003), four of the sub-programmes mentioned in the knowledge strategy paragraph should explicitly contribute to this: 1) recruiting or founding a top knowledge institution; 2) setting up at least three lectorships/professorships in ICT/logistics and industry (2003), metalelectro, process chemistry, urban/architecture (2004) and healthcare/medical/life sciences (2005); 3) founding knowledge circles with stakeholders from private firms, education and government to strategically cooperate in the area of innovation (R&D) in process chemistry, audio-visual, metalelectro, architecture and healthcare/medical/life sciences, among others; and 4) establishing a centre of excellence in trade, transport and logistics. Recruiting the top knowledge institution is aimed to take place in 2005. The lectorships/professorships are currently being filled. Starting subsidies have been granted to several knowledge circles, including high-opportunity clusters (medical, ICT, logistics, AV, new media, metalelectro and industrial chemistry), a local innovation platform (the Economic Development Board Rotterdam) and a knowledge circle SME. The academic centre for trade, transport and logistics is still in the formation stage. Besides these initiatives, cooperation between EUR and TUD should improve, according to the PKE.

9.4.3 Applying Knowledge/Making New Combinations

In line with the Dutch situation, in Rotterdam the linkages between firms and educational institutes are rather weak. Only 15 per cent of companies in the Rotterdam region have regular contacts with universities, polytechnics and other knowledge institutions; 38 per cent interact every now and then. This implies that almost half of all firms do not keep up any contact with knowledge institutes. There is room for improvement: over one-third of the firms say they want to increase the number of such contacts. Most of the interaction between knowledge institutions and companies takes place on a regional scale: about 75 per cent of these contacts concern institutes in South-Holland. No more than 4 per cent of the firms think that the knowledge and expertise of the knowledge institutions are well tuned to the companies' needs; 40 per cent thinks the match is very bad. In addition, some 20 per cent of the companies do not know where they should go for answers to specific knowledge-content questions. The majority of the firms (70 per cent) believe that the main use of contacts with knowledge and education institutions is to obtain trainees; 40 per cent mention the buying/exchanging of knowledge, and 22 per cent cite the joint development of goods or services (Etin Adviseurs – OBR, 2003a). The municipality and the province (the Knowledge Alliance among others) have acknowledged this. A multitude of initiatives aim to improve this situation: several are discussed below.

In 2003 the municipal policy to support start-up firms (by means of consultation, courses, accompaniment and funding) helped 510 companies, a a few more than the intended goal of 500. The investor 'Ondernemers Fonds Rotterdam' (Entrepreneurs Fund) will be expanded with a young firms fund. This plan is being elaborated and participants are being recruited. In 2004 the municipality, together with Knowledge Alliance South-Holland, organised the 'Zakenfestival' (Business festival) in Rotterdam; here, big firms, SMEs, network organisations, knowledge institutes and city districts were able to exchange information and extend networks (OBR, 2003b).

Port infolink The Port infolink initiative (www.portinfolink.com) provides an example of applying knowledge. Port infolink – which is included in the municipal Knowledge Economy Programme – was set up to improve communication in the harbour complex by using ICT. The processes in the logistic chains that run through the harbour have to be optimised by using online ICT services that increase customer efficiency and enhance product quality. It is mainly funded by the port authority, but is also expected to obtain funds from shipping companies, the Chamber of Commerce and local government.

Techno centres Techno centres are initiatives that aim to increase innovation in Rotterdam by stimulating the application of new knowledge and products in firms and by strengthening the educational output to companies.

A techno centre that is already running is Digital Port Rotterdam; it is related to Port infolink. Digital Port is an ICT expertise centre that aims to stimulate the effective use of ICT in SMEs. It focuses on firms in trade, transport and logistics. Together with start-up supporter Area 010, the Digital Port will accompany start-up firms to the market. Besides municipal funding, businesses also contribute financially.

The process industry and metalelectro techno centres recently received starting grants. The Knowledge Centre Mainport Rotterdam (Kenniscentrum Mainport Rotterdam; KMR) is responsible for management; this is a representative organisation of companies in the harbour and industrial complex. The techno centre for process industry aims to realise a practising factory where process operators can be educated. The techno centre metalelectro has started a 'technology and labour market' centre that aims to increase the number of people who study Industrial Technology and to help graduates find suitable jobs.

Media schools In the media schools (this includes a polytechnic and several vocational colleges) about 130 participants attend labour market-oriented classes in the AV sector (make-up, location management, sound/camera/light courses, etc.). A project that was about to start at the time of writing is the Transferium, directed by the W. De Kooning Academy. The Transferium is a platform for media and ICT education and labour market. Here, cooperative ties should be created between firms and schools; it should also distinguish Rotterdam as a media and culture city. The media schools are meant to contribute to the AV sector in the city.

Knowledge Circles The Knowledge Circles mentioned previously also play an important role in the application of knowledge. They are in an early development stage: having received starting grants, the business plans have to be elaborated further. The leading parties are the Erasmus Medical Centre (see Incubator EMC) and the polytechnic Hogeschool Rotterdam.

Lectureships in ICT and logistics, and architecture Also at an early stage of development are lectureships in ICT and logistics, and architecture. Both lectureships aim to increase knowledge transfer and innovation. According to the Knowledge Economy Programme, this should be achieved through applied research, involvement in product development and facilitating start-up firms. The lectureship in ICT and logistics aims to have 50 participants in a trainee project each year. Both lectureships are directed by the Hogeschool Rotterdam.

Start-ups: Foundation Incubator Erasmus MC A Knowledge Circle is planned in the medical sector, to be set up in 2004–2005. This should enhance cooperation between the Erasmus Medical Centre (EMC), companies and government. Recently EMC became really involved in applying knowledge that is an academic research spin-off. Under the old regime, when medical faculty employees were more restricted by university regulations and policy, only a limited amount of medical scientific knowledge was used in private business activities. The EMC loosened its ties with the Erasmus University in 2002–2003: now the university only has influence through the Board of Supervision and consultation meetings. The new situation implies more entrepreneurial freedom for the EMC. In 2003 the EMC set up the department for knowledge transfer: this was badly needed since the EMC was lagging behind other Dutch academic medical centres. The Foundation Incubator Erasmus MC was founded at the beginning of 2003. The incubator is financed by the Ministry of Economic Affairs (SIT subsidies by Senter and Technostarter funding), the Rotterdam municipality (via OBR development company), the EMC and Erasmus University. At the time of writing four start-ups were running, but another 11–13 firms were about to be launched. The start-ups are supported via subordinated loans (not through shares). The start-ups can be placed in the EMC Holding which presently contains some 20 firms. Most of the start-up companies are active in the field of life sciences. There are enough ideas to potentially start-up 30–40 firms within the incubator in the next two to three years. Yet the search for such ideas is at present not yet optimal. The newly-discovered ideas for medical products are first investigated by the Bureau for Industrial Ownership (www.bie.nl): they examine whether the idea is really new and unclaimed. If this test proves successful, the idea is judged by an external board. If an idea has sufficient potential, a patent is applied for. Finally, it is decided whether to sell the patent or to develop it into a product inside the EMC.

Besides the Incubator, the EMC is involved in several other innovative activities where knowledge is applied: there is a virtual clinic for Hepatitis B that advises patients in, for instance, Poland, via ICT communication. Another high-tech EMC activity is the Skills Lab.: this concerns post-academic education at a distance, using ICT communication and robots (e.g., doctors physically present in the Rotterdam hospital operating on people in Bosnia). Also, the EMC financially supports professors who set up companies outside the incubator: they receive support for three years from the EMC.

Provincial start-up policy The province of South-Holland also aims to stimulate start-up firms. Its ambition is to achieve a 20 per cent growth in the number of technostarters by the end of 2007. The policy should be geared to the nine pillars (see Section 9.2) and will build upon current activities such as the Information Point Techno Starters. Currently, the provincial starters policy is not yet linked to plans at other governmental levels (the municipalities and the state). Therefore, South-Holland will develop an action plan in the first half of 2004. Within the 'Nieuwe bedrijvigheid' (New business activity) projects the province aims to accompany 90 start-ups in the municipality of Rotterdam and the region The Hague-Leiden until 2005; these firms should generate about 500 extra jobs. The Area 010/015 project aims at stimulating start-up activities in Rotterdam and Delft. The South-Holland Investment Fund will partly finance the start-ups: it aims for participation in 15 firms and to support 50 start-ups up to 2007. This fund is financially supported by the province, the municipalities of The Hague and Delft and the Rabobank (Province of South-Holland, 2004).

New combinations: the audio-visual sector The audio-visual (AV) sector in The Netherlands is concentrated in Amsterdam and Hilversum. In recent years Rotterdam has tried to kick-start economic development in this sector. The municipality did not have any specific policy aimed at innovation. The city board did, however, support the set up and growth of AV education in schools in Rotterdam. The educational AV policy is part of the Knowledge Economy Programme. Although the current economic contribution of the AV sector in Rotterdam is quite limited, there are opportunities to develop the sector in the direction of new media, special effects montage, etc. These fields are relatively young and might be interesting areas to establish 'new combinations' with fields like ICT. Nevertheless, there have to be some conditions for such development: besides government support and a solid supply of education, the demand side has to develop. Currently, the large firms present in Rotterdam are not particularly interested in acting as local buyers of new media (for instance, in the field of advertising).

9.4.4 Developing Growth Clusters

The Rotterdam municipality uses a cluster approach to broaden the scope of economic activities. The city focuses on clusters that contribute to an increased diversification of the economy, that bind knowledge to the region, increase the quality of the city and improve the innovative climate. Cluster development includes network stimulation, focused acquisition and offering specific facilities to entrepreneurs. The spearhead clusters are discussed below.

The ICT cluster According to a cluster policy evaluation (Policy Research Corporation, 2001) the information and communication technology cluster belongs to the fastest growing sectors in the Rijnmond region. The cluster offers good prospects for economic diversification and embedding knowledge in the region. Moreover, it contributes to the innovative climate. What is lacking is sufficient attention on the demand side of the market. The ICT cluster is seen as an important binding factor for other economic sectors such as the harbour and industrial complex (including petrochemistry and maritime services). The municipality adjusted its ICT cluster policy in 2002: entrepreneurship regarding ICT applications was to be strengthened, the relations with other sectors (especially SMEs in the harbour complex) were to be enhanced, the ICT cluster would be broadened to include multimedia (a 'new combination' with the AV sector), ICT infrastructure was to be improved and knowledge-intensive start-ups were to be supported (OBR, 2002). Since then, several measures have been taken, including Port infolink (upgrading the harbour complex by using ICT), I-Portal (a virtual meeting place for ICT firms) and Digital Port (stimulating ICT in SMEs) were started, a lectureship ICT and logistics is about to commence, two pilot projects in ICT infrastructure (Lloydkwartier and Nesselande) are being executed, and redevelopment of the 25kv building and Schiecentrale (to house ICT and AV businesses) has begun. Still waiting to happen are increased ICT education and implementation of the ICT Fund that aims to stimulate innovation through financial incentives (OBR, 2003b).

The AV cluster The audio-visual cluster's contribution to the Rotterdam economy is rather limited. The biggest spin-offs are found in the field of improving the cultural climate and the image of the city. The AV cluster has been mainly supported by the municipality which has provided subordinated loans, and offered specific facilities in combination with low rents (Policy Research Corporation, 2001). The Rotterdam Movie Fund ('Rotterdams Filmfonds') is a key instrument: the OBR development company lends money to movie producers if an amount of 150 per cent is spent in Rotterdam. In 2002, €2.5m in loans resulted in almost €7m in turnover (OBR, 2003b). The contribution of the AV cluster to diversifying the Rotterdam economy is limited

– the cluster is too small and relies too heavily on government money. In 2002 the municipality decided to continue supporting the cluster, but some policy adjustments were made: the focal point was more directed to including new media and the recruitment of AV institutes and firms was intensified. ICT and new media were jointly strengthened in educational programmes at the vocational institutes and polytechnic Hogeschool Rotterdam (OBR, 2002). The redevelopment of the Lloydkwartier, 25kv building and Schiecentrale which is part of stimulating the ICT cluster also concerns enhancing the AV cluster.

The International Film Festival Rotterdam (IFFR) spin-off regarding city image is well-known: abroad the IFFR is considered to be one of the main film festivals in the world (IFFR, 2004) and many foreign people place the IFFR in the top three when thinking about Rotterdam's most striking features.

The medical cluster Several years ago, van den Berg and van Klink (1993) and Policy Research Corporation (2001) mentioned the considerable potential spin-off of the medical cluster towards the city, but identified the lacking entrepreneurial climate as an inhibiting factor. Several areas (such as biotechnology and life sciences) are still in a rather elementary phase globally; this offers good opportunities for successful active intervention (founding an incubator). The municipality has acknowledged the opportunities for the medical cluster. Recent liberalisation developments concerning the Erasmus Medical Centre have already been discussed above; they indeed seem to have a very positive influence on the economic utilisation of medical knowledge (e.g., through the EMC Incubator). The medical activities in Rotterdam have recently gained momentum, but there is a long way to go: some other Dutch cities and several cities abroad are more advanced in this field. There are advanced plans to greatly extend the medical complex; the geographical focal point is the Hoboken area where the EMC is currently located. The complex is planned to be extended to include businesses that have intense linkages with the hospital and the medical faculty; this will include more incubator space and facilities. The EMC construction plans also offer a good opportunity to create positive spin-offs for poorly-served neighbourhoods nearby.

Although the medical cluster in Leiden (which includes the pharmaceutical multinational Crucell) is geographically quite nearby (about 40 minutes by car or train), there are no current plans to build up cooperative linkages. This has to do with the fact that the Rotterdam actors who are involved in the medical cluster do not consider such cooperation paramount. Besides, Leiden is probably too far away for really intense collaboration. According to actors in the medical cluster such intensive teamwork requires travelling distances to be about ten minutes at most. In addition to Hoboken, some other geographic locations are considered for medical activities: Marconiplein/Veerhavens (ten minutes by metro from Hoboken) and Schieveen (this site is further away when transportation time is considered and therefore less appealing).

Cooperation between polytechnics (Hogeschool Rotterdam) and the EMC is still largely absent. This is an opportunity since these institutes complement each other: while the EMC is active in academic medical research and product development, the polytechnics are involved in innovation in health and elderly care.

Other clusters The clusters evaluation (Policy Research Corporation, 2001) included several other clusters. The recycling cluster that was set up in 1994 had been mostly oriented towards recruiting some companies. Broadening this policy to other cluster activities (like increasing interaction between network actors and stimulating innovations) could eventually enlarge the (export-oriented) market.

The leisure cluster has been mostly focused on improving the quality of life in Rotterdam. A vibrant city is expected to boost economic activities indirectly (e.g., by attracting talented knowledge workers). The clusters evaluation recommended focusing more on entrepreneurs and merging the institutions that are active in the leisure field. This is in line with other research (van den Berg et al., 2000). The municipality adjusted its policy according to this evaluation: among other things, by centralising the coordination of leisure activities and defining product market combinations (OBR, 2002).

Architecture is also present as a cluster in Rotterdam. The city is at the architectural forefront at the global level: various important architecture companies are established in Rotterdam. The cluster is clearly recognised and several institutions play a driving role by organising a broad range of activities (e.g., The Netherlands Architecture Institute and the Berlage Institute). This implies that the role of the municipality can be kept limited in this respect. However, the city is able to contribute to the image of Rotterdam as a city of architecture through its role as real estate developer and spatial policy coordinator (Policy Research Corporation, 2001). Architecture can contribute extensively to Rotterdam's image (van Holsteijn, 2004).

Rotterdam hosts several insurance companies. Considering the scale of these firms, the cluster is operational on a national scale. These enterprises would not gain much from being regionally embedded. Therefore, municipal cluster policy is not needed in this sector. However, since a development of concentration/centralisation is going on in this sector, it is recommendable for the municipality to maintain regular contacts with these firms as a form of general economic policy: taking the insurance companies' needs into account can contribute to their staying in Rotterdam (Policy Research Corporation, 2001).

In addition to the clusters discussed above, the municipality has continued to identify (potential) economic clusters. Some potential clusters mentioned are trade, food/fruit, certain forms of chemistry and maritime business services. The municipality has established relationship managers for these activities. The managers together with entrepreneurs identify the specific needs of economic sectors and judge whether the municipality should play an active role (OBR, 2002).

Geographical focal points: 'landmarks' Up to the time of writing the municipality has been taking the lead in the cluster policy: the input of private parties was limited. The Knowledge Economy Programme (PKE) wants to change this focus more towards the businesses. Virtually all the cluster activities mentioned above are part of the PKE. The PKE and the draft economic vision for 2020 both want to link the creation and development of clusters and economic sectors to geographical focal points ('landmarks'). This comprises Hoboken (medical cluster), Lloydkwartier (AV and ICT cluster), the new central railway station area (high-grade international services), the second Maasvlakte port area (harbour and industry), the 'Stadshavens' (City Harbours) district (focus of economic activities still to be decided) and the R&D zone. The latter area is discussed below.

R&D zone The R&D zone (Rotterdam and Delft zone; research and development zone) centres on the the area of Schieveen next to Rotterdam Airport within the Rotterdam municipality, and the Technopolis Innovation Park next to the Technical University in Delft. Both Schieveen (75 ha.) and Technopolis (70 ha.) are in the development stage. These business parks should become high-grade areas embedded with much vegetation and water. Technopolis is focusing on (technical) research and product development; Schieveen will be more oriented towards business activities. Both sites are to be fully occupied by around 2020; Technopolis has already begun to house some firms, while Schieveen is expected to start in 2007. Technopolis' grounds are fully owned by the TU Delft, the Schieveen land is owned by the Rotterdam municipality. A problem for both these areas is road congestion on the A13 highway. The R&D zone is expected to play a key role in the cooperation between the Erasmus University and TU Delft: e.g., the Academic Centre for Trade, Transport and Logistics could lead to business activities in the

R&D zone that benefit from knowledge and experience gained at Erasmus University and TU Delft.

Provincial cluster policy The province of South-Holland has made cluster policy a task of the Knowledge Alliance South-Holland. Its management will follow the nine 'knowledge pillars' described in Section 9.2: for each pillar a trigger and an ambassador will be appointed. The trigger will organise the necessary activities within the cluster, while the cluster ambassador operates as an external focal point and representative. The cluster management will also involve organisations like Syntens, Senter (both are agencies working for the national Ministry of Economic Affairs) and TNO (a national applied science organisation). SME activities are also expected to be included in the cluster approach. The provincial cluster activities are currently still at the planning stage (Province of South-Holland, 2004).

9.5 Conclusions and Perspectives

This section contains conclusions and unfolds perspectives for Rotterdam's position in the emerging knowledge economy.

9.5.1 Foundations of the Knowledge Economy

The foundations of Rotterdam's knowledge economy are weak to moderate. The importance of the harbour is still considerable for the city from an economic point of view. Although the harbour is by far the largest in Europe, in recent years it has lost market share to other European ports (Antwerp, Hamburg, Amsterdam): while trans-shipment in Rotterdam increased in absolute numbers, growth in competing ports was greater. Moreover, harbour and industry employment is declining as a result of increased productivity. Probably even more important is the shift towards more knowledge intensive economic activities: the share of (business) services in Rotterdam's GRP is further increasing. Compared to the other G4 cities, Rotterdam has relatively many economic activities in which employment is declining: the port and industry especially contribute to this.

An indication of the value of the knowledge economy foundations in Rotterdam is given in Table 9.4. It can be concluded that the city has a rather weak overall score. The city's *knowledge base* can be characterised as rather good. The Rotterdam labour force has a low education level compared to the Dutch average, but the percentage of people with third-level education is similar to other European cities. The region contains two well-performing universities; the Erasmus University Rotterdam (EUR) and the Technical University in Delft (TUD) contain over 35,000 students. The TUD and the medical and economic/management fields at the EUR are of high quality. The Rotterdam labour force is relatively young. There is, however, a shortage of knowledge workers and those with a technical background are particularly in demand. The *economic base* is moderate: the city has comparatively many jobs in sectors with declining employment (industry and port). Even so, employment in services has grown considerably in recent years. Productivity is high but its growth is cause for concern. A positive side of the economic base is the good economic dynamism: Rotterdam has a high number of start-up firms. The *quality of life* is rated neutral. The housing conditions and living environment are valued low. Trash, vandalism and (feelings of) insecurity are considered important. The city contains too few medium-priced and expensive homes. Leisure facilities are good and plentiful. Although Rotterdam is increasingly becoming known as a nice, vibrant place, the city needs further image improvement. *Accessibility* is judged as very good: car access and public transport are fine, while international accessibility is good (with Schiphol Airport quite nearby) and further improving (soon to be connected to the European high-speed train network). *Urban diversity* can be considered rather good: large ethnic minority groups contribute to significant cultural diversity (restaurants and leisure events). Furthermore, these groups rejuvenate the working population. On the other hand, they have low education levels and a high percentage is unemployed. The lack of integration of some ethnic

Table 9.4 An indication of the value of knowledge economy foundations in Rotterdam

	Foundations	Score in Rotterdam
1	Knowledge base	+
2	Economic base	□
3	Quality of life	□
4	Accessibility	++
5	Urban diversity	+
6	Urban scale	+
7	Social equity	-

-- = very weak; – = weak; □ = moderate; + = good; ++ = very good.

minorities has caused problems throughout the years which have negatively affected the city image. Regarding *urban scale*, Rotterdam is a secondary city in Europe. Within The Netherlands it is the second largest city and part of an urban network (Randstad) that seems to be integrating further. The *social equity* in Rotterdam is quite weak as a result of high unemployment compared to Dutch figures. The city houses many low-income households.

9.5.2 Activities in the Knowledge Economy

In accordance with our framework of reference, Table 9.5 considers four types of activities in a knowledge city, and attaches scores for Rotterdam. The overall score is moderate. In *attracting knowledge workers* the performance is average. The number of students in third-level education increased significantly in previous years; this includes foreign exchange students. Now the city has to try to retain the talented students in Rotterdam; policy efforts in this field have just begun. Other recent developments are the (joint) efforts by companies, knowledge institutes and municipality to increase the number of students who advance to higher education levels after their initial graduation. The current upgrading policy in housing construction is positively contributing to enhancing the quality of life. Rotterdam performs rather well in *creating knowledge*. The Erasmus University Rotterdam and Technical University Delft both are in the top 20 of European universities with the biggest impact in scientific publications. R&D expenditures compared to GRP are quite low in South-Holland and R&D intensive sectors are not abundant. The number of patents is substantial. The Rotterdam municipality has recently started strengthening knowledge creation capacity by, among other methods, recruiting a top knowledge institute and setting up lectureships/professorships. *Applying knowledge* is moderate in Rotterdam: few companies have regular contacts with knowledge institutions. However, there are multiple initiatives to improve this situation (e.g., the incubator at Erasmus Medical Centre and several techno centres); they are supported by the municipal Knowledge Economy Programme. Chances for *new combinations* are present in combining ICT with traditional harbour and industry firms, the medical cluster and the audiovisual cluster. In addition to this, the *development and growth of economic clusters* is also taking place with the support of the municipal Knowledge Economy Programme. As well as the clusters in ICT, AV and the medical field, some other clusters offer considerable potential. The insurance cluster does not seem to need specific municipal cluster support; the leisure and architecture clusters, however, would probably benefit considerably from municipal cluster support.

The last two clusters contribute significantly to the city image. Finally, there are plans to cluster R&D activities on the verge between science, management and high-grade business in some areas between Delft and Rotterdam. Although the number of cluster initiatives is extensive, Rotterdam has a head start in virtually none of the sectors compared to other Dutch cities: only architecture and port-related activities seem to have such a position.

Table 9.5 An indication of the value of knowledge economy activities in Rotterdam

	Activities	Score in Rotterdam
1	Attracting knowledge workers	□
2	Creating knowledge	+
3	Applying knowledge/making new combinations	□
4	Developing (new) growth clusters	□

-- = very weak; – = weak; □ = moderate; + = good; ++ = very good.

9.5.3 Perspectives

Traditional economic activities like the port and industry still make up a large part of Rotterdam's GRP: these activities should be maintained by upgrading effectiveness and efficiency. This implies that certain sectors will show further decreases in employment. Therefore, Rotterdam will have to develop further its economic activities in more knowledge-intensive fields. The current position of Rotterdam concerning knowledge economy foundations and activities is moderate. The plans and programmes that were and are started to improve these assets are extensive. A strong facilitating role by the Rotterdam municipality to involve all relevant parties and coordinate their activities is necessary: the setting up and execution of the Knowledge Economy Programme is expected to be crucial in this respects Rotterdam municipality's activities, however, need to be geared to initiatives in neighbouring cities: Delft in particular and perhaps The Hague and Leiden should be more involved. Currently interaction and cooperation with these cities and with the province is low. However, cooperation between Rotterdam and Delft has been intensified recently: the universities and cities signed an agreement to strengthen collaborative efforts.

Organising capacity (van den Berg et al., 1997) plays a crucial role in successfully developing Rotterdam towards a knowledge economy. In previous years many municipal policies were created to enhance the urban economic development, but relatively few were executed. Since 2002 the municipal policy implementation seems to have increased considerably: many of the plans created in the recent past are now being carried out (e.g., the medical complex extension). Another positive aspect of the current situation is the interactive aspect of the policies: instead of organising matters top-down, a bottom-up approach is now prevalent. This can be witnessed, for example, in the Economic Development Board, the coordinated policy efforts with the Erasmus University and the Technical University Delft and also in increased inter-urban cooperation between the Randstad cities. Such cooperative efforts are important to influencing decision-making processes at higher levels like the Dutch government and the European Union. Furthermore, the strengthening of the link between creating and applying knowledge is being improved in Rotterdam. The city is now involved in coordinating knowledge creation in the important economic activities; for instance, with regard to harbour activities and research.

Summarising the Rotterdam case, it can be stated that, although several of the knowledge foundations are rather weak to moderate, the prospects for the city are looking bright. The city has put together an integral plan to enhance and extend knowledge-driven activities and currently a vast number of such activities are being implemented.

Note

1 Clusters are defined here as sector-crossing (regional) networks, horizontally and/or vertically mutually differing and complementary enterprises and/or organisations (including education and knowledge institutes), that bring about a specific economic dynamism in a specified geographical area.

Discussion Partners

Mr Cees Jan Asselbergs, Manager, Deltalinqs (harbour network organisation).
Mr Hans Beekman, Manager, Department for Education (DSO) Rotterdam.
Mr Steef Blok, Financial Manager, Faculty of Medicine, Erasmus Medical Centre.
Mr Joop van Boven, Development Corporation Rotterdam (OBR).
Mr Wim Dik, Technopolis Innovation Park – Technical University Delft (TUD).
Ms Sandra den Hamer, Director, International Filmfestival Rotterdam.
Ms Merel Heimens Visser, Manager, Kennisalliantie Zuid-Holland (Knowledge Alliance).
Mr Sander de Iongh, Manager, Development Corporation Rotterdam (OBR).

Ms Miranda Janse, Manager, Development Corporation Rotterdam (OBR).
Mr Bart Ronteltap, Manager, Kennisalliantie Zuid-Holland (Knowledge Alliance).
Mr Huibert Tjabbes, Manager, Incubator MC, Erasmus Medical Centre.
Mr Jasper Tuytel, Director, Hogeschool Rotterdam (polytechnic).
Mr Jan van 't Verlaat, Development Corporation Rotterdam (OBR).

References

Berg, L. van den and H.A. van Klink (1993), *Naar een 'Medisch Centrum West' in Rotterdam? De betekenis van medische complexen voor de stedelijke economie*, Rotterdam: Euricur.

Berg, L. van den and P.M.J. Pol (1998), *The European High-Speed Train and Urban Development*, Aldershot: Ashgate.

Berg, L. van den and A.P. Russo (2004), *The Student City – Strategic Planning for Students' Communities in EU Cities*, Aldershot: Ashgate.

Berg, L. van den, E. Braun and J. van der Meer (1997), *Metropolitan Organising Capacity*, Aldershot: Ashgate.

Berg, L. van den, E. Braun and A.H.J. Otgaar (2000), *Sports and City Marketing in European Cities*, Rotterdam: Euricur.

Bureau Louter (2003), 'De Economische Hittekaart van Noordwest-Europa', http://www.bureaulouter.nl.

COS (Centre for Research and Statistics) (2003a), 'Factsheet Prognose Bevolkingsontwikkeling Rotterdam 2003–2017', http://www.cos.nl.

COS (2003b), *De staat van Rotterdam 2003 – een pilot.*

COS (2003c), *De G4 in de peiling* (citizen survey).

COS (2003d), *Key Figures Rotterdam 2003.*

Ecorys (2003), *Monitor Toerisme G4.*

Etin Adviseurs – OBR (2003), *Economische Verkenning Rotterdam 2003.*

G4 (2002), *Problemen en kansen in de G4.*

Het Financiële Dagblad (2001), *Gespleten Randstad*, 26 October.

Van Holsteijn, J. (2004), 'Economische aspecten van architectuur – De bijdrage van architectuur aan de stedelijke economie met een toepassing op Rotterdam', MSc thesis, Rotterdam.

IFFR (2004), 'Beleidsplan International Film Festival Rotterdam 2005–2008', http://www.filmfestivalrotterdam.com.

Kloosterman, R.C. and B. Lambregts (2001), 'Clustering of Economic Activities in Polycentric Urban Regions: The case of the Randstad', *Urban Studies*, Vol. 38, No. 4, pp. 717–32.

Knowledge Alliance South-Holland (2003), 'Handvest Kennisalliantie Zuid-Holland', http://www.kennisalliantie.nl.

Knowledge Alliance South-Holland (2003/4), *Werkplan Kennisalliantie Zuid-Holland.*

Ministerie van Economische Zaken (2002a), *Benchmark Regionaal Investeringsklimaat.*

Ministerie van Economische Zaken (2002b), *Benchmark Gemeentelijk Ondernemersklimaat.*

OBR (OntwikkelingsBedrijf Rotterdam) (2002), *Kwaliteit werkt door – Economische Visie Rotterdam 2002–2006.*

OBR (2003a), *Economische Verkenning Rotterdam 2003.*

OBR (2003b), *De Economische Programmamonitor*. Rotterdam.

Office of the Deputy Prime Minister (2004), 'Competitive European Cities: Where do the Core Cities stand?', London, http://www.odpm.gov.uk.

Pol, P.M.J. (2002), 'A Renaissance of Stations, Railways and Cities – Economic Effects, Development Strategies and Organisational Issues of European High-Speed-Train Stations', PhD thesis, TRAIL, Rotterdam/Delft.

Policy Research Corporation (2001), *Clusterbeleid in Rotterdam*.

Port of Rotterdam (2003), *Haven- en Industrie Monitor*, No. 3/4.

Province of South-Holland (2004), *Ontwerp Innovatiebrief kenniseconomie Zuid-Holland*.

Stichting Nederland Kennisland (KnowledgeLand Foundation) (2003), 'Time to Choose – Knowledge Economy Monitor 2003', http://www.kennisland.nl.

Stuurgroep Kenniseconomie (Steering Group Knowledge Economy) (2003), *Programma Kenniseconomie – Rotterdam maakt werk van de Kenniseconomie*.

TU Delft (2003), 'Ingeschreven studenten TU Delft, studiejaar 2003/2004', http://www.tudelft.nl.

VROM (Ministry of Housing, Spatial Planning and Environment) (2001), *Vijfde Nota over de Ruimtelijke Ordening 2000/2020*.

Winden, W. van and P. Woets (2003), 'Urban Broadband Internet Policies in Europe – A Critical Review', paper presented at ERSA Congress, Jyväskylä, Finland, 27–30 August.

Chapter 10

Zaragoza

10.1 Introduction

The city of Zaragoza counts 628,401 inhabitants (2003), making it the fifth city of Spain. Geographically, Zaragoza is strategically situated between Barcelona and Madrid, Valencia and Bilbao. It is the administrative and economic capital of the Aragon region, a large and sparsely populated area of some 1.2m inhabitants. In the last few decades, the city has maintained its strong position as an industrial region – it is the fourth largest local economy in Spain – and has benefited from the generally favourable economic developments in Spain.

In the years to come the city will be confronted with major changes. One of the most important developments is the improved rail link with Barcelona and Madrid. In a few years time, Zaragoza will be connected to the high-speed link that connects Spain's two urban powerhouses. This will affect the geo-economic position of the city and offers a number of new opportunities, but also threats. This chapter aims to give an analysis of Zaragoza's position and opportunities in the knowledge economy. After this introduction, Section 10.2 briefly describes the knowledge strategy. We will describe and analyse the city's 'knowledge foundations' (as defined in our analytical framework), in Section 10.3, and the 'knowledge activities' in Section 10.4. Section 10.5 concludes.

10.2 Knowledge Strategy

There are 'knowledge economy strategies' relevant to our case study at the levels of the national state, the autonomous Region of Aragon, and the City of Zaragoza.

The *Spanish central government* has committed itself to meeting the Lisbon Challenge, which implies an increase in spending on (higher) education and R&D. Furthermore, on a national level, the España.es programme is relevant. This is a major action plan for the development of the information society in Spain. It aims to give strong impulses to the ICT sector, to raise awareness of the importance of ICTs in the society, to improve ICT education and to ensure

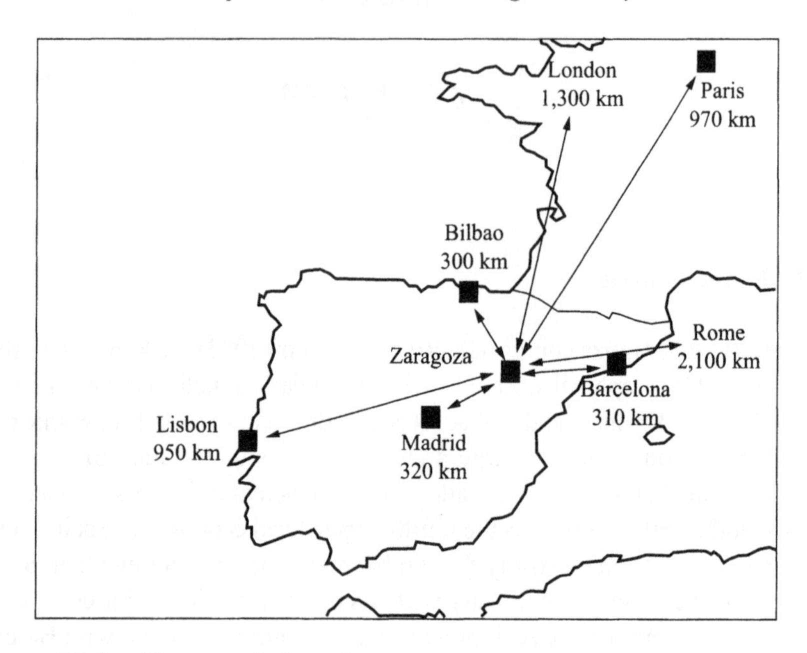

Figure 10.1 Zaragoza's location

Source: Ebropolis (1998).

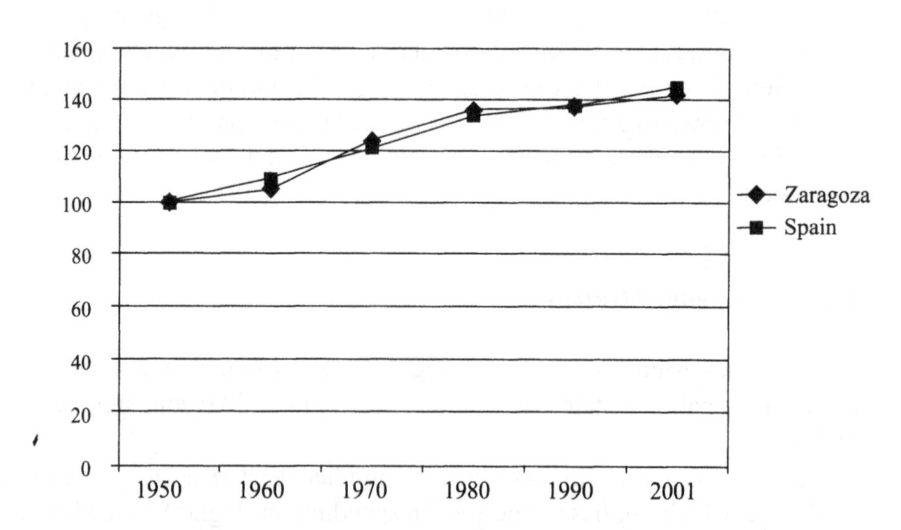

Figure 10.2 Index of population growth in Zaragoza and Spain (1950 = 100)

Source: IAE (2003).

access to ICT for all citizens and SMEs. However, at the time of writing a new government has just been inaugurated. It remains to be seen what the knowledge economy strategies of this government will be.

In Spain, the regions enjoy a high level of autonomy. They are important players and are responsible for education policies, innovation policy and lately also for higher education policy. The Region of Aragon – of which Zaragoza is the undisputed political and economic capital – has several action plans to promote the transition to the knowledge economy. Furthermore, it develops a range of policies. Two initiatives are worth mentioning:

- the WALQA initiative. This is a set of plans to promote e-business, create ICT centres and invest in telecom infrastructures;
- InnovaAragon. This project aims to improve links between science, technology and companies to promote innovation. It supports innovative cooperation between firms and universities, and supports the exchange of good practice.

Between the regional government (Aragon) and the municipalities (Zaragoza), there are two more government layers: the province (there are three provinces in Aragon) and the 'comarca'. Both have limited capabilities. Furthermore, it is worthy to note that the municipalities obtain some 80 per cent of their budget from the state. Their own tax-raising capacity is limited.

The City of Zaragoza has recently become active as well. In 2003, the municipality presented its comprehensive knowledge economy strategy named 'Zaragoza towards the knowledge society'. Key elements of the strategy are economic diversification – notably towards a more services and information oriented economy – and an improvement of the utilisation of the existing knowledge base. This strategy contains an extensive programme of actions directed at fulfilling these aims.

This strategy is an elaboration of Zaragoza's more general 'Strategic Plan'. This plan points out the main directions for Zaragoza's future development. A first version of the plan was drawn up in 1998 by Ebropolis, the Association for the Strategic Development of Zaragoza and its Metropolitan Area, headed by the City Council of Zaragoza. Ebropolis is a network organisation with members in all sectors of urban life, from civic organisations to companies to public entities. The plan has full backing from all these sectors and was drawn up in close collaboration with them. Several elements of the strategic plan refer to the need for a shift towards a knowledge economy: Zaragoza should offer an attractive environment for firms, based on innovation, quality and environmental sustainability, and should seek to create an effective infrastructure to link markets, firms, technology and research throughout the region.

To elaborate on the theme of the knowledge economy, in 2002 a working group on new technologies and knowledge society was set up, comprising 50 experts from the industry, knowledge institutes and public entities. This group had the aim of developing a strategy to incorporate Zaragoza and its metropolitan area to the knowledge society in a balanced way, taking advantage of this opportunity as a tool for economic and social development.

The strategy was developed by a wide group of experts and key people in the city. The ideas were widely discussed and validated in workshops with a number of stakeholders and participants, making use of the Ebropolis network. The strategy has become more practical and concrete, and has been linked to the newly-emerging urban space in and around the old and new station areas. It was decided to promote the creation of a city of knowledge in the new multifunctional urban spaces produced by the intermodal station, boosting the development of technological centres, with special attention to the promotion and production of audio-visual content.

It is here that the Digital Mile project enters the picture. This project has the objective of creating in Zaragoza an area of innovation and knowledge, located in the city centre. Its development has been stimulated by the arrival of the high-speed train, the urban land that will become available due to the construction of a new station and the Expo 2008 project: Zaragoza is competing to organise the Expo in 2008 and if successful, the Expo will be held in the areas to the right of the meander in the Ebro river (see Figure 10.3).

The area under consideration for the Digital Mile consists of three zones, pictured as the darker shaded areas in the figure: 1) the railway land to be recovered after the arrival of the high-speed train; 2) the Expo 2008 area (north of the river); and 3) the area in between (the Almozara district). The first is the most important and concrete, the last two could be included in a later stage.

The Digital Mile is intended to be an urban place in which the possibilities of new information technologies can be used at any time, place and position. This requires an excellent fixed (optic fibre) and mobile electronic infrastructure.

The objectives of the Digital Mile are as follows:

- to be a laboratory for new ways of living, learning and working. The area will be very mixed in terms of urban functions. As such, this is nothing new. But it will also be a location for pilot projects and massive testing of new technologies and concepts. This would make the area attractive

Figure 10.3 The Digital Mile area

for innovative people and companies, and give it a competitive edge. This can only be realised with a superior digital infrastructure;

- to improve Zaragoza's image as an innovate city. Currently the city does not score highly in the ranks of 'technology cities'. But being a latecomer, it has the opportunity of taking new approaches;
- to strengthen the knowledge-based economy of the city. The project will try to boost the number of high-skill jobs by attracting or nurturing companies that are active in advanced services, technology development and innovation;
- to attract national and international companies. These companies are now under-represented in the Zaragoza economy. They should be pulled by the 'accessibility jump' of the high-speed train, the offer of high quality premises at a good price, and the availability of highly skilled workers, in a high-tech setting.
- to create a cluster of innovation and professional creativity. Companies and R&D facilities should reinforce each other and interact intensively to create innovation. It is recognised that a certain degree of specialisation should be aimed for, in order to reach relevant critical mass. Two specialisations are being considered: digital convergence at home/ubiquitous computing, and the creation of content for new electronic hardware and software;

- to boost the establishment of research centres of national and international standing;
- to create a community of residents with access to the most advanced technology: this can boost the area as pilot area or test-bed for new developments.

With its knowledge economy agenda in general, and more concretely, the Digital Mile project, Zaragoza hopes to achieve the following goals:

- to make a qualitative jump in order to provide the city with better national and international exposure and image;
- to acquire the characteristics of big cities – creativity, innovation, self-confidence and independency – but without 'big city problems' such as excessive urban sprawl, insecurity, destruction of cultural heritage and social segregation and disintegration. Put even more strongly, the aim is to '... bring us closer to the Aristotelian ideal of the perfect city: a city with a size that does not prevent everybody from knowing each other and taking part in the forum' (Digital Mile project memorandum, p. 23);
- to expand its tertiary sector, notably business services, as this sector is likely to generate high growth and jobs for the highly skilled, thus contributing to income and productivity increases.

For more details of the plan, see Ayuntamiento de Zaragoza (2004).

10.2.1 The Policy Process

The Digital Mile project currently (2004) only exists on paper. At the time of writing, a Masterplan is being developed, which should be finished by the end of October 2004. The ideas have been (and are to be) discussed with many actors and specialists inside and outside the city, among whom are some well-known experts such as William J. Mitchell. Reference projects in other cities have been studied as well.

At the time of writing, consultations are being held with large private companies, most of them in high-tech industries, to involve them in the elaboration of the plans and to evoke their interest in and awareness of the new area. Initial agreements have been reached with Telefónica (Spains's largest telecom operator), Indra (the largest IT company in Spain) and multinational companies such as Samsung and Siemens. The university is involved in the policy process as well: it will carry out technical and feasibility studies that should result in a general vision for the project concerning its form, management, finance and infrastructure. Some speed in the planning process is needed, as the free spaces in the railway area need to be filled up with new developments in order to finance the displacement of railway facilities.

10.3 Knowledge Foundations

In this section, the knowledge foundations – as discussed in the theoretical chapter – will be assessed for Zaragoza. They include the knowledge base, the economic base, the quality of life, the accessibility, the urban diversity, the urban scale, and the social equity in the city.

10.3.1 Knowledge Base

Zaragoza's knowledge base is relatively strong, taking into consideration both the educational level of the population and the quality of the knowledge infrastructure. The educational level of the population is at about the Spanish average, and it has risen strongly in the last decades. Zaragoza is an important university centre in Spain, with some 30,000 students. The university is very old, and offers education in a wide range of academic disciplines. For a long time it had faculties only for sciences, philosophy/humanies, and law. In 1974 a school of engineering was created. At first, it only offered mechanical and electrical engineering. Later, new disciplines were added: telecommunication technology (in 1989), informatics (1994) and chemical engineering (1996).

Of the 30,000 students, 9,000 are studying engineering – 50 per cent each at master's level and bachelor's level. The university in general enjoys a good reputation. Its strongest areas are chemistry, materials, optical and laser technology, and mechanical and chemical engineering.

10.3.2 Economic Base

Table 10.2 shows some basic data on the development of the region. By the end of 2003 the unemployment rate was 6.2 per cent, well below the Spanish average (11.2 per cent) but with an upward tendency (Fundear, 2004). Estimates from the City Council indicate that GDP per capita in Zaragoza is 112 per cent of the Spanish average. For 2003 this figure was €20,085.

The economy of Zaragoza is dominated by services activity, but less so than many other cities of a similar size. The city has a relatively strong

Table 10.1 University of Zaragoza and polytechnic: faculties and number of students, 2002–2003

Faculty/School	Number of students
Faculty of Sciences	2,726
Faculty of Economics	3,519
Faculty of Law	3,289
Faculty of Philosophy/Humanities	3,070
Faculty of Medicine	1,322
Faculty of Veterinary Medicine	1,264
Higher Polytechnic Centre	3,950
Faculty of Education	2,027
School of Engineering	3,654
School of Health Sciences	972
School of Commerce	1,626
School of Social Studies	2,009
School of Tourism	655
Total	30,083

Source: IAE (2003), p. 42.

Table 10.2 Population and GDP per capita, 2001

	Population (2001)	GDP/capita (2000), €
Aragon	1,204,215	16,206
Zaragoza	622,602	20,085
Spain	40,847,371	15,248

Source: IAE (2003).

manufacturing sector. In 1999, over 32 per cent of employment was in manufacturing, and the share even increased between 1995 and 1999 (see Table 10.3), which is remarkable from the perspective of international comparison.

Within manufacturing, major changes have taken place in the last few decades. The textile industry has suffered strong losses, but the automotive sector has gained in importance. General Motors has had a large plant (producing Opels) near the city since 1982. GM came to Zaragoza for several reasons: it considered the quality of workers to be good, and the

Table 10.3 Structure of the Zaragoza economy (province): employment in 1995 and 1999

	1995	1999
Agriculture	7.3%	5.3%
Industry	30.1%	32.4%
• Energy	0.3%	0.4%
• Manufacturing	23.4%	25.5%
• Construction	6.3%	6.6%
Services	62.6%	62.3%
• Business services	41.0%	41.0%
• Non-business services	21.6%	21.3%
Total	100.0%	100.0%

Note: data for Zaragoza city itself are not available. Authors suggest that the city probably has a higher presence of services, especially business services.

Source: FUNDEAR 1/2003, cuadro 8.

labour climate in the region was quiet, with few conflicts; furthermore, the company was offered substantial incentives in the form of tax reductions and cheap premises. During the 1980s, a sizeable automotive cluster emerged with many suppliers. The arrival of GM has played a significant role in the upgrading of the industry: suppliers of the plant had to meet high technological standards, and other industrial companies in the region were eager to have the same technological level as well. Currently, the Opel plant has over 8,000 employees, with a lot of additional employment with the suppliers. Other strong manufacturing subsectors are machine building, metal industry and food processing.

Apart from GM in 1982, there has been little greenfield investment by foreign multinationals in the region; however, many multinationals have taken over local firms in the last few decades.

The service sector is relatively underdeveloped for a city of Zaragoza's size, with a percentage of only slightly over 60 per cent; the share has even declined between 1995 and 1999. It is estimated that some 50 per cent of the business services used/purchased by Zaragoza-based companies is imported from other regions, notably from Madrid and Barcelona. Major business service companies in those cities consider the market in Zaragoza too small to open up a branch office. An important and emerging branch in Zaragoza is the logistics and transportation industry (more on this in Section 10.4.4).

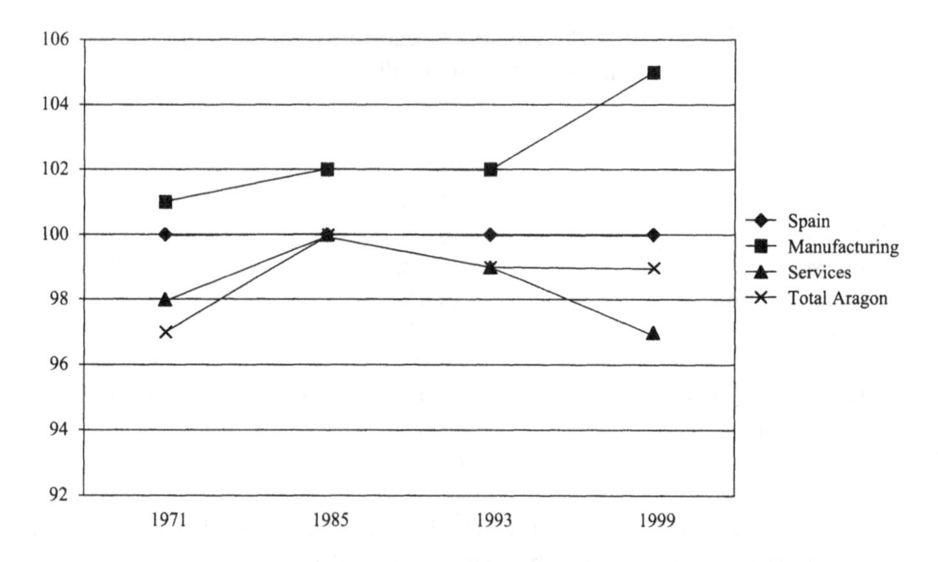

Figure 10.4 Productivity of economic sectors in Aragon, compared to Spanish average, 1971, 1985, 1993 and 1999

Source: Pardos and Gomez Loscos (2003), p. 16.

Table 10.4 shows the economic performance of Aragon compared to other Spanish regions. In the period from 1995 to 2002, Aragon's share has declined slightly.

10.3.3 Quality of Life

The Spanish consider Zaragoza an attractive city. It offers a rich variety and high quality of urban amenities (cultural, culinary, etc.) without the disadvantages of the major Spanish metropolitan areas of Barcelona and Madrid, such as heavy traffic congestion, high crime rates and extremely high real estate prices. The city has an image of being quiet and safe.

In terms of leisure, both the city and the surrounding region have a lot to offer. The mountains are not far away: there are ski stations within a one and a half-hour drive. An increasing number of inhabitants of Zaragoza have a second home in the Pyrenees, and many have one at the Mediterranean coast, which is two hours' drive away. The city is seeking to improve what it has to offer in terms of leisure and shopping through projects such as 'Puerto Venecia'.

Table 10.4 Gross regional product at current market prices, shares of the Spanish regions, 1995 and 2002

Region	Share in 1995	Share in 2002	Change in %
Andalucia	13.41	13.65	+0.24
Aragon	3.27	3.10	–0.17
Asturias	2.42	2.23	–0.19
Balears	2.30	2.49	+0.19
Canarias	3.80	4.02	+0.22
Cantabria	1.25	1.28	+0.03
Castilla y Leon	6.10	5.67	–0.43
Castilla la Mancha	3.53	3.40	–0.13
Cataluna	18.90	18.42	–0.48
Comunidad Valenciana	9.45	9.79	+0.34
Extremadura	1.72	1.71	–0.01
Galicia	5.61	5.31	–0.30
Comunidad de Madrid	16.79	17.32	+0.53
Region de Murcia	2.29	2.41	+0.12
Navarra	1.70	1.70	0.00
Pais Vasco	6.32	6.37	+0.05
La Rioja	0.76	0.75	–0.01
Ceuta y Melilla	0.28	0.28	0.00
Spain	100.00	100.00	0.00

Source: INA (2004), Espana en Cifras 2003–2004.

One advantage of Zaragoza is its low tax level. Property tax (levied by the city) is one of the lowest amongst Spain's cities: it is only one-third of the levels in Barcelona and Madrid.

The city is not particularly known as a tourist destination. Although it has lots to offer in terms of cultural and shopping amenities, this holds much less for its surrounding region. The relatively poor external accessibility (hardly any air connections) may now be a factor that negatively affects the number of tourists, but here also the HST may bring improvement. In particular, day trips from Barcelona and Madrid could sharply increase. Accesibility by road is quite good in Zaragoza: the city is connected by motorways to Madrid, Barcelona, Bilbao/France, Huesca, Pamplona and, soon, Valencia. The railway connection to Madrid already offers an excellent service: just one hour forty-five minutes. The main problem is the airport, but by the end of 2004 new European connections will begin operations.

The Spanish real estate market has been booming for some years now; Prices per square metre in Zaragoza increased by 61 per cent between 1999 and 2002. This is in line with the Spanish average. The sharp price increases make it increasingly difficult to find accommodation at a reasonable price in the city.

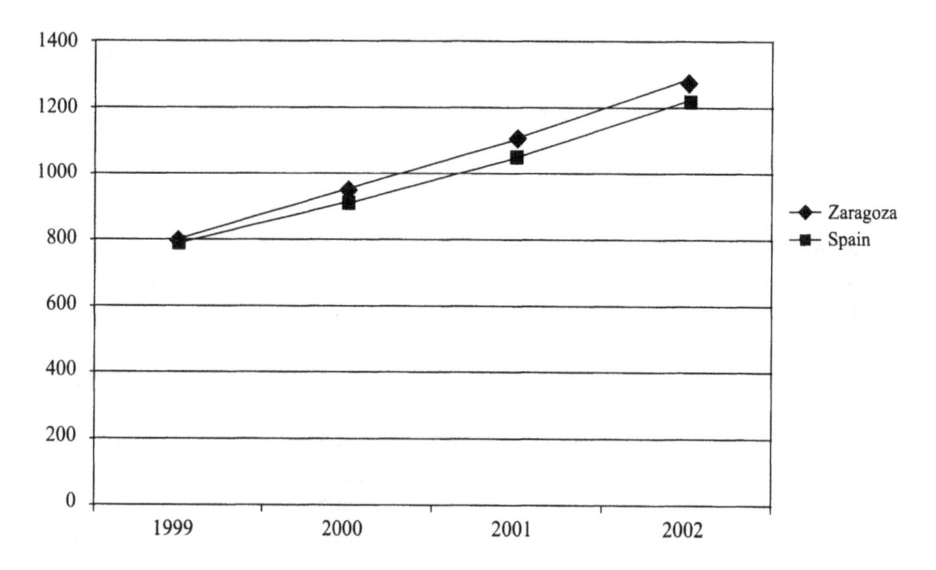

Figure 10.5 House prices: development of medium price (in €) per square metre, 1999–2002

Source: IAE (2003).

10.3.4 Accessibility

Zaragoza is located on the major rail and road axis that runs from Madrid to Barcelona. Transport connections to France are relatively poor, as there is no highway that crosses the Pyrenees in the middle. Rail connections are poor, and road traffic always has to pass either through the west side (Irun) or the east side of the Pyrenees.

The city has an airport, but it offers very few destinations. In 2001, the airport carried only 218,750 passengers, of which some 18,000 were on international flights (IAE, 2003). At the time of writing, talks are being held with low-cost carriers that could start operations at the airport: this would significantly increase the city's external accessibility for short trips.

As mentioned already, Zaragoza is to be connected to the high-speed rail network which is currently being rolled out in Spain. This process started with the inauguration of the Madrid-Seville line when Seville was hosting Expo 1992. Since then work has begun to create a Madrid-Zaragoza-Barcelona line, and further extend the line to the French border.

By the end of 2004, the link with Madrid will be completed. This will reduce travel time from two hours to 75 minutes. The link with Barcelona will be completed by 2007: here, travel time will decrease from three hours to 75 minutes.

The high-speed train not only affects the city's external accessibility but also offers scope for urban transformation in the city centre. A new station has already been built in the city centre. The old station area and areas adjacent to the rail track – the land of which is owned by the government – can now be fully developed, and this opens up new opportunities for urban transformation. In the Digital Mile project, plans are being developed to utilise these spaces for new economic functions.

Public transportation within the city is now solely operated by a bus company. The city has advanced plans to construct a metro tube (15 km): creating a tramway is another option.

Figure 10.6 The high-speed train in Zaragoza's new station

10.3.5 Urban Diversity

One criterion for urban diversity is the composition of the urban population. In total, 5.27 per cent of the population is of foreign origin. For a European city of Zaragoza's size, this is a low percentage. Figure 10.7 shows the origin of the foreigners who live in Zaragoza. It shows that Latin Americans form the largest proportion, followed by Africans. It is important to note that the proportion of foreigners in the population is rising rapidly, from 1.35 per cent in 1999 to 5.27 per cent in 2002. Many of the newcomers are poor and unskilled immigrants from underdeveloped countries.

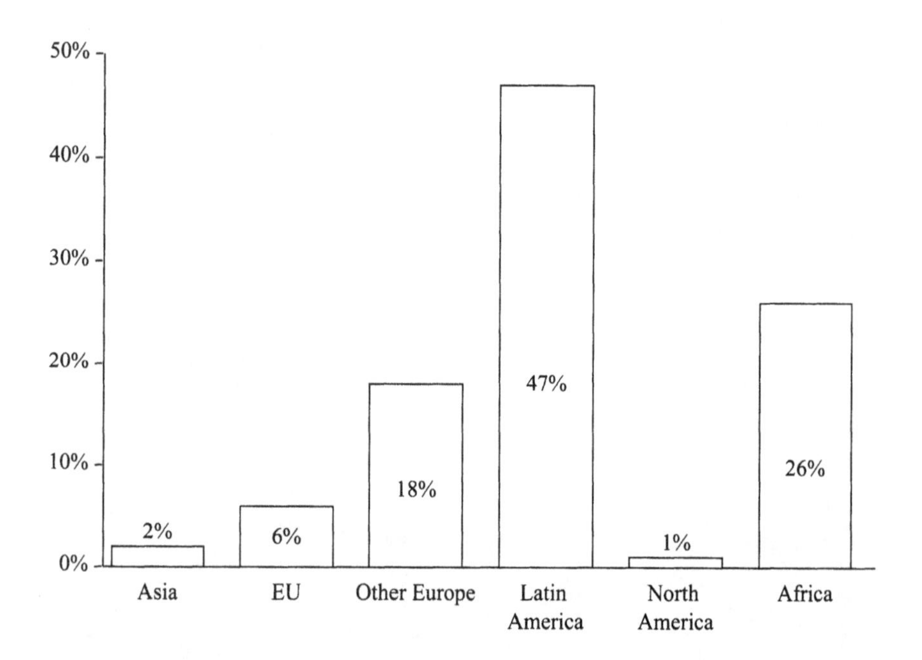

Figure 10.7 Foreigners in Zaragoza in March 2003: total = 37,263 (5.27% of the population)

Source: PHM Ayuntamiento de Zaragoza (2003).

10.3.6 Urban Scale

With some 600,000 inhabitants, Zaragoza can be considered as a medium-sized European city. However, it is not part of a larger urban conurbation, like many similar sized cities in northwestern Europe: it is the capital city of a very large and very sparsely populated region. The city has a central function for that region and it offers a number of amenities and facilities that reflect this function. However, for some types of companies – for instance branches of consultancy firms or other specialised industries – the size of the market is too small to justify opening an office in the city. These functions are operated from Madrid and Barcelona.

Also, there is not much commuting to and from Zaragoza, as there are no real suburbs or towns within commuting distance. The labour market in the city is therefore relatively small. The arrival of the HST may have an impact here: as Madrid and Barcelona will be at commuting distance, the labour potential of Zaragoza will increase significantly.

10.3.7 Social Equity

The social situation is Zaragoza is relatively favourable. There are no ghettoes or no-go areas in the city. However, since the late 1990s the city has attracted poor immigrants from North Africa and Latin America (notably Ecuador). These people often lack education and skills to fully participate in the labour market. However, there are also positive signs. Most notably, during the late 1990s the unemployment rate dropped sharply, from a peak of almost 20 per cent in 1994 to 5.9 per cent in 2002 (IAE, 2003, p. 65). The number of people dependent on unemployment benefits has dropped accordingly.

10.4 Knowledge Activities

In this section, we discuss Zaragoza's 'knowledge activities'. We discern several types of knowledge activities: 1) attracting and rataining knowledge workers; 2) creating new knowledge; 3) applying knowledge/making new combinations; and 4) developing new growth clusters.

10.4.1 Attracting and Retaining Knowledge Workers

How well does the city manage to attract and retain knowledge workers? Evidently, the university is an important magnet for talented young people. It attracts students from all over Spain, although the majority come from the Aragon region. If we look at university graduates, only 50 per cent of them work in the region of Aragon: over 40 per cent of the alumni work in either Barcelona or Madrid, and 10 per cent elsewhere (source: interview).

The university has a somewhat inward-looking attitude as far as academic staff is concerned: most of them have made their careers within the walls of the Zaragoza university, and there are no programmes or incentives to attract top researchers from outside.

In general, the city struggles to attract and retain skilled workers. Many people are educated in Zaragoza but leave the city because of the lack of suitable jobs. On the whole, the region of Aragon is losing skilled people, while at the same time it is attracting low-skilled people from other Spanish regions. Figures from 2002 support this: in that year, the region lost 1,842 highly skilled people, and attracted 7,696 low skilled people from other Spanish regions (Baguena, 2003).

As already mentioned, the city is seeking to change its economic structure by promoting economic diversification and, more specifically, with the Digital Mile project, in order to create more high-level jobs. But this will be a long-term process.

The fact that highly-skilled people leave the city does not imply that they can be of no value to it. Many of the 'emigrants' remain socially and emotionally attached to the city; they return frequently to maintain their social relationships. The connection of the HST may reinforce this process. Furthermore, many Zaragoza alumni have high positions in leading Spanish companies, which can benefit Zaragoza as well.

Given this, what seems to be lacking is an active alumni policy. Such a policy could help to create further benefit from existing social, emotional and economic linkages between the city/university and its alumni.

10.4.2 Creating Knowledge

Knowledge is created in the private sector and in universities. If we look at R&D expenditures, the region of Aragon scores somewhat below the Spanish average (there are no data on the urban level) but well below the EU average: only 0.8 per cent of GDP is spent on R&D (see Figure 10.8). In absolute numbers, in 2002 this comes to €160m.

When we look at the growth rate of R&D expenditures, Aragon is somewhat above the Spanish average, with an average annual growth rate of 13.5 per cent between 1997 and 2002. The private sector takes the highest share (63 per cent) of R&D expenditure and this share has also grown over the last five years. In Spain, only the Basque Country (76 per cent), Navarra (69 per

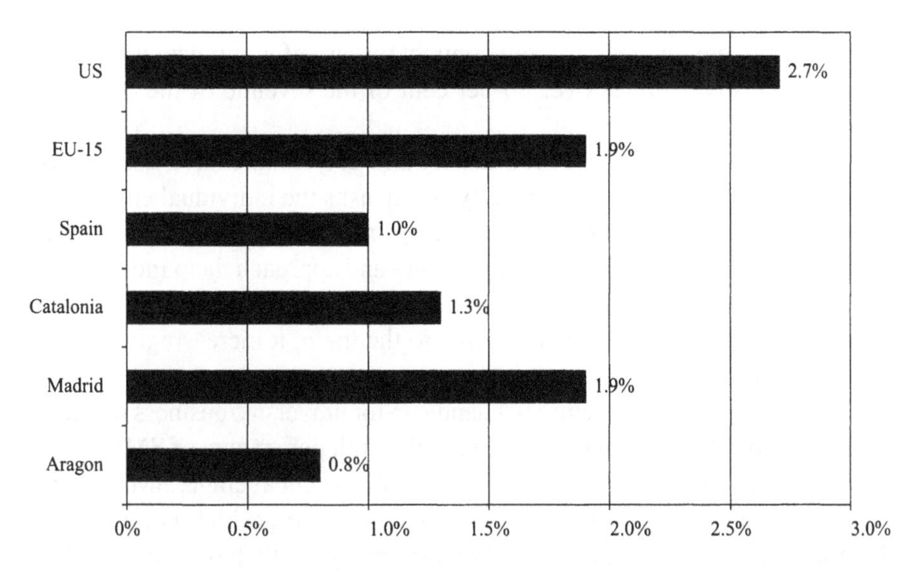

Figure 10.8 R&D expenditures as percentage of GDP

Source: Gobierno de Aragón (2004), 'Boletín Trimestral de Coyuntura No. 5, Marzo 2004', Dpto de Economía, Hacienda y Empleo, Zaragoza.

cent) and Catalonia (68 per cent) have higher shares. This high share can be regarded as positive for the economy, as private sector R&D is usually more easily translated into business growth than is public sector R&D.

Public R&D expenditure in Aragon is mainly channelled into higher education institutes. The university's strongest areas are in chemistry, materials, optical and laser technology, and mechanical and chemical engineering. However, despite apparent quality differences, officially the university does not have a policy to strengthen specific spearheads or science fields.

10.4.3 Applying Knowledge/Making New Combinations

Here we make a distinction between knowledge transfer (from research institutes/universities to the private sector) and entrepreneurial activity in the city.

Knowledge transfer activities and initiatives The University of Zaragoza has a renowned Faculty of Sciences, and a good School of Engineering with many contacts in the industry, not only in Zaragoza but also in other industrial regions in Spain.

University employees carry out contract research for companies for some €9m p/a. This constitutes over 60 per cent of the revenue of the School of Engineering.

Most of the relations between industry and polytechnic have a short-term (six to 12 months) character. Typically, a firm asks the individual engineer or research group to solve a specific problem for a fixed budget. Normally, the businesses know the experts in the university and approach them individually. The polytechnic has recently started a shift to a more proactive approach: instead of waiting for the firms to come to the them, it increasingly markets itself to the companies in the region.

The local economic structure is a handicap for university/business relations. A particular problem for the university is the high percentage of SMEs in the local economy and the low number of large firms. On average, SMEs prefer short-term contracts to solve very practical problems, which is not always the prime interest of the university. Larger firms typically have more need for strategic scientific cooperation.

Most of the cooperative projects are short-term, but there are longer-term partnerships as well: for instance, industrial groups finance two chairs at the polytechnic.

In rare cases, the university develops an innovation that can be patented. There are several possibilities concerning the ownership of the patent: either the university can become the full owner and reap all the potential economic benefits of it, or the research group that developed the patent can claim a certain percentage of the revenues. But in cases where a professor has a part-time appointment, he/she can become an individual owner of the patent and the university can be bypassed. This occurs frequently, and implies that the university gets no benefit.

Industry is not the only relevant partner for the university. Recently, the university has started to work closely with the city of Zaragoza in the Digital Mile project: the university is one of the key partners in the highly ambitious Digital Mile Consortium. The city makes use of the university's technical and economical expertise and involves it in the generation of new ideas. This seems to be a natural cooperation, given the educational and technology components of the project.

Promotion of start-ups Officially, promoting start-ups is an important objective of both the university/polytechnic, the city and the region. However, several interviewees indicated that the entrepreneurial spirit in the region leaves much to be desired and the number of start-ups is low. Figure 10.9 supports this view.

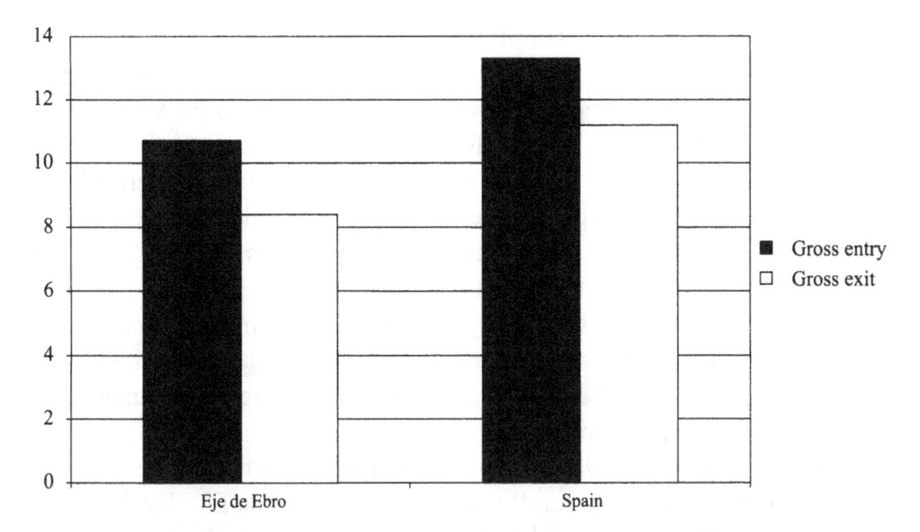

Figure 10.9 Creation and destruction of companies in Eje de Ebro and Spain

Source: Pardos and Gomez Loscos (2003), p. 351.

Note: The gross entry rate is calculated as the number of new companies in year t divided by the total number of companies in period t–1. The gross exit rate is the number of companies that went out of operation in year t divided by the total number of companies in year t–1.

University spin-offs (professors or students who start their own company) are rare: in the last ten years, there have only been four spin-offs from university professors. This is at least partly due to the fact that professors tend to have a 'public sector' attitude: often they have lifetime employment and are generally risk-averse. This situation can only be changed with a different incentive structure: one solution could be to stop the lifetime contracts; another to reward not only scientific publications, but also successful and innovative cooperations with industrial partners. It is important to note that this is not a situation specific to Zaragoza or Aragon: rather it is endemic in Spain. There are also positive signs of change: young students and staff seem to be more inclined to take risks and become entrepreneurial.

The Region of Aragon also has an active policy to promote and support start-ups. The CEEI (European Centre of Companies and Innovation), an incubator centre for starting firms, has been located in Zaragoza since 1992 The Centre offers not only space for companies (20 rooms), training and conference rooms, but also support for finance, marketing and business plan writing. The Centre only accommodates industrial companies, but in its definition ICT and biotech companies are also included. The focus is on metal, plastics and automotive suppliers; in other words, it plays to the strengths of the urban economy. There is demand for this facility: 90 per cent of the Centre is occupied. In the 12 years between 1992 and 2004, a total of 69 firms have started up in the Centre; 83 per cent have survived and still exist. Companies can stay there for five years. After that period, their rent is increased by 5 per cent per month, giving them an incentive to leave.

The CEEI, together with the university, runs a programme to promote start-ups. Staff who start their own company receive a guarantee that in the case of failure they can retain their university position. Despite this arrangement, the number of start-ups is low.

The Centre started in 1992 as a partnership between the region, the city and the University of Zaragoza. It was co-financed by the EU. By 2004, 61 per cent of the shares were held by the region, 20 per cent by the city, and the rest is shared among the university, the Chamber of Commerce and the Ministry of Economic Affairs. The region carries 50 per cent of the annual operational costs of the centre, some €600,000 p/a.

Despite CEEI's relative success, the city of Zaragoza considers the start-up rate too low. The Digital Mile project should be a catalyst to promote entrepreneurship.

Promotion of ICT use One of the ways to promote innovation is to help companies to adopt new technologies. In the province of Aragon, ICT-use among companies, especially among SMEs, is low, even lower than the Spanish average. In 2001, the percentage of companies that used PCs stood at less than 80 per cent in Aragon against 85 per cent in Spain (CEPYME Aragon, 2003, p. 41), and the percentage of companies with an Internet connection (for 2002) was 60 per cent, against a Spanish average of 68 per cent (ibid., p. 47). There are no data for Zaragoza, but it can be safely assumed that the figures for the city are a little better than those of the region.

The regional government of Aragon is very active in promoting ICT use, believing that it can substantially raise companies' competitive position. Its actions include raising firms' awareness of possibilities and convincing firms of the need to have ICT qualified staff. It also promotes educational facilities in ICT in the region (for regular educational programmes but also for education of the elderly); facilitates the exchange of knowledge and experience on ICT adoption; helps firms to analyse and reconsider their work processes and invest in ICT accordingly; is involved in the development of e-security systems for firms; and has plans to promote the roll out of high-quality electronic infrastructure and provide access for all SMEs. Also, the region invests in its own ICT facilities and plans to expand and upgrade its offer of electronic services to its firms and citizens. Finally, the region wants to promote the development of the ICT sector itself, spatially focused in the cities of Zaragoza and Huesca, where a special park was established for ICT firms.

Unfortunately the electronic infrastructure in Aragon and Zaragoza is a bottleneck (although this is a problem for Spain as a whole): there are frequent drop-outs, and prices for broadband are high from a European perspective. The incumbent telephone company Telefonica still has a dominant market position. However, companies do not seem to have a problem with communications infrastructure; they can get what they need at a reasonable price.

10.4.4 Developing New Growth Clusters

Zaragoza wants to develop new economic growth clusters in the years to come. Two ambitions stand out. The first is the creation of a logistics and transport cluster, near the airport; the second is the development of a cluster of ICT companies and advanced services in the new station area, in the 'Milla Digital' (Digital Mile) project.

Creating a logistics cluster By 2000 the regional government of Aragon had taken the initiative to make more out of the region's strategic position on transport axes in Spain and Europe. The idea was launched to create an intermodal logistics platform near the airport, which would be easily accessible by rail, road and air. In March 2000, a company named 'PLAZA' was founded to realise those ambitions. It is a joint stock company of the region, the city of Zaragoza and two banks, with a total capital of over €18m. The company, though largely publicly owned, has been set up as a professional and relatively independent organisation, in order to increase its effectiveness and efficiency.

PLAZA is developing an area of over 1,000 ha. exclusively devoted to transport and logistics. This also has implications for the development of the inner city of Zaragoza: many rail freight activities that were previously scattered over the city and the region are now being concentrated in the area. This will free up space in the city, notably near the old station, for new developments.

The PLAZA company is preparing the area for logistics use by investing in road and rail access infrastructure and intermodal facilities, a power station, sewage systems, security, etc; considerable investment is needed to move

Figure 10.10 Aerial photograph of the Plaza area.

Source: www. Plazalogistica.com.

rail installations to the new area. Some 5 km^2 is destined for profit-making operations; the rest is being developed as either open space or transport infrastructure. The revenues for PLAZA come from the sale and rent of land, and the renting of facilities.

It is an explicit policy to reserve the commercial areas for transport and logistics use only: manufacturing or other companies are not allowed to buy or rent land.

By April 2004 some 50 per cent of the available land had already been sold, at prices of around €100/m^2. Most of the logistics operations are in the fashion and the food industries; 80 per cent of the companies are Spanish. The share of foreign firms is expected to increase. US companies especially (Dell, Motorola) are interested in basing logistics operations in Zaragoza, because of the high security standards that are implemented.

There are several arguments why the area could have potential to become a major logistics hub. First, the logistics park is at a strategic intersection of

Spanish transport nodes and could develop as a complementary hub in the national transport system. It is situated at a point equidistant from the main Eastern ports of Spain (Barcelona, Tarragona and Valencia) and the Cantabrian ports (Santander, Bilbao), making it a good 'dry port' location.

Another key asset is the low price of land compared to the port cities of Barcelona, Bilbao and Valencia: land is far from scarce in Aragon, and prices are half those of Barcelona or Madrid. The logistics cluster could be a good area for land-intensive activities such as container storage and redistribution. The PLAZA organisation is developing strategic relationships with seaports to exploit complementarities: for instance, Tarragona Port Services has opened a terminal in Zaragoza to collect sea freight.

Thirdly, the high-speed train will use a new track, which implies that more capacity will become available for freight traffic in the future. Finally, there will be more competition in the Spanish rail freight market as new rail operators enter the market from 2004 on. This could improve the quality of rail services to and from the PLAZA area, and provide more flexibility and lower prices.

The location near the airport is considered crucial, even though the airport is a very small one, because it enables logistics companies to fly in missing parts or emergency deliveries. Some expect traffic at the airport to grow with the growing demand from the logistics park.

In the long run, the logistics cluster is expected to generate between 7,000 and 10,000 new jobs in the city. Interestingly, these are not only low-skilled jobs. The MIT (Massachusetts Institute of Technology) from Boston will open a centre in the PLAZA as a base for logistics-related education and research. A high-level postgraduate course in logistics will be set up, admitting some 40 top students a year. Some teachers from MIT will be involved, but other professors (from Spain and Europe) will be hired as well. In addition, MIT will open a logistics laboratory to conduct logistics research, in cooperation with companies located in PLAZA. This initiative is a clear example of how seemingly 'ordinary' economic activities can have a high level spin-off and can be built upon in terms of education and research. For more information on PLAZA see to www.plazalogistica.com.

Services and ICT: the Digital Mile The high-speed rail link to Barcelona and Madrid will strongly affect Zaragoza's position in Spain's urban grid. The sharp decrease in travel times could have a catalytic effect on the city's attractiveness to business services and technology companies, and could put the city more firmly on the (mental) map of large companies. But in addition, the high-speed train is the catalyst for the transformation of a large area in the city centre. A brand new station has been built, and a large amount of land is waiting to be redeveloped. The land is largely owned by the Spanish government. In 2002, a company – 'Zaragoza Alta Velocidad 2002' – was founded by the major stakeholders: the Spanish government, the Aragon region, the city and the railway company RENFE. The aim of this company is to carry out the necessary infrastructure works and to develop urban projects in the area.

The Digital Mile project aims to develop plans for the area. The ambition is to attract services industries and ICT companies: this would redress Zaragoza's relatively poor position in these sectors (see Section 10.2).

10.5 Conclusions and Perspectives

Zaragoza is at the crossroads of a new era. In the last vew decades, the city has adapted itself gradually to changing economic circumstances, by modernising its industry and keeping pace with the favourable development of the Spanish economy as a whole. The city has a lot to offer in terms of *quality of life*: it is a real urban place, of a sufficient *urban scale*, that creates a market for numerous urban amenities; the city centre is beautiful and there are a number of cultural facilities. At the same time it is relatively quiet, there is no severe congestion, the city is relatively safe, and there is no large-scale social exclusion either. This, in principle, makes it an attractive location for knowledge workers.

However, compared to the major Spanish metropolises (especially Madrid and Valencia) the city has lost economic weight. In particular, it did not develop new urban growth sectors – notably business services, headquarter functions, creative industries and tourism – as rapidly. One of the prime reasons for this is the relatively poor external *accessibility* of the city, with long travel times by train to large agglomerations, and poor air connections (although car accessibility is fairly good). But this will change when the HST is completed.

The city's *knowledge* and *economic bases*, however, are still quite strong: the city has a good and large university, and a well-developed economic fabric with a modern industry that can compete because it is innovative. Threats to local industry come from new EU member states and Eastern European countries, as there is a general tendency for manufacturing to locate there, but up to the time of writing there are no strong signals that the urban economy will suffer too much. Nevertheless, the trend gives local industry an incentive to be even more innovative, to cut costs and improve quality. Competing on (relatively) low wages is no longer an option, so this upgrading means that industry's knowledge intensity has to improve. That calls for intensified cooperation between the university and local business. Although many collaborations already exist, there is ample room for improvement in this respect. One of the key impediments is the incentive structure at the university: university staff are rewarded for the scientific publications, not for cooperation with the industry. In this respect, the industrial mix in Zaragoza also forms an impediment: small- and medium sized firms dominate the economy, and it is difficult to forge fruitful links between the university and those companies: they have a different time horizon (long term vs. short term), different interests (scientific research vs. solving practical problems) and the culture of both worlds is very different.

The service sector – notably business services – is relatively under-developed in Zaragoza. The city is considered too small to justify the opening of a branch office for some types of business services. We found indications that the same holds for new growth sectors such as the ICT sector and the creative industries. One of the consequences is that Zaragoza, overall, can offer relatively few jobs that require a high (university level) qualification. The city thus loses talent to other regions, notably to the larger Spanish agglomerations. At the same time, the city attracts students to its university.

The key catalyst for change in the years to come is undoubtedly the arrival of the high-speed train. The potential effects of this relative change in accessibility are manifold. In the first place, the HST will bring the cities of Madrid and Barcelona into commuting distance. This may bring new residents to the city, and opens the large job markets of those agglomerations to existing residents. Secondly, the HST will improve the opportunity for business contacts with companies in Barcelona or Madrid: they can be expected to intensify, and also, it may boost tourism, as it will be much easier to reach the city. Thirdly, space will be freed up in the centre of the city because the old railway facilities will be removed. The Digital Mile project aims to fill these spaces in an innovative way: it promotes the growth of new urban sectors, thereby diversifying and upgrading the urban economy. It may raise the number of high-skilled jobs in the city, which helps to keep and attract talent. Also, it will create up-market housing facilities with top electronic infrastructure provisions at a very strategic location in the city (i.e., at walking distance from both the HST station and the attractive city centre). Fourthly, the high-speed track will free up rail capacity on the old track that can be used for rail freight. This is a key factor to developing Zaragoza as a logistic node, as is currently being undertaken by the Plaza organisation.

The two key projects in Zaragoza (the Digital Mile and the Logistics Platform) aim to attract companies from outside. Another way to develop business is to promote indigenous growth. From this perspective, we note that, overall, the entrepreneurial climate of the city could be improved. The number of start-ups is not impressive, especially from the university. The university culture is still rather risk-averse.

The city's future perspective to developing growth clusters is rosy. The logistics platform that is already under development is a source of growth and will create many new jobs. It is highly interesting and also innovative that this development is being linked with the creation of a new 'centre of expertise', with MIT from Boston setting up an advanced logistics Masters' course opening a research facility in the area. The Digital Mile project, if elaborated and implemented adequately, is an important catalyst for R&D, the ICT sector, business services and the creative industries. It is an interesting feature that the university will be closely involved in the further development of the plan.

At least two conditions should be met to make the Digital Mile a success. Firstly, although the project has political and societal backing and companies are interested in the concept, it is still mainly something 'on paper'. A very strong and professional project organisation should be set up to elaborate and implement the project. It is a very complex project from a financial, planning and conceptual point of view, with many actors and stakes involved. Only a strong and experienced manager and a high quality team can manage the project. A second condition concerns the quality of electronic infrastructure and services. Without a reliable infrastructure and operations, the power of the concept will be lost. This offers a challenge to the project, given the relatively sorry state of the Spanish telecom sector: line drop-outs happen frequently and broadband charges are relatively very high from an international perspective. Moreover, it is a problem that broadband take-up is low in Zaragoza, both among firms and citizens. Higher ICT access rates in the city would make Zaragoza more attractive for companies that want to test new products and services. Therefore, ICT adoption policies – for SMEs and for citizens – should receive priority.

There is much at stake: if the Digital Mile is a success, it could broaden the economic base and increase the knowledge intensity. It could even raise international attention. It is also a chance for Zaragoza to distinguish itself from other (Spanish) cities and do something focused and specific. It could give the city much more of an image as a 'knowledge city'. There could be a positive spin-off to the quality of life in the city as well, as the type of people

Table 10.5 An indication of the value of knowledge economy foundations in Zaragoza

	Foundations	Score
1	Knowledge base	+
2	Economic base	□
3	Quality of life	+
4	Accessibility	□
5	Urban diversity	□
6	Urban scale	□
7	Social equity	+

-- = very weak; – = weak; □ = neutral; + = good; ++ = very good.

Table 10.6 An indication of the value of knowledge economy activities in Zaragoza

	Activities	Score
1	Attracting knowledge workers	□
2	Creating knowledge	+
3	Applying knowledge/making new combinations	□
4	Developing new growth clusters	+

-- = very weak; – = weak; □ = neutral; + = good; ++ = very good.

who would work in the Digital Mile area would probably contribute to the liveliness of the city.

Discussion Partners

Mr Jose Carlos Arnal, Digital Mile Project, City of Zaragoza.
Mr Victor Calvin, Factory Director, B/S/H (Bosch and Siemens Home Appliances Group).
Mr Ricardo Cavero, Councillor of Science and Technology, City of Zaragoza.
Mr Javier Celma Celma, Technical Director, Ebropolis.
Mr Manuel Doblare, Director, Institute of Engineering Research Aragon, University of Zaragoza.
Mr Fernando Fernandez, Managing Director, CEEI incubator centre Aragon.
Mr Ricardo Garcia Becerril, Managing Director, PLAZA.
Mr Alberto Lafuente, Alderman of Economic Affairs, City of Zaragoza.

Mr Jose Felix Munoz, Coordinator, APTICE (association for Internet and e-commerce promotion).
Mr Carlos Oehling, President, Going Investment (Venture Capital).
Mrs Eva Pardos, Lecturer of Applied Economics, University of Zaragoza/Fundear.
Mr Jose Miguel Sanchez, Managing Director, Chamber of Commerce Zaragoza.
Mr Javier Subias, General Coordinator, Ebropolis.

References

Ayuntamiento de Zaragoza (2004), Zaragoza Digital Mile Project, Project Memorandum.
Baguena, J.A. (2003), in *Heraldo de Aragon*, 31 August.
CEPYME ARAGON (2003), 'Integratic 2003: Plan para la integracion de las Empresas Aragonesas en las Tecnologias de la Informacion y el Conocimiento'.
Ebropolis (1998), 'Plan Estrategico de Zaragoza y su Area de Influencia'.
Fundear (2004), *Boletin Trimestral de Coyunture*, No. 5, p. 35.
IAE Instituto Aragones de Estadistica (2003), *Datos basicos de Aragon, 2003*.
INA (2004), 'Espana en Cifras 2003–2004'.
Pardos, E. and A. Gomez Loscos (2003), 'Aproximación a los servicios a empresas en la economia aragonesa', Fundear Working Paper 1/2003.
Plaza (2004), 'The Largest Intermodal Logistics and transport Complex in the South West of Europe'.

Chapter 11

Synthesis

11.1 Introduction

In this comparative research we have tried to find out how cities are transforming towards the knowledge economy, and to identify some typical elements by charting and analysing the development paths of a number of cities. Moreover, a very important question in this research was: what can the local government do to upgrade the local economy and guide it towards greater knowledge intensity? For each participating city we have investigated what the local government has done in the past few years to raise knowledge intensity, and how the stakeholders have been involved.

The research has been carried out in two stages. Desk research was central to the first stage. Much has already become known about the knowledge economy and the role of cities in it. In an extensive literature study, we described the 'state of the art', and derived the building blocks for a theoretical framework.

In the second stage we have probed the subject more deeply in nine European case studies: Amsterdam, Dortmund, Eindhoven, Helsinki, Manchester, Munich, Münster, Rotterdam and Zaragoza. In each city we held some 10 to 12 in-depth interviews with key people from government and enterprise. For each city we made a generic analysis of the development of its knowledge economy in the last decade. How knowledge-intensive is the local economy? How can that be measured? What are the city's prominent assets and resources? What does the economic structure look like? What generic economic/spatial policy has been conducted?

The findings are useful for many other European cities as well.

Nine cities were involved in the empirical part of this study. The results of this study, however, can be used within many more cities than this sample. In principle, all urban regions are operating within the knowledge economy. The global economy has changed into an economy in which knowledge-intensive activities become increasingly important. All urban regions which want to offer a competitive location environment have to adapt to this new economic reality. Thus, no choices can be made whether to be or not to be within the knowledge economy. But the starting positions of cities to become successful within this knowledge economy are different. Some of the cities (still) have a comfortable position for attracting and retaining knowledge economy companies and workers; others, conversely, have a hard job in attracting and retaining them.

We have developed a theoretical framework for the role of cities in the knowledge economy and used this framework to assess the situation of nine European cities. The cases also provided feedback to refine the framework. Furthermore, the nine participating cities can draw general and specific lessons from this comparative study. Because of the generalised character of the study, other cities that want to become (more) active within the knowledge economy can also learn from the outcomes of the research.

A high level of government involvement is typical of European cities. From every €100 earned within European countries, about €40 to €50 are spent by governmental institutions. These organisations can therefore have a large influence on the competitiveness of economies. If they operate in a very innovative and efficient way, they will support the economy. Conversely, if these governments are not innovative and efficient they may seriously hamper the economy. This will become even more important in a further globalising economy, in which knowledge-intensive activities are looking for the most efficient places to locate.

The most important findings of this comparative urban research are presented in this synthesis. In the next section, we will discuss the findings about the knowledge foundations. Then in Section 11.3 knowledge economy strategies will be addressed. Section 11.4 will present the findings of the knowledge economy activities. Finally, in Section 11.5, concluding remarks and perspectives will be given.

11.2 Knowledge Foundations

This section deals with the position and development of the knowledge foundations in the nine case study cities. We confront the outcomes of the cases with the developments described in the introductory chapter regarding the literature overview and our research framework.

In the first chapter we identified seven knowledge foundations: knowledge base, economic base, quality of life, accessibility, urban diversity, urban scale and social equity. All these foundations contribute to the overall point of departure for succeeding in the knowledge economy. However, the degree of importance differs: the knowledge and economic bases can be considered as fundamentals, since cities without sound scores in these fields will find it very difficult to successfully build up and maintain a knowledge economy. The other five factors can be characterised as supportive: they add extra strength to the fundamentals. Table 11.1 summarises the case study outcomes regarding the knowledge economy foundations.

11.2.1 Knowledge Base

The creation of new knowledge is important for cities. The quality and quantity of knowledge institutes such as universities, polytechnics, public and private R&D organisations determine to a large extent the starting position and development of a city in the knowledge economy. Cities with a strong knowledge base have better chances for economic growth than places lacking such a foundation (OECD, 2002; Mathiessen, 2002; Castells and Hall, 1994). One of the indicators contributing to the knowledge base is the number of people with a third level education: see Figure 11.1.

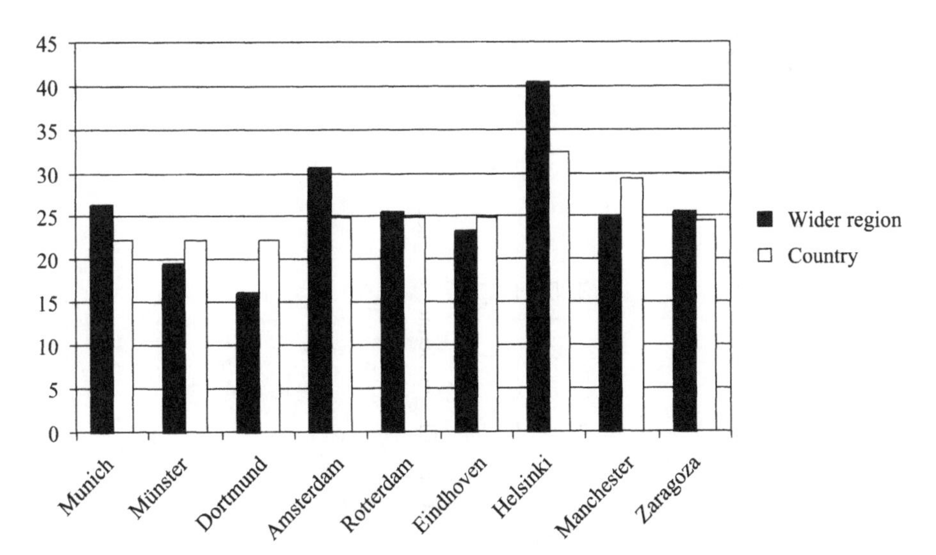

Figure 11.1 Percentage of inhabitants with third level education in 2002

Source: EU – European Innovation Scoreboard 2003.

Table 11.1 Knowledge foundations in the case study cities

Foundations	Manchester	Munich	Amsterdam	Helsinki	Eindhoven	Rotterdam	Münster	Dortmund	Zaragoza
Knowledge base	++	++	++	++	++	+	+	+	+
Economic base	□	++	+	+	+	□	+	+	□
Quality of life	+	++	++	+	+	□	+	+	+
Accessibility	+	++	++	+	□	++	+	+	□
Urban diversity	++	□	++	□	□	+	□	□	□
Urban scale	+	+	+	□	□	+	□	+	□
Social equity	-	++	□	+	+	-	+	□	+

-- = very weak; – = weak; □ = moderate; + = good; ++ = very good.

Our case study research identified the following cities as having a predominantly strong knowledge base: Eindhoven, Helsinki, Manchester, Munich, and Münster. Manchester and Münster have particularly high ratios of students per inhabitant: this is the result of the large number of higher education facilities in these cities. Helsinki has by far the highest educated population, which is in line with country comparisons where Finland scores significantly higher than the UK, The Netherlands, Germany and Spain. Cities with relatively many low-educated inhabitants are Dortmund, Rotterdam and Manchester. This education backlog is the result of these cities' industrial past. In some cases the industrial background can also have advantageous effects on the knowledge base. In Manchester, for instance, specialist courses and research in creative industries take place in the combined field of fashion and textile production. Rotterdam and Dortmund are strengthening their knowledge base regarding electronic logistics to enhance the logistics sector that already makes up a big part of these cities' economic foundations.

The relative size and specialisation of the knowledge base focus differs strongly among case study cities.

Some cities have a strong focus: Eindhoven is strongly focused on technology (electrical and mechanical engineering, industrial design etc) that is mostly concentrated in private companies; Münster – which has a large generic university – is concentrating on biotech and nanotech (the Münster knowledge base mainly consists of the university and public research institutes); Dortmund focuses on ICT and micro electronic mechanical systems (MEMS); and Rotterdam has clear focal points in medicine and transport and logistics. Other cities have a more broadly developed knowledge base: Amsterdam, Manchester, Zaragoza and Munich. Helsinki also has a broad knowledge base, but at the same time it houses the most important technical university in Finland.

11.2.2 Economic Base

The economic base is the second main determinant for the success of a city in the knowledge economy. The economic base depends on the economic sector composition and size and the level of productivity. In the literature review it was stated that cities that used to specialise in traditional industry and port activities perform less well than cities that have a more diverse economic base.

The case studies confirm that former industrial cities have a weak economic base.

Manchester, Rotterdam and Dortmund were heavily dependent on industrial and port activities that have seen large employment decreases in previous decades. Today, these cities are still burdened with this legacy: they are confronted with high unemployment levels, a less well-educated population (which is an important part of the knowledge base), lower quality of life and housing stock and they suffer from a poor image that is heavily based on activities that mostly belong to the their past. Cities that are more diversified (e.g., Amsterdam and Munich) and cities that depend on economic sectors that have not shown major declines (technology in Eindhoven, science, services and administration in Münster) have a more favourable economic base. Also, the more diversified cities have certain economic sectors that stand out. Amsterdam has many jobs in financial and business services, port and related industry, airport and related activities, and ICT and new media. Munich has considerable employment in modern industry (Siemens, BMW), banking and insuranceand and ICT. Helsinki is dominated by (financial and business) service sector jobs, while government and ICT also provide considerable employment. Compared to the other case study cities, Zaragoza has relatively much employment in the manufacturing sector (especially the automotive industry).

An indicator for the strength of the economic base is the cities' GDP per capita compared to the national average (see Figure 11.2). Indeed, the cities of Manchester, Rotterdam and Dortmund show rather low numbers compared to the national GDP per capita figures. Cities with high figures are Munich, Amsterdam, Münster and Helsinki. Figure 11.2 should be read with caution: in some cases the regional GDP per capita is considerably higher than the urban figure because the more affluent people live outside the city. This is for instance the case in Eindhoven and Rotterdam.

11.2.3 Quality of Life

The urban knowledge economy depends on talented people who create and apply new knowledge. A key determinant to attracting and retaining such people is quality of life (Florida, 2000). Quality of life is a difficult concept, because it has a very subjective element to it: it includes many dimensions and diverse groups of people value such aspects very differently. Knowledge workers (especially the young and the more creative kind) value the presence of

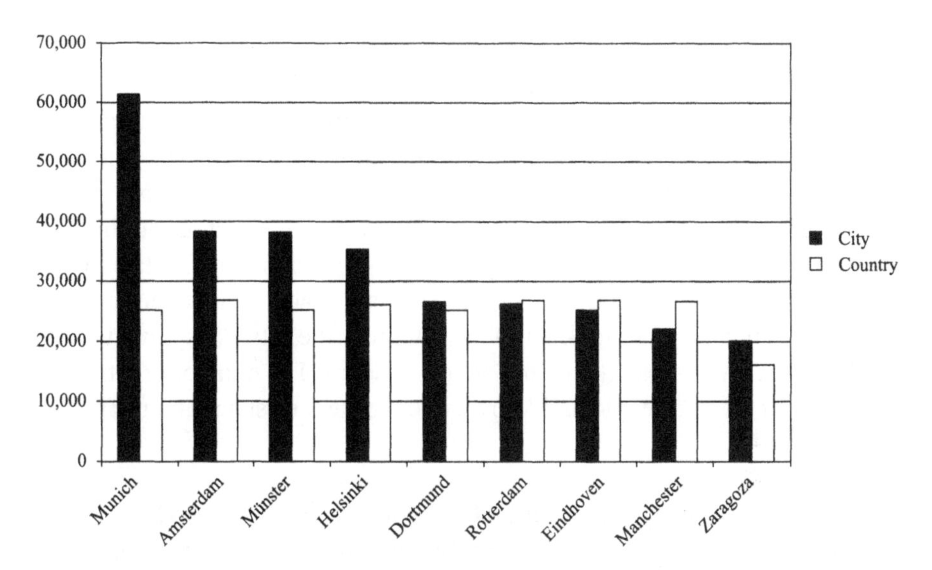

Figure 11.2 GDP per capita in 2001

Source: Office of the UK Deputy Prime Minister (2004); Zaragoza case study data for 2003.

cultural activities and amenities highly. Other quality of life dimensions include quality of the houses, (architectural) attractiveness of the built environment and natural surroundings/parks.

The urban quality of life in capital cities is often considered attractive.

The urban quality of life is judged particularly high in Amsterdam. This city not only has many cultural facilities, but also the vibrant, lively image that reflects such a quality of life. Also, Munich and Helsinki offer a high urban quality of life: compared to Amsterdam, however, they lack the 'sexy', vibrant city image. Rotterdam and Manchester both have developed an attractive quality of life in recent years that is accompanied by a rather dynamic, lively city image. Nevertheless, these cities have large dilapidated neighbourhoods and the outward city image is negatively influenced by crime. The quality of life in Dortmund has increased significantly in the previous decades: old industries have shut down, resulting in an increasingly green, less polluted city. The supply of cultural facilities and amenities is good but not extraordinary. Dortmund's main problem concerning quality of life is its lagging image to the outside: many people in other places still think of Dortmund as a grey, dirty industrial place, while this image no longer matches reality. Quality of life in Münster can be typified as being that of a rather small town. The city is very green and rather quiet. Münster has a varied supply of cultural activities and facilities, but it does not have an extensive, exciting nightlife. It seems that Münster's quality of life matches its image quite well. Zaragoza has a good urban quality of life: a rich variety of urban amenities, without serious big-city problems like traffic congestion, high crime rates and high real estate prices. The city has an image of being a quiet and safe place. Eindhoven too does not have a metropolitan look and feel, although the city does have various facilities such as a concert hall and an art museum. The number of attractive facilities for young people and knowledge workers (restaurants, pubs, etc.) is fairly low.

Our research suggests that technological knowledge workers in general do not seem to find a 'vibrant' city quality of life very important; they seem to like relatively quiet, easy-going places.

For example, the less vibrant quality of life in Eindhoven is currently not a main issue, because the city is very much technology oriented. However, this might change in the (near) future if the technological focus of Eindhoven changes.

A general finding from the case studies is the importance of English-language facilities. This part of quality of life has to do with the knowledge activity of attracting and retaining international knowledge workers. Facilities like international schools, but also entertainment in English are key facilities in this respect. Cities like Amsterdam, Helsinki and Manchester are well endowed with English-language facilities.

11.2.4 Accessibility

Innovative activity is highly concentrated in some metropolitan and regional capital cities. Critical factors for international knowledge transfer are international linkages conducted by face-to-face contacts that are facilitated by international hub airports. Cities that are most successful are those that combine rich local knowledge spillovers and international exchange (Simmie, 2002). This implies that both internal and external accessibility are important. In our research we based the cities' performance mainly on the external accessibility, which is determined by the presence of international airports and, to a lesser extent, high-speed train (HST) networks.

Case study cities with relatively good accessibility are those located near an international airport: Amsterdam, Helsinki, Manchester, Munich and Rotterdam.

Case study cities that are connected to an HST network are Amsterdam, Dortmund, Munich, Münster and Rotterdam. Helsinki has plans to build several HST links, including a line to St Petersburg. This should help it become a gateway between Europe and Russia. Zaragoza is to be connected to the HST network currently being rolled out in Spain. The strategic geographic position of the city between Madrid and Barcelona implies that the HST connection will offer a multitude of development opportunities for Zaragoza. Eindhoven lacks both a large international airport and an HST link.

Besides face-to-face contacts, the knowledge economy depends on the exchange of codified knowledge via ICT networks. Several case study cities have deployed activities to enhance digital access. Amsterdam and Münster have Internet exchanges that facilitate large scale data traffic: being close to such an exchange can lead to cost reductions. Most case study cities have good access to broadband Internet (xDSL, cable, fibre, etc.).

Several case study cities are busy implementing projects to further enhance ICT infrastructure and access.

Amsterdam is constructing a neighbourhood where all buildings are connected to optic fibre, Manchester has established a wifi (WLAN) network in the deprived district of East Manchester, Eindhoven runs a project where houses and offices are connected to fibre, Rotterdam is connecting two districts to fibre and also Zaragoza has plans to improve Internet access in parts of the city.

11.2.5 Urban Diversity

Cities with a high urban diversity are better positioned to attract (creative) knowledge workers. Places with diverse groups of people (by ethnicity, nationality, etc.) have an environment that is easy to plug into: they attract talented people (Florida, 2000). Furthermore, in the first chapter we stated that diversity fosters innovation in cities (Duranton and Puga, 2000). However, urban diversity also has a downside: several ethnic minorities consist of many low-educated people; unemployment among these groups is relatively high.

Also, urban diversity can imply tensions between different groups of people because of differing life styles and their according values. Which case study cities are most diverse? The percentage of foreign-origin inhabitants gives a good indication (see Figure 11.3).

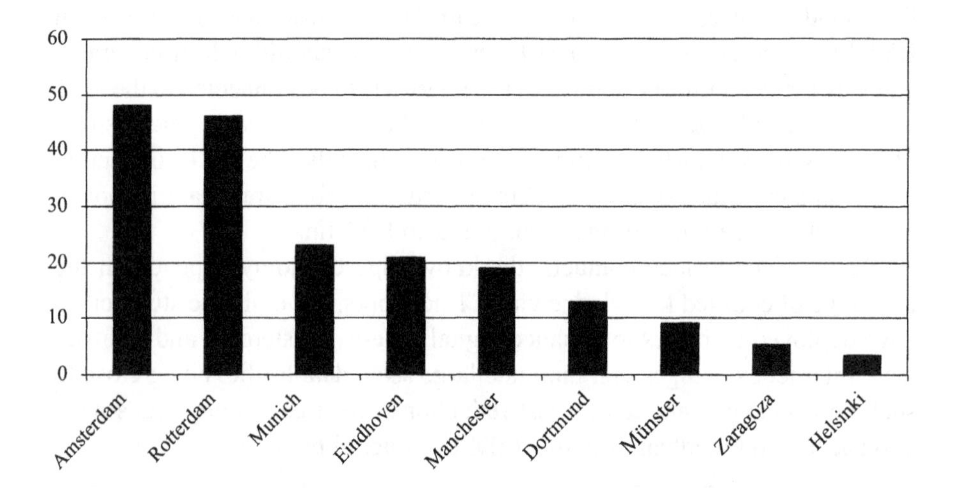

Figure 11.3 Percentage of inhabitants of foreign origin

Note: These figures are only a rough indicator, since measurement techniques among the cities differ. The figures for the Dutch cities would be lower (although still high) if the same definitions were used as in other case study cities.

Source: case studies.

Amsterdam, Rotterdam, Munich and Manchester have the highest figures: this outcome is in line with our evaluation of the overall urban diversity in the case study cities.

Besides high numbers of foreigners, these three cities have a rich blend of cultural activities (see 'Quality of Life') and residents. Amsterdam, Rotterdam, Manchester and Munich all have a very diverse mix of inhabitants – young urban professionals and dynamic student populations – but also significant groups of deprived people. Although the foreign origin inhabitants add to the cultural diversity – and enhance possibilities for economic development (e.g., through 'new combinations' – at the time of writing several ethnic groups throughout most of the case studies also pose a problem for these cities. Various foreign-origin inhabitants are relatively poorly educated, which has resulted in weak labour market positions and high unemployment levels. This touches upon the social equity foundation.

A specific group of foreigners that contributes to urban diversity are migrants from industrialised countries.

In general this involves highly educated people from the EU and Western hemisphere countries who go somewhere to get advanced education and/or work in high-grade jobs. Most of these people have a high economic potential. They look for places where sufficient English-language facilities are offered. Relatively large 'Western' immigrant communities (the number of Western immigrants in the total foreign population) can be found in Amsterdam and Eindhoven. Münster has a rather large, well-educated foreign population consisting of students and researchers originating from Southeast Asia, Eastern Europe and Turkey. Zaragoza is rather special: it has a low number of foreigners, most of whom come from Latin America, because of the Spanish language connection. Most of these Zaragoza immigrants are poorly educated.

The main policy point concerning urban diversity for most case study cities is to improve the labour market position of several ethnical minorities. This will enhance the knowledge base and economic position of the cities and will strengthen the social equity dimension.

11.2.6 Urban Scale

The knowledge economy is a network economy: rapid developments in knowledge make that no single person or company can master all disciplines or even monitor all the latest developments; this also applies to cities. Cities – especially those that have a smaller urban scale – need to engage in strategic networks to be able to timely respond to rapidly changing markets and technologies. Every city in the network needs to develop its own specialisation (Castells, 2000). In the knowledge economy, to a large extend, big is beautiful. For companies it is easier to find specialist staff in larger cities and for citizens big cities are attractive because of the large variety of jobs (Glaeser, 2000). The scale of large cities offers scope for bigger airports and rather specialised facilities such as international schools.

Munich, Amsterdam, Manchester, Helsinki, Dortmund and Rotterdam have an extensive urban scale.

Rotterdam and Dortmund are medium-sized cities from a European perspective: their current urban scale is not remarkably large, but they have many opportunities for increasing the scale by improving cooperation with nearby cities, since such linkages are currently quite limited. Rotterdam is part of the dense urban Randstad region (7 m inhabitants) which also includes The Hague, Utrecht and Amsterdam. Dortmund is part of the industrial Ruhr region (5.4 m inhabitants), which includes cities like Essen and Duisburg. Münster and Eindhoven have a smaller urban size and are not particularly close to other big cities: their possibilities for increasing the urban scale by enhancing interurban cooperation are more limited. Finally, Zaragoza itself is a medium-sized city (bigger than Münster and Eindhoven), but it is not part of a larger urban conurbation: the city is located in the middle of a sparsely populated area. Table 11.2 gives an indication of the case study cities' urban scale.

Table 11.2 Number of inhabitants in the city and region, 2003

City		Region	
Amsterdam	735,000	Greater Amsterdam consisting of 17 municipalities	1,400,000
Dortmund	588,000	Eastern Ruhr Region (Dortmund, Unna and Hamm)	1,200,000
Eindhoven	206,000	Greater Eindhoven Area (Southeast Brabant)	670,000
Helsinki	560,000	Helsinki region comprised of 12 municipalities	1,200,000
Manchester	440,000	Greater Manchester	2,500,000
Munich	1,300,000	Urban region comprised of 8 'Landkreize' (districts)	2,500,000
Münster	280,000	Münsterland	1,500,000
Rotterdam	600,000	Rijnmond region	1,100,000
Zaragoza	600,000	Aragon region	1,000,000

Source: case studies.

Most case study cities are aware of the importance of urban scale. They try to strengthen their position concerning this knowledge foundation in several ways. In Manchester, two universities have merged to attract more and better professionals and students. In the Randstad region, currently most interaction takes place within the northern (Amsterdam) wing and southern (Rotterdam) wing: activities including both wings are limited. The construction of the new HST rail links currently under way is expected to significantly reduce travel times between the two cities and increase interaction and cooperation. Helsinki tries to extend its urban scale by looking abroad: cooperation with Tallinn is set up and interaction with St Petersburg is expected to improve when a new HST train connection is in place. Eindhoven acknowledges that its limited size implies a relatively low degree of influence on national government decisions. Therefore it cooperates with some other cities in the province to increase their political influence. Eindhoven is also involved in cross-border cooperation: it works together with Aachen and Leuven to jointly become a top technology region. Münster is involved in international cooperation as well: these efforts mainly run through the university and polytechnics that participate in EU projects. Cooperation between cities in the Ruhr region (including Dortmund) is limited and up to the time of writing few activities have been deployed to increase such linkages. For Zaragoza, the urban scale is expected to increase when the HST train connection is finished.

11.2.7 Social Equity

In the knowledge economy a two-way relationship exists between economic performance and social equity. On the one hand, growth of knowledge-intensive sectors increases the number of jobs in personal services, hotel and catering industry and retail: such jobs often require low levels of education. On the other hand, reducing poverty and inequality can stimulate economic growth by increasing (the perception of) safety (Hall and Pfeiffer, 2000) and by enhancing the purchasing power, which will strengthen the demand side of the economy. In their national contexts Munich, Münster, Eindhoven, Helsinki and Zaragoza perform well. These cities do not have significant social equity problems.

The former industrial cities of Manchester, Dortmund and Rotterdam are performing rather poorly. Amsterdam, too, has a high unemployment figure. As we have already identified in the urban diversity section, a significant part of the unemployment problems applies especially to ethnic minorities. These communities started their formation decades ago with the large inflow of migrant workers that came to Dutch and German cities, among others, to fill low-skill, low paid job vacancies. Today, cities like Dortmund, Rotterdam and Amsterdam have large groups of foreign-origin inhabitants from countries like Turkey and Morocco. The social problems are mostly concentrated in certain urban districts.

The case study cities try to mend the social divide – whose main cause is low levels of education – in several ways.

In Manchester not only does renovation of houses take place, but also the university is involved in solving the problems: a university institute is located in East Manchester and several academic fields (e.g., urban development and education) try to apply their knowledge to improve the neighbourhood. Amsterdam wants to increase social inclusion by improving the quality of education. In Eindhoven specific education programmes are aimed at unemployed people who lost their jobs in the economic downturn: they can now strengthen their knowledge and skills to obtain a more favourable position in the job market once economic circumstances improve.

11.3 Knowledge Economy Strategies

In the preceding section the knowledge foundations of the nine participating cities were discussed. In this section, we will discuss their knowledge strategies. In our study we have noticed that, increasingly, cities are formulating comprehensive strategies that bring together several knowledge-related issues such as education, research, technology transfer, but also housing policies and city marketing. This indicates that cities are becoming increasingly aware of interactions and inter-relations between formerly separated policy fields. But first, we will discuss national and regional knowledge economy policies.

Some national/regional governments have developed knowledge economy strategies, but there are large differences.

More explicit knowledge economy strategies at other government levels can be found in Germany, Spain, The Netherlands and Finland. Within Europe, Finland was one of the first countries to develop explicit 'knowledge economy strategies'. During the 1980s the national government of Finland was already halfway along the transition process towards the knowledge economy. The STPC – the national council for the knowledge economy – was set up in 1985, and Tekes – the investment fund – in 1983. Furthermore, there is a national *Centres of Expertise* programme, aiming to support dedicated knowledge economy clusters within Finnish regions, which can operate in an internationally competitive way. Thus, the strategy has a regional component. In

Germany, the regions (Länder) are quite active regarding these strategies, as we found for the regions of Bavaria and North-Rhine Westphalia. The Länder are largely responsible for R&D policy and education policy and they have a large influence on some economic sectors, such as the media sector and the health and biotechnology sector. In Spain, the autonomous Region of Aragon appears to play a similar role to that of the German regions in supporting education and innovation policies. The national government and regional governments in The Netherlands have recently set up innovation platforms to develop strategies to improve Dutch performance within the knowledge economy.

Halfway through the 1990s, a number of cities participating in this study developed early knowledge economy strategies. In 1993 Eindhoven started the Stimulus programme. In 1994 Manchester published its City Pride prospectus, which was renewed in 1997.

At the start of the new millennium, more cities came up with knowledge economy strategies. Dortmund had started the Dortmund Project by 2000. In 2003, Rotterdam, Münster and Zaragoza published their latest strategies. Obviously, because of the newness of these strategies, their results could not be studied within this research project.

Some cities have very explicit and comprehensive knowledge economy strategies.

Quite explicit and comprehensive strategies could be found in the cities of Manchester, Rotterdam, Dortmund, Zaragoza and Eindhoven. The other four participating cities, Amsterdam, Munich, Münster and Helsinki, have less explicit strategies.

The focus of Dortmund and Rotterdam is to diversify the urban economy. In Zaragoza, the key ambition is to develop the service sector and benefit more from the HST connection. Eindhoven's strategy aims to strengthen its technological profile in many respects. Manchester has developed a vision of its role in the knowledge economy as a secondary city within Great Britain: 'Manchester Knowledge Capital'. An important question in this respect was what the role of Manchester in the global knowledge economy in relation to the dominant London region could be. This vision states that Manchester wants to make better use of its strengths (i.e., four universities, international airport, Metrolink system and growing industry clusters) to attract and retain knowledge economy activities. Another important element of the strategy is improving the urban equity (by supporting disadvantaged people). This is considered essential to achieving a sustainable regional economic growth.

There are different catalysts for developing a strategy.

For some cities, an economic crisis was the key catalyst for the development of a knowledge strategy. For the city of Eindhoven, for example, the economic crisis in the early 1990s was an important catalyst for setting up the economic stimulation programme 'Stimulus'. This programme has contributed to increased cooperation between firms and education institutes and to attracting research institutes. The desire to further strengthen the structure of the regional economy has been a catalyst for a new programme – 'Horizon' – which includes a number of projects supporting knowledge economy activities, such as stimulating the flow of students from lower via intermediary to higher education, increasing the number of start-up companies and setting up a high-tech campus. For Helsinki, the economic crises of the early 1980s and the collapse of the Soviet Union in the early 1990s (previously Finland's most important trade partner) were major incentives to taking adequate steps to strengthen the local economy. Amongst other things, this led to an internationalisation strategy (for instance, by cooperating with cities outside Finland), and in particular to much more attention being paid to education, science and research. With respect to the latter elements, the national government played a very active and stimulating role. Furthermore, for the city of Helsinki, cultural and environmental factors were considered to be important to improve its role in the knowledge economy. Representatives of cultural organisations are even involved in initiatives to strengthen the regional economy.

For Rotterdam and Dortmund, the *steady decline of the economic base* was a key reason to design a knowledge economy strategy. Desired structural economic changes have stimulated them to come up with active strategies. In 2003 the Development Corporation Rotterdam developed a Knowledge Economy Programme. Within this programme, three tracks to support the regional knowledge economy are identified: developing the regional labour potential; reinforcing the innovation power; and improving knowledge infrastructure and facilities. In addition, for Dortmund the perceived need for regional economic structural changes was the catalyst to develop a strategy. It was set up by the City of Dortmund in close cooperation with private business companies. This 'dortmund-project' has a number of goals: stimulating new specific knowledge economy clusters (ICT, e-logistics, micro electronic mechanical systems); strengthening local companies; improving education institutes; upgrading the urban quality of life; and better administrative services. Besides this public-private partnership, the help of the regional government (North-Rhine Westphalia, see below) and the EU (Urban 2) are also considered important.

For Zaragoza, the *connection to the HST* was a key catalyst to thinking about the future of the city. 'Zaragoza towards the knowledge society' is the name of the knowledge economy strategy presented in 2003 by the City of Zaragoza. Major elements of the strategy are economic diversification and better utilisation of the existing knowledge base. The strategy includes a large number of projects that contribute to these aims, but acquired a clear focus in and around the station areas of the high-speed train. In these areas, the 'Digital Mile' project is envisaged: the creation of a highly innovative urban environment, which should become a laboratory for new ways of living, learning and working. It is hoped to attract knowledge-intensive activities to these areas, with major spread effects on the surrounding regional economy.

Other cities have less explicit and comprehensive knowledge economy strategies.

But in all cases, they are very active in separate fields of knowledge economy policy. The City of Amsterdam had few codified policies on the knowledge economy. Within the city, there is however a knowledge network, in which the city, universities and companies are involved. Until recently, this network has focused on networking and information exchange. It is expected, however, that it will become more proactive in bringing relevant actors together to create and apply relevant knowledge. Within the urban region of Amsterdam there is a wide variety of knowledge institutes and knowledge-intensive business companies. There is, however internally as well as externally, insufficient awareness of the existence of these knowledge elements. In order to change this, the City of Amsterdam supports better city-marketing strategies.

The City of Munich does not have a comprehensive strategy, but it undertakes a number of activities that promote the regional knowledge economy, such as promoting promising economic sectors, promoting innovation, helping SMEs in innovation and supporting starters. In addition, the Freestate of Bavaria plays a very active role in stimulating the knowledge economy.

The City of Münster does not have an explicit and comprehensive knowledge strategy, but has developed several related policies. The two most important aims of these policies are to increase interaction between regional education institutes and companies and to better promote the knowledge activities of the city. The first aim is considered important because the education institutes are still operating too much as islands. More interaction is expected to lead to more applied knowledge creation. Because the knowledge activities of the city of Münster are not very well known, within as well as outside the city, promotion activities are also considered important in order to create more awareness of the strengths and attractiveness of the urban region.

In their strategies, some cities take a regional perspective.

In the knowledge economy, administrative boundaries matter less. Some cities realise this and develop their strategies accordingly. Rotterdam, for instance, wants to improve cooperation between the Technical University and relating business activities of the neighbouring city of Delft. Both cities are located so close to each other that it can be argued that they form one coherent economic region. Manchester and Eindhoven came up with clear comprehensive regional strategies. Other cities even look beyond the direct urban region and engage in partnerships with cities in other regions. Examples are Münster (part of the Scientific Region, together with Enschede (NL) and Osnabrück) and Eindhoven (linked with Aachen and Leuven). It must be noted, however, that it is not always clear how the 'strategic' cooperations work out in practice.

Our study suggests that strong knowledge economy foundations lead to fewer incentives for active urban management.

We found that the stronger the knowledge economy foundations are, the less explicit the knowledge economy strategies are (see Figure 11.4). This was particularly noticeable in the urban regions of Amsterdam, Munich and Helsinki. All three cities have relatively strong knowledge economy foundations (see previous section). Because of these strong foundations, they will attract knowledge-intensive activities more easily than regions with weaker foundations. They thus have fewer incentives to develop explicit policies to strengthen these foundations and/or to have adequate knowledge economy activities (see next section). In the case of Helsinki we could also see that the national government plays an active role in stimulating the knowledge economy; in the case of Munich, the regional government of Bavaria played such a role. It can be argued that the more active other governmental levels are, the less necessary an active local authority will be. In the case of Amsterdam, however, other governmental levels are not yet really active in stimulating the knowledge economy. For this urban region, therefore, a more active public involvement in stimulating the knowledge economy is considered to be necessary in order to improve and guarantee Amsterdam's role in the knowledge economy in the long term.

Conversely, we found that cities with weaker foundations are quite active in developing explicit and comprehensive strategies. Because of their weaker starting position they have to be proactive to catch up with their competitors. For these cities, several catalysts could be noticed for formulating explicit and comprehensive strategies to strengthen the regional economy.

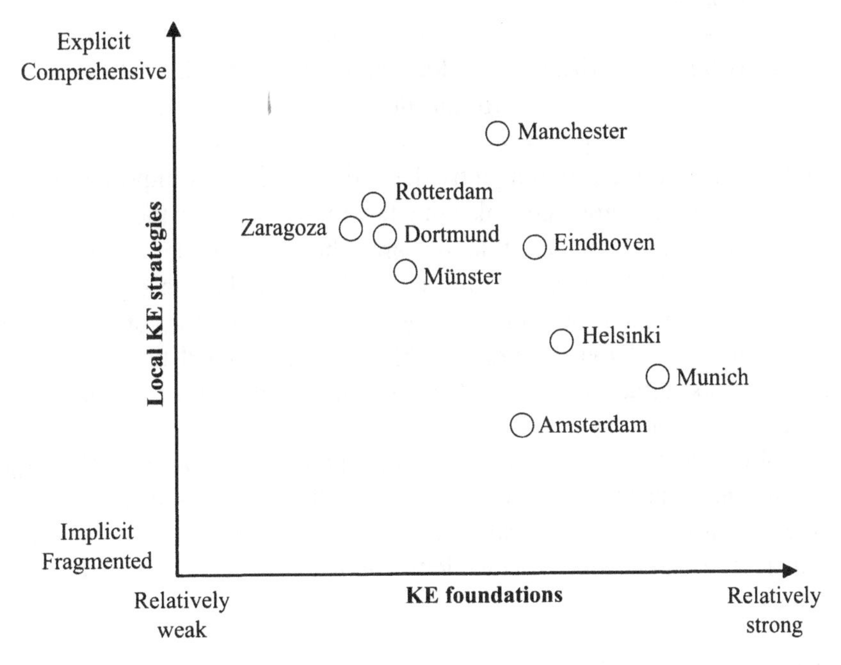

Figure 11.4 Local knowledge economy strategies and foundations

11.4 Knowledge Activities

What can cities do to become stronger in the knowledge economy? This was one of the key questions in this research project. In Chapter 1, we make a distinction between four types of 'knowledge activities':

- creating new knowledge. This can be pure scientific knowledge, but also other types of knowledge;
- applying new knowledge. For this category, we analysed how scientific knowledge is transferred into business, and, more generally, how the knowledge infrastructure and the business sector cooperate;
- attracting knowledge workers. Here, we looked at the cities' ability to attract and retain highly educated people, not only students but also workers in the creative industries;
- developing new growth clusters. For each city, we have described and analysed how it tries to strengthen the knowledge foundations by developing new growth clusters.

Knowledge activities, in the long run, change the knowledge foundations.

Before treating the different activities individually, it is important to see their inter-relations, and the link with the foundations of the knowledge economy. In Figure 11.5 these relations are schematically represented.

In the figure, the knowledge activities are linked to the two key economic foundations: knowledge activities change the knowledge and economic bases of a city. In the previous section, we have already argued that cities with relatively weak initial foundations have the highest incentive to change their knowledge and economic base.

The four knowledge activities are in the centre of the figure, in the large box. After analysing the nine cases, we conclude that the activities are mutually related: the last category – creating new growth clusters – is a function of the others. For the creation and development of a growth cluster, all the other activities are relevant.

11.4.1 Creating New Knowledge

The cities are very different in their ability to create knowledge. Knowledge is created in the public sector (universities and public research institutes) and in the private sector. It is far from easy to measure the degree of knowledge creation. One rough indicator is the number of patents. Although we do not have fully comparative data, Munich, Helsinki and Eindhoven lead the pack in this respect. This is because they have relatively many strong technology companies that submit patents. In most cities, private companies are far more important as patent holders than universities (in Munich, universities less than 2 per cent).

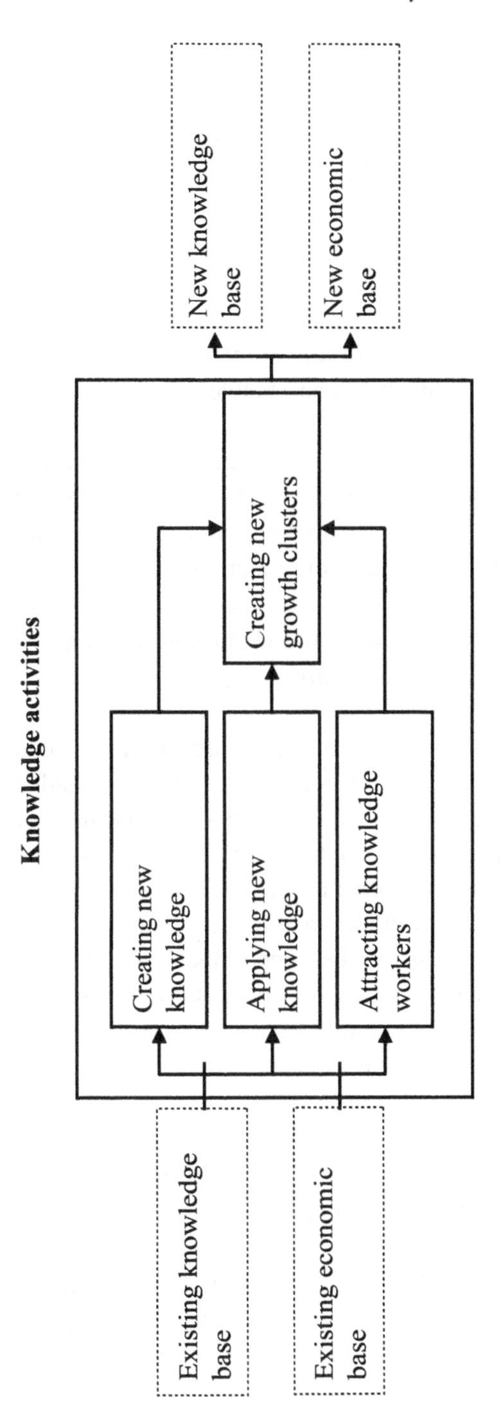

Figure 11.5 Knowledge activities of cities

Patent data should be interpreted with great caution. First, the patent data refer to the location where the patent was registered; this is not necessarily the location where the invention took place. This may overestimate Munich's innovativeness, with its headquarters of BMW and Siemens. Furthermore, innovations in services (which make up over 70 per cent of the economy) are normally not patented, although they make a major contribution to productivity and quality improvements. Third, it is not the number of patents registered that is important for the economic development and innovative character of a city, it is the number of patents which are transformed into production or new processes. This is the important difference between invention and innovation.

Another indicator for new knowledge creation is R&D expenditure (by firms and governments). Finland is the leading country here (and much of it is spent in Helsinki), followed by Germany (with Munich as a focal point) and then The Netherlands. The UK and Spain score much lower.

Innovation and new knowledge creation is too often measured as high-tech innovation only.

High-tech indices (R&D expenditures, patents, etc.) are often regarded as a measure for a region's innovativeness and its strength in the knowledge economy. Many types of policies – national, local and European – also heavily focus on high-tech companies and promote technological innovation. This underestimates the importance of innovations in services, concepts, design and other areas that are crucial activities in the knowledge economy. Cities such as Amsterdam (financial and business services) and Manchester (creative industries) create a lot of new knowledge of this type but nevertheless receive a low score in the common rankings of innovative regions (for instance, the EU's Top 50 European Innovation Index). In a recent article, Den Hartog, Broersma and van Ark (2003) confirm this point.

Local governments have a limited degree of *direct* influence on knowledge creation in the city.

The creation of new knowledge takes place in universities, research institutes and private companies. In theory, it would be beneficial for a city if the knowledge infrastructure and the business sector were complementary. However, in practice, local governments have very little *direct* influence on this. Universities are relatively autonomous, and decide their own research and education priorities. University *policy* is designed at the national level (Netherlands, Finland, UK) or regional level (Germany, Spain). Most science funds also operate on a national or regional level. In many cases, we found that governments face a trade-off between equity and equality. In Finland, where most of the national research funds end up in the capital city of Helsinki, a debate is going on as towhether these resources should be spread more equally over the country. The Freestate of Bavaria, where Munich is a natural 'magnet' for these funds, has an explicit policy to spread research and education spending equally over all Bavarian cities. Nevertheless, Munich has benefited greatly from the relatively high investments of the Freestate of Bavaria in knowledge, science and education in the last decades.

Cities can do something to improve their knowledge base. One option is to attract research institutes. Eindhoven, for instance, managed to attract a highly innovative research institute (TNO Industry) to the city, offering support of €7m.

The direct competences of local governments to steer the local knowledge creation are limited: they cannot do it alone.

However, city governments can achieve a lot more when they engage in partnerships with other actors.

We found several good examples of cities that use their organising capacity to influence knowledge creation.

- Develop research institutes. The city of Münster invested €2.3m in a centre for nanotechnology that offers optimal conditions for research in this field, in a strategic cooperation with the university and a local bank.
- Promote cooperation between universities. For cities with several universities, making them cooperate is a challenge. Manchester has ambitions to coordinate the activities of the four universities in its region: two universities have already decided to merge, with the aim of becoming stronger vis-à-vis top science cities in the UK (London, Cambridge and Oxford). The city invests substantially in the campus. Rotterdam promotes the cooperation of higher education institutes in the fields of media and transport and logistics.
- Establish chairs/professorships. Rotterdam also has concrete plans to establish professorships in fields that it considers important for the local economy: logistics/ICT, process chemistry, architecture, and life sciences.

- Promote and support cooperation with universities in other regions. In particular the smaller cities (Eindhoven, Münster) see this as a viable option. Eindhoven has started to cooperate with the cities of Aachen (Germany) and Leuven (Belgium). The regions share a strong R&D profile, and want to fine-tune their specialisations. Interestingly, it was Philips, the electronics company, that initiated this cooperation: the company has important research institutes in each of the three cities. Münster works with Osnabrück and Enschede/Hengelo to create a 'knowledge region' to strengthen the networks between the research centres in those cities.

11.4.2 Applying Knowledge/Making New Combinations

How is knowledge translated into business, and how can this be improved? These are crucial questions for urban policy makers in all our case study cities. In our study therefore, *applying knowledge* is the second type of knowledge activity. Under this heading, we described and analysed university/business cooperations and activities to improve this, as well as the promotion of start-ups. What have we learnt from the cases?

In general, we found that many policy initiatives (on local, regional and national levels) aim to improve the cooperation between firms and universities, in order to make more out of the regional knowledge base. Universities and companies too often inhabit separate worlds: they have different drivers (academic prestige versus profit making), different time horizons, different work attitudes, and in practice, find it difficult to cooperate. However, there are many potential synergies and interdependencies: some knowledge developed at universities is commercially very valuable, or may become so in the longer run; university students and researchers are potential entrepreneurs; universities are the source of new staff for companies; university laboratories may be used by start-ups; universities may benefit from applied research assignment from the industry, and vice versa.

Declining direct and unconditional funding for universities gives them an incentive to cooperate with the private firms.

It is a general trend in Europe that universities are urged to cooperate more with businesses: they receive less direct funding, and have to increase their earnings from cooperation with industry. This may improve collaboration. In Münster there is an extra reason: the polytechnic school there obtains less structural funding from the region of North Rhine Westphalia, but more when it is more successful in attracting financial means from industry. This example shows how regional governments can create incentives to improve business/university collaboration.

In order to improve the willingness of scientists to work with industry, incentive structures within Europe's universities need to change.

Despite the trend of decreased funding, working with industry is still considered inferior by many (top) researchers. In the current situation, in all our case studies researchers are mainly judged according to their scientific publications. Their careers depend on it, and the allocation of research grants and funds is also based on researchers' publication track record. This reduces their willingness to work with industry. If the aim is to promote university/business cooperation, this incentive structure needs to change. This is something that cannot be achieved by local governments, but they could join forces and lobby to modify the policies of national governments and research funds.

In cities with a strong and diversified economy and big firms, cooperation between companies and universities/research institutes is easier.

In our case study cities, we found that larger and innovative firms are more willing to cooperate with universities. This is an advantage for cities with a diverse and innovative economic structure, such as Munich and Helsinki, and also Eindhoven: they form a very 'natural' environment for cooperation. In some other cities (i.e., Münster and Zaragoza), the economic structure is much less favourable. Münster has a lack of business activity: top research teams in the Münster university see this as a disadvantage, because if they want to work with industry they have to travel to other cities. Münster tries to attract knowledge-intensive business with its excellent knowledge infrastructure. In Zaragoza, the business sector is dominated by medium-tech SMEs that are less inclined to work with the university. We conclude that cities with *both* a strong knowledge base and a strong economic base have the best perspectives for fruitful university/business interaction.

All the case study cities want to improve university/business cooperation, but they do so in different ways.

In our case study cities, a lot of policies are directed to improve university/business interaction. In such policies, cities have to work together with universities and companies. Here again, the degree of *organising capacity* is crucial: the success of policy depends heavily on the quality of the local networks and the ability of the partners to work together and arrive at concrete results.

There are many instruments to improve university/business cooperation, and for each instrument there can be a different role for the city. For some activities, cities can take the lead. In others, leadership is more naturally in the hands of the other partners, and cities can play a more supportive or enabling role. Below, we list some of the policy options.

Activities where cities take the lead

- Cities can create networks by providing financial rewards for cooperation. In Eindhoven, the Stimulus project financially rewarded innovative cooperations between business and knowledge institutes.
- Cities may co-invest in joint facilities and technology centres where firms and research institutes can cooperate. German cities are very active in this respect. One example is CeNTech in Münster, a centre for nanotechnology research. It provides high-level facilities to university research groups, and encourages the creation of new business out of that research. Similar centres are to be found in Dortmund and Munich.
- Cities may create technology parks: special zones dedicated to high-tech business. Several cities have created technology parks near university establishments. The largest we found was in Dortmund where, in 1985, the city created a technology park, together with the Fraunhofer Institute, the Chamber of Commerce and a bank. By 2004 it had 200 companies, most in high-tech industries. They contribute substantially to the Dortmund economy.
- Cities can involve the university in urban projects. In Zaragoza, the city deeply involves the university in the design of its largest urban project, the 'Digital Mile' near the central station area. Here, the local government takes the role as third-party funder.

Activities where universities take the lead

- Set up of transfer agencies. In most of our case study cities, higher education institutes have transfer agencies that create links with business and help administrative and contractual issues.

- Proactively approach companies and invite them to participate in projects. In Zaragoza, the polytechnic no longer waits for companies to ask for their help, but rather markets itself to the companies in the region.
- Enable firms to use university facilities. In Munich, for instance, young firms are allowed to use laboratories of the Technical University, and may use the university as their address.

Activities where firms take the lead In general, larger firms are more likely to cooperate with the local knowledge infrastructure than small ones are. It is a trend that companies are becoming more open to the outside world and engage in activities that are new to them. Philips, in Eindhoven, has initiated cooperation between Eindhoven, Leuven and Aachen (see previous section) to fine-tune knowledge activities and strategies. Also, it is building a technology campus in the city. Siemens (in Munich) plans to construct a campus in the city centre, and become more open to the city than it is now.

There is a substantial role for regional and national governments in promoting business/university cooperation.

Although cities have some instruments to contribute to improve interaction, they depend on others in many instances. Much depends on the administrative organisation of the country. In the federal countries (Germany and Spain) the regions are the most important; in the others, it is the national government. In Germany, the regions (Länder) are the main actors to promote university/business cooperation. Bavaria deploys an impressive variety of policies. One example is the provision of financial incentives for projects in which the university cooperates with a company. The so-called Bonus Programme (a programme encouraging research financed by sponsors from commerce and industry) pays financial rewards for the acquisition of funds from third parties for carrying out specific applied research assignments. In Spain, the regions are also important players in innovation policies. In Finland, there are strong financial incentives from the *national* government to make universities and business cooperate. Tekes, the national technology agency, funds cooperative R&D projects of universities and businesses.

Ideally, national or regional policies should support local strengths.

In many countries, there is a gap between national innovation policies, which are very general, and local competences and ambitions, which are more specific and concrete. Too often national innovation policy is non-spatial, i.e. it does not take regional differences in economic and knowledge specialisations into account (although it must be stressed that this holds less for regional policies in Bavaria and North-Rhine Westphalia). With this in mind, Finland has designed its national Centre of Expertise Programme, launched in 1994. The idea is to focus local, regional and national resources on the development of internationally competitive fields of know-how in specific places. The programme pays special attention to SMEs to develop selected internationally competitive fields of expertise and stimulate technology transfer from universities to firms. The programme covers the whole country and it is carried out in regional Centres of Expertise, appointed by the Council of State, that work closely with universities and companies in their respective sectors. The Netherlands recently set up a national innovation platform, chaired by the prime minister, to promote the knowledge economy. However, there is little explicit recognition of or attention on the regional and local dimension of the knowledge economy, and there is little integration of national strategies and local/regional strategies.

In Germany, the national government (and sometimes the regions) organises national competitions, in which cities or regions can win financial support. By doing this, cities are encouraged to develop innovative strategies in certain fields. For instance, cities could apply for a subsidy to create a Biotech region. Another example is the 'City of Science' competition, where cities can submit plans to strengthen the link between their university and other urban actors. Often, the city benefits even when it does not win the competition: actors have cooperated to draft the proposal and have developed ideas that they want to implement anyway. In The Netherlands, this 'competition approach' is also frequently used, for instance in the Twinning initiative.

Promoting entrepreneurship and spin-outs from universities is a priority in all the cities.

Cities are highly aware of the importance of entrepreneurship for their economic future. Every city in our study has invested in starters' centres, where start-up companies can obtain cheap business space and administrative support. In many cases, the centres organise meetings where the companies can meet potential new clients, partners or experts, visits to trade fairs, etc. It is a trend to have specialised starters' centres rather than generic ones: Munich has five starters' centres, more or less specialised in specific branches; Zaragoza's centre focuses on industrial companies; Münster's centre focuses on ICT and life sciences; Eindhoven's Twinning centre is dedicated to ICT. Typically, starters' centres are set up jointly by the city, the university, the region and the Chamber of Commerce.

Another instrument is the business plan competition, which is common in German cities. The entrepreneur who submits the best business plan wins financial and other support. Dortmund probably has the most aggressive programme: starters from all over Germany can participate in the business plan competition and win the money even if they do not locate in Dortmund. But if they do locate in that city, they can win even more.

Many cities hope to increase the number of spin-out companies from universities (successful researchers who develop a commercially viable product or idea). Some cities support chairs in entrepreneurship (Münster, for instance). In Munich, researchers can get support if they want to register a patent (which can be quite complicated). Some cities have agencies to spot commercially viable ideas in the university.

Venture capital goes where the business is, and proximity matters.

In general, since the collapse of the dot.com boom, it is a lot more difficult for companies to obtain venture capital, as the providers have become much more cautious. However, there are spatial variations. Venture capital providers (official VC companies or private investors) tend to locate where the number of investment projects is the highest: the larger cities. For the smaller cities, this can be a problem. In Münster, for example there is no active venture capital company. This makes it more difficult for start-ups or SMEs to obtain finance compared to firms in Munich, where the offer is large.

To increase the supply of VC, many local and regional governments in our case study cities have public or semi-public venture capital funds. One example is the Stimulus Venture Capital Fund in Eindhoven which provides VC for high-tech SMEs for a period of 5–7 years. The Freestate of Bavaria also has numerous VC provisions, most of them in the role of 'accelerator funds': it is easier to obtain more capital if you already have one investor.

11.4.3 Attracting/Retaining Knowledge Workers

In our framework we have argued that the knowledge economy hinges on the availability of qualified staff. As a general trend, the demand for workers with tertiary education is rising. One of the key questions for cities is, therefore, how to attract these people to the city, or how can they be retained. In this respect, it is important to make a distinction between different types of knowledge workers, for instance, between top scientists, engineers, managerial staff, people in the creative industries (artists, architects, etc.), or specialists in the service sector. Another category is students. All of these groups have different location preferences.

Capital cities easily attract knowledge workers.

In all our case study countries the capital regions had operated as a magnet for the higher educated since the mid-1990s. In Finland, Helsinki is the focal city for the highly skilled: by 2004, 50 per cent of Finland's academics lived in the Helsinki region and there is still an upward tendency. In the UK, London is still a strong magnet for talent from other cities. In The Netherlands, Amsterdam has had a high percentage of academics and knowledge workers in recent years. In Spain, Madrid and Barcelona attract a disproportionately high proportion of the highly educated. In Germany, Munich (not a national capital but the important regional capital of Bavaria, with more than 10m inhabitants) is in a strong position, albeit it a much less dominant one than Helsinki or London.

The reasons for this tendency are manifold. One key factor is that large cities offer more job opportunities than small ones do. This makes it easy to change jobs (many knowledge workers are relatively mobile) but also easier for the partner of the knowledge worker to find a job. Secondly, these cities are well connected internationally (through airports and HST lines), a factor that has become more important with internationalisation of the economy; thirdly, these cities offer an international infrastructure (schools, social life) that makes it easier to attract talent from other countries. This is a key asset given the increased international labour mobility in high-skilled jobs. Fourthly, these cities are natural 'cultural capitals', with subsequently strong cultural and creative industries. Their cosmopolitan feel attracts creative people.

But their success may also drive out some categories of knowledge workers.

In Munich, for instance, costs of living are so high that many students choose to live outside the city, or decide to study in smaller towns. Also, there is a lack of affordable space available for artistic/creative people who are not (yet) commercially successful. The same problem applies to Amsterdam. Furthermore, for some categories of workers such as teachers, nurses or police, whose salaries are determined on a national level, these cities are relatively expensive places. Those cities may need specific policies to keep these vital workers in the city: they could consider subsidising student housing, creating specific housing projects for public servants, or subsidising locations for artists.

Smaller (secondary or provincial) cities lack many of the advantages of the capitals but may offer other assets to attract talent.

Smaller cities tend to be quieter, greener, and have less congestion and crime, and some of them have specific strengths. Eindhoven, notwithstanding its small size, manages to attract top engineers and many foreign workers. They appreciate the green surroundings and the quietness of the city. The strong technology-oriented economic base of the city generates interesting jobs for these people. Münster (also a quiet and green city) attracts top scientists because of the good reputation of its university. Both cities have a relatively high share of Western immigrants. Zaragoza currently has problems retaining its highly educated workforce, and many students move away after graduation. When the city is connected to the HST, it may become a commuter town for Barcelona and Madrid and capitalise on its high quality of life. More generally, the lack of international accessibility may hamper the ability of some smaller cities to attract talent. This is a problem for Eindhoven, which will not be connected to the HST networks and will experience a relative loss of accessibility.

Overall, smaller cities have to compensate for some disadvantages they have. There are several options for this:

- generic city marketing strategies: increase the visibility and profile of the city. This can be achieved by organising events, etc.;
- targeted city marketing actions: make sure the city is part of the 'mental map' of the target group. This should be done in cooperation with local business leaders and research institutes;
- investment in short-stay accommodation. This helps to attract temporary 'expats';
- development of active supportive policies to help foreign workers, for instance, to find housing, or a job for the partner;
- setting up or subsidising international schools, perhaps in cooperation with neighbouring cities;

- engagement in strategic partnerships with other cities (Münster and Eindhoven do this).

Again, organising capacity is crucial: these cities do not obtain the fruits of the knowledge society automatically, so they have to be proactive.

Cities with a traditional economic base (Rotterdam, Manchester, Dortmund) have specific problems attracting talent, but also specific strengths.

Some of these cities have an image problem that makes it more difficult to attract highly-educated knowledge workers. This is a particular problem for Dortmund: although this city is no longer dominated by smokestacks and heavy industry, many people in Germany and abroad still perceive it as such. Changing the image is therefore a key priority for Dortmund. Secondly, given their sector structure, these cities offer fewer jobs for which high skills are needed. Rotterdam has a relatively poor housing stock: there is a lack of housing for middle-class families. Therefore, upgrading the housing stock is a key priority for that city.

This class of cities not only has drawbacks but also specific opportunities. One is the abundance of (cheap) space, which can be a factor in attracting creative talent and provides room for experimentation. Many cities are converting their industrial heritage in order to develop creative industries. In the last few years, Manchester and Rotterdam have built up a young and trendy image in certain subcultures, which may offer promising economic perspectives.

For top scientists, the reputation of the university and the quality of the specific research group is the most important location factor.

Recent studies show that Europe loses many scientists to the US, where several top universities attract the brightest people from all over the world. In Europe, only the UK has any universities in that league. But there are several activities in our case study cities to change this. Some of our case study cities are trying to build up strong competence centres. Manchester, for instance, wants to be on a par with Oxford and Cambridge by promoting a merger between its universities. In Germany, the Max Planck institutes are setting up top-level international research schools together with local universities (one is in Munich), to attract top talent. This is an important trend, because top universities/research groups may attract and generate private research activity.

Cambridge in the UK is an outstanding example of a city whose excellent knowledge base has attracted many high-level private research institutes (Microsoft, for instance).

11.4.4 Developing Clusters

The last knowledge activity is the promotion of cluster development. As outlined above, this is a crucial policy to influence and modify the economy and knowledge base of a city, and it involves all the other knowledge activities as well. What can we conclude, based on our case studies?

Many cities choose similar spearhead clusters.

We must make a distinction between growth clusters and spearhead clusters: *growth clusters* are activities that grow in the local economy regardless whether they are the specific target of policies or not. *Spearhead clusters* are activities that receive special policy attention.

Table 11.3 shows the list of sectors/clusters that receive policy attention in our case study cities. It is quite striking that many cities choose the same spearheads: ICT and biotechnology, with some variations, appear in almost every city; nanotechnology is becoming popular as well.

Table 11.3 Spearhead sectors/clusters that receive policy attention

City	Spearhead sectors/clusters
Amsterdam	ICT/new media, life sciences
Dortmund	MEMS, ICT, logistics
Eindhoven	Medical technology, automotive, ICT, mechatronics
Helsinki	Materials and microsystems, gene technology and molecular biology, medical technology, logistics, ICT/media
Manchester	ICT, cultural/creative industries, biotechnology and health, nanotechnology, finance/business services
Munich	ICT, biotechnology, media, environmental technologies, new materials
Münster	Biotechnology, nanotechnology, ICT
Rotterdam	ICT, audio-visual media, health, medical technology, transport and logistics
Zaragoza	ICT, logistics, business services

Many cities try to add a 'knowledge component' to traditional economic sectors.

As stated in our theoretical introduction, the knowledge economy affects every economic sector. 'Traditional' sectors are also part of the trend. From this perspective, we found interesting examples of cities that design policies to upgrade traditional sectors and make them more knowledge intensive, or add new knowledge-intensive activities. Rotterdam wants to improve its knowledge base in the field of transport and logistics. To that end, it invests in an Academic Centre for Trade, Transport and Logistics, which is a cooperation between two universities and the port community. Zaragoza is developing a logistics park, but at the same time it has attracted a subsidiary of MIT (the Massachusetts Institute of Technology, from Boston) which will conduct research and offer top-level education in the field of logistics.

The selection of target clusters should be based on existing and distinctive strengths of the city.

It is easy to state that a cluster should be developed or promoted: it is less easy to actually make it happen. Many factors play a role in cluster development and market forces are very strong. The direct influence of policy is limited. However, based on what we have seen in our case study cities, we are able to draw some conclusions. Policy makers should be aware of the market potential for the cluster in their specific city: in the end, market forces strongly dominate corporate location decisions and the future development of sectors, and policy can only support and encourage certain developments. A careful analysis of current strengths and weaknesses is therefore crucial.

Cluster policy can be built on several types of strengths:

- *academic* strengths in the city (Münster, Rotterdam, Helsinki). Münster's nanotechnology cluster policy, for instance, is built around the strength of a very specific research group at the university. Other good examples are Rotterdam (the medical cluster policy is built on strong academic competence in the medical school) and Helsinki;
- *economic* strengths. Eindhoven's choice for mechatronics is based on its strong business profile in this sector; the same holds for Manchester's choice for finance and business services;

- *cultural* strengths. This holds most true for Manchester's choice of the creative industries as a growth sector, but also for Amsterdam's potential in the new media sector;
- strengths in *accessibility*. Zaragoza's ambition to develop a logistics cluster and a business services hub near the new station is directly related to the connection of the city to Europe's high-speed train network.

The chances for a target cluster to flourish are the highest when the cluster is based on *several* strengths of the city. One good example is Munich's biotechnology cluster, which is linked to both academic strengths (several university departments have a good reputation in biotech research) and to economic strengths – the city already had a cluster of pharmaceutical companies and related business.

Growth cluster policy should be comprehensive.

Promoting a growth cluster is a complex challenge. Therefore, many partners should be involved and it should combine a number of policy instruments. Ideally, promoting a growth cluster includes all the knowledge activities in our study:

- starters' policy: specific starters' facilities and support for start-up firms in that particular sector;
- knowledge creation policy: cities should try to influence the research infrastructure in a way that supports the targeted growth cluster;
- knowledge transfer policy: make sure that universities and the firms in the target cluster cooperate;
- location policy: for some clusters, where face-to-face interaction is crucial, it can be wise to co-locate the various cluster actors in one building or in a technology/business park;
- policy to attract talent. Clusters need qualified staff, and sometimes, targeted actions are needed to ensure that such staff becomes available.

This implies that organising capacity is crucial.

Cities cannot design and implement cluster policies on their own: all the relevant actors should be involved. Coalitions are needed that include the private sector, the university and the relevant public actors. Eindhoven is a good example of such coalition formation: cooperation in the 'triple helix' (government–knowledge institutions–business world) fuels the drive to excel as a top technology region. The actors jointly set up and implement policy efforts; they have a strong feeling of responsibility towards their part of the activities.

Rotterdam is also a good example of how cluster policy can be structured. That city has established an economic development board whose members are high-level local actors from the private, semi-private and public sector. Each member is responsible for one target cluster and develops the policy with all the stakeholders involved.

There is a lack of inter-regional coordination in cluster policies.

In their cluster policies, cities hardly look beyond their administrative borders. However, cluster actors (firms, universities, workers) operate on a different scale level and do not mind administrative borders. With improving transportation systems (the high-speed train network is particularly important), functional regions are still growing: commuting distances will increase and the relevant region for clusters will continue to grow accordingly. This is not reflected in policies, however. We found that neighbouring cities often develop identical clusters and compete instead of cooperate. In the Randstad, for instance (of which Rotterdam and Amsterdam are part), there are at least three cities that want to develop a health/biotech cluster. Similar situations prevail in the Ruhr area (where Dortmund is situated). Eindhoven is a rather positive exception: this city defines and promotes its clusters on the regional level and also seeks to coordinate its efforts with other cities (Leuven and Aachen). Rotterdam also has recently included the technical university of the neighbouring city of Delft in its cluster policy.

It would be more efficient if cluster policies were designed at least on the level of functional urban regions (commuting regions) or even higher spatial levels. To achieve this, different incentives are needed. Regional and national governments should obtain a stronger coordinating role in order to prevent the current fragmentation and create bigger and stronger clusters in specific places. This is particularly relevant for clusters where 'critical mass' is an important growth condition, as is the case in biotechnology clusters. This is not only important for the cities but for the economic competitiveness of Europe as a whole. More research is needed into the societal costs of the current cluster fragmentation, and to develop efficient policy frameworks to make the fragments into a whole.

11.5 Concluding Remarks and Perspectives

We hope that this highly ambitious study sheds more light on the development and policy options of cities in the knowledge economy. It has demonstrated that the knowledge economy is a trend that affects every city, but in different ways depending on the specific local situation. Therefore, there is no single set of policy options that applies to all cities. Rather, cities should carefully analyse their strengths and weaknesses and define their policies accordingly. Our frame of analysis can be of help in this respect. We have described a number of policy options that can improve cities' positions in this new environment.

A key conclusion for all cities is that sitting back and waiting is not an option. Even cities that currently do very well and seemingly have all the ingredients to thrive in the knowledge economy face longer-term threats that require an answer. History has shown that success is temporary and the winners of today can be the losers of tomorrow.

Another key conclusion is that, whatever the situation of the city, organising capacity is crucial to develop and implement appropriate strategies. Key drivers behind this are the increasingly networked character of the knowledge economy, and the increased interactions between actors who used to live in separate worlds. Therefore, the current society looks for joined-up policy making, which implies the involvement of the relevant actors, from public to private. This requires new governance structures and changes the role of local governments. They should break out of their ivory tower and become part of the urban network, as network leaders, supporters, orchestrators or initiators, depending on the situation. This puts high demands on the abilities of the local management. In fact, in the knowledge economy, urban management itself becomes more knowledge intensive!

Our study shows that the knowledge economy is global, but also highly localised. Urban regions are the focal points of the knowledge economy, and they have their specific strengths and weaknesses. From this perspective, we conclude that many national and EU policies are not sufficiently adapted to this reality. One of the key bottlenecks is the aspatial character of national and European innovation and R&D policies. In order to make Europe the most competitive economy in the world (an ambition formulated in the EU Lisbon Challenge), this should change. Policies should take into account the power and potential of local clustering, and capitalise on agglomeration economies. New policy frameworks are needed that create incentives for scale enlargement of local clusters, and to fight the current fragmentation of efforts among cities and nations.

References

Castells, M. (2000), 'The Information City, the New Economy, and the Network Society', in *People, Cities and the New Information Economy*, materials from an International Conference, Helsinki, 14–15 December.

Castells, M. and P. Hall (1994), *Technopoles of the World*, London: Routledge.

Duranton, G. and D. Puga (2000), 'Diversity and Specialisation In Cities: Why, where and when does it matter?', *Urban Studies*, Vol. 37, No. 3, pp. 533–55.

European Union (2003), 'European Innovation Scoreboard 2003', http://trendchart.cordis.lu/scoreboard2003/html/pdf/eis_2003_tp3_regional_innovation.pdf.

Florida, R. (2000), *The Economic Geography of Talent*, Pittsburgh: Carnegie Mellon University, September.

Glaeser, E.L. (2000), 'The New Economics of Urban and Regional Growth', in G. Clark, M. Gertler and M. Feldman (eds), *The Oxford Handbook of Economic Geography*, Oxford: Oxford University Press, pp. 83–98.

Hall, P. and U. Pfeiffer (2000), *Urban Future 21: A global agenda for twenty-first century cities*, London: E&FN Spon.

Den Hartog, P., L. Broersma and B. van Ark (2003), 'On the Soft Side of Innovation: Services innovation and its policy implications', *De Economist*, Vol. 4, pp. 433–52.

Matthiessen, C.W., A.W. Schwarz and S. Find (2002), 'The Top-level Global Research System, 1997–99: Centers, networks and nodality. An analysis based on bibliometric indicators', *Urban Studies*, Vol. 39, Nos 5–6, pp. 903–27.

OECD (2002), *Territorial Review of Helsinki*, Paris.

Office of the UK Deputy Prime Minister (2004), 'Competitive European Cities: Where do the Core Cities stand?', London, http://www.odpm.gov.uk.

Simmie, J. (2002), 'Knowledge Spillovers and Reasons for the Concentration of Innovative SMEs', *Urban Studies*, Vol. 39, Nos 5–6, pp. 885–902

Index

Note: Tables are indicated by a 't' after the locator, e.g. 'Manchester 175t'. Page numbers referring to figures are followed by 'f', e.g. 'GDP, comparison between cities 325f'.